Ro

RINEHART PRESS SERIES IN ELECTRONICS TECHNOLOGY

BASIC ELECTRON DEVICES *by E. George Griffith*
VOLTAGE AND POWER AMPLIFIERS *by Robert E. Sentz*
SPECIAL ELECTRON DEVICES *by Walter Sowa and James Toole*
ELECTRONIC POWER SUPPLIES *by Joseph Grabinski*
PULSE AND DIGITAL CIRCUITS *by John Canavan*
FEEDBACK AMPLIFIERS AND OSCILLATORS *by Robert E. Sentz and Robert A. Bartkowiak*
INFORMATION TRANSMITTAL AND COMMUNICATING SYSTEMS *by John Paul Froehlich*
ROTATING MACHINERY *by Harold J. Herbein*
TECHNICAL CALCULUS *by H. Richard Stillwell and Dennis H. Price*
TRANSMISSION LINES AND ANTENNAS *by Stephen Roman*

Rotating Machinery

HAROLD J. HERBEIN
Pennsylvania State University

UNDER THE EDITORSHIP OF JOHN M. DOYLE

RINEHART PRESS
SAN FRANCISCO

Library of Congress Catalog Card Number: 79–126531
SBN: 03–084675–7
PRINTED IN THE UNITED STATES OF AMERICA
1 2 3 4 9 9 8 7 6 5 4 3 2 1

Preface

This book has been written primarily for students of electronics technology in a two-year college, junior college, community college, or technical institute to provide an understanding of electrical rotating machines. Electronic control circuits are employed more and more for automatic control, and the advance of such circuits by solid-state devices has increased the demand for rotating machines for energy conversion. In the application of electronic circuits, it is necessary that the electronics technician understand the operation and characteristics of the rotating machine. In industrial production systems, rotating machines may be a part of the control system.

This book can also serve as a text for any nonelectrical or nonelectronics technology student. It would be especially useful for the industrial or production technology student, or the mechanical technology student. The industrial or production technician needs to know electric motor characteristics and applications in plant operation and layout. The mechanical designer needs to know motor characteristics for the machines he designs.

The electrical technology student can use this book as a text with a small amount of extra study from the references. References are noted; especially in the more advanced study of synchronous machines, induction motors, control circuits, and motor applications. The use of these references with the given subject matter greatly expands the book's use as a text.

This book does not include a discussion of transformers since they are not rotating machines. The study of transformers appears in most of the electronic texts and discussions of their applications can be found in texts on power plants.

The sample solutions of the examples and the problems that are included with each chapter, except the last one, assume that the student has completed courses in algebra and trigonometry and basic direct-current and alternating-current theory. For the nonelectronic or nonelectrical students, sections on power-factor correction with synchronous motors and the mathematical circuit solution of the induction motor can be omitted since these sections require the use of vector algebra.

The material is presented in logical order, starting first with a general, unified approach to motors and generators. Next, the construction and theory of direct-current motors and generators are taken up as a group because each is a dynamo, with the generator converting mechanical energy to electric energy and the motor converting electric energy to mechanical energy. The study of the characteristics of direct-current generators and motors follows. The discussion of alternating-current synchronous machines, both generators or alternators and motors (synchronous motors) follows. Alternating-current induction motors or alternating-current asynchronous machines are covered in the largest chapter of the book (Chapter 5). Direct-current machines as amplifiers appears in Chapter 6 and a chapter on motor applications concludes the book.

Harold J. Herbein

Oley, Pennsylvania
January 1971

Contents

Rotating Machinery

Electric Motors and Generators

1

Rotating machines may be classified as *motors* or *generators* and as either *direct current* (dc) or *alternating current* (ac). They are further classified as to principle of operation and type of construction. Examples such as a dc *shunt* generator or a three-phase ac *squirrel-cage* induction motor are typical rotating machines. They can be very small such as an electric watch motor or very large such as a 900-megawatt (MW) turbogenerator unit in a large power generating plant. Basically, motors and generators are the same type of machines, the only difference being in the energy conversion process. A dynamo refers to either a motor or a generator and when electric energy is converted into mechanical energy, the dynamo is called a motor. When the energy conversion is from mechanical to electric, the machine or dynamo is a generator. Figure 1-1 shows an electric watch whose input of about 8 microwatts (μW) comes from an aspirin size mercury-oxide power cell. The tuning fork is driven electromagnetically to vibrate 360 times per second. The vibrations are translated to rotating motion and turn the hands of the watch. Figure 1-2 shows an artist's picture of a large turbine generator. The cutaway sections on the left show steam turbine blades. The electrical generator is shown on the right side with part of the casing cutaway. The output of this unit is from 500,000 to 750,000 kilowatts (kW) or almost one hundred billion (10^{11}) times the input to the watch.

In the study of fundamentals of electricity and electric circuits, many of the new books have a unified approach and refer to direct current (dc) as a special case of alternating current (ac). In the study of rotating machinery, however, the unified approach is not used; dc machines, ac synchronous machines, and ac induction motors are the general classifications considered in this book.

This book concerns itself mainly with the conversion of electromechanical

FIGURE 1-1. **Transistorized electric watch. (Accutron by Bulova Watch Co., Inc.)**

energy in rotating machines. The control, operation, characteristics, and applications of such machines will be advanced in the material that follows.

1-1 PHYSICAL PRINCIPLES[1],[2]

Electric motors convert electric energy into mechanical energy and electric generators convert mechanical energy into electric energy. In both cases some energy is converted into heat energy, but in rotating machines heat energy is not useful and is considered as a loss. The principle with which this book is concerned is the reversible electromechanical energy conversion through the medium of a magnetic field.

The principle involves a magnetic field that can be set up by permanent magnets or electromagnets and a conductor, usually copper, situated perpendicularly in the magnetic field. If force is supplied to cause motion of the conductor in a perpendicular direction to both the field and the conductor, an electromotive force (EMF) will be induced in the conductor. This action

FIGURE 1-2. Artist's conception of a 3600-rpm, tandem-compound, four-flow, four casing, reheat large steam turbine-generator unit; 500,000 to 750,000 kilowatts. (General Electric Co.)

is *electromagnetic induction*. If in the same field and conductor an electric current is made to flow through the conductor, a force will act on the conductor in a direction perpendicular to both the conductor and the field. This action is *electromagnetic force*. Figures 1-3(a) and 1-4(a) show this principle. Figures 1-3(a), (b), (c), and (d) illustrate electromagnetic induction; (b) shows a cross-sectional view of the magnetic field ϕ between a north and south pole. Figure 1-3(c) shows a cross-sectional view of the conductor just having moved from position *a* to position *b*; (d) shows the conductor in the field of (b). Since magnetic lines stretch much like rubber bands and break when stretched beyond their elastic limit, in the position shown in Figure 1-3(d), the bottom line of flux is about to break. Since this hypothesis causes semicircular lines of flux around the conductor, an EMF is induced toward the reader by applying the right-hand rule for the relationship of flux around a current-carrying conductor. This phenomenon, discovered by Michael Faraday in 1831 and is known as *Faraday's law*, is stated as follows:

The magnitude of voltage induced in a conductor is proportional to the rate at which the conductor cuts magnetic lines of force.

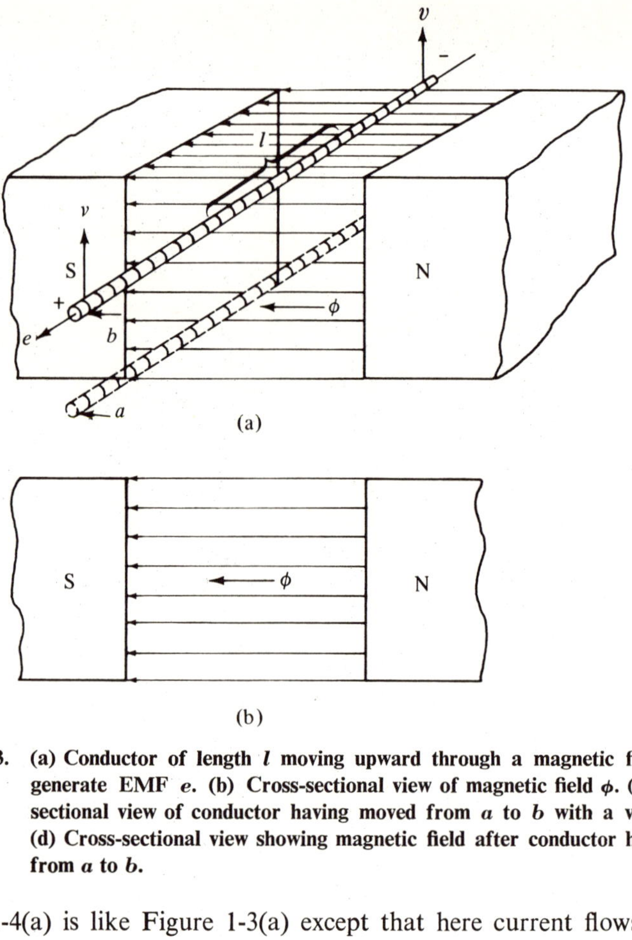

FIGURE 1-3. **(a) Conductor of length *l* moving upward through a magnetic field ϕ to generate EMF *e*. (b) Cross-sectional view of magnetic field ϕ. (c) Cross-sectional view of conductor having moved from *a* to *b* with a velocity *v*. (d) Cross-sectional view showing magnetic field after conductor has moved from *a* to *b*.**

Figure 1-4(a) is like Figure 1-3(a) except that here current flows in the conductor that is placed in the magnetic field ϕ. Figure 1-4(b) shows the magnetic field as in Figure 1-3(b). In Figure 1-4(c) the field around the current-carrying conductor is shown for conventional current flow and the field is clockwise as can be verified by the *right-hand rule* for the field around a current-carrying conductor. Figure 1-4(d) shows the interaction of the field of Figure 1-4(b). Because the semicircular lines around the conductor are compressed below the conductor, an upward force acts on it. The principles may be summarized as follows:

(1) A magnetic field must be established, either from permanent magnets or from electromagnets (usually electromagnets).

(2) A conductor is situated perpendicularly in the magnetic field.

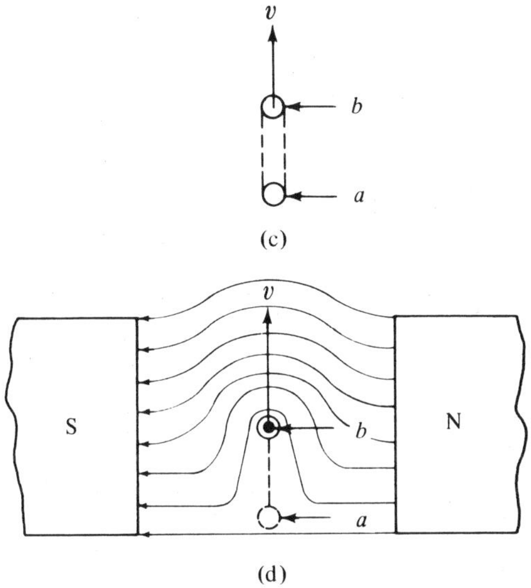

FIGURE 1-3. *Continued.*

(3) (a) If motion is produced by mechanical force on the conductor perpendicular both to the field and to the conductor, a voltage is induced in the conductor (generator action); or (b) if current is made to flow through the conductor, a force results that is perpendicular both to the conductor and to the field. A sufficient force will produce motion in the direction of the force (motor action).

1-2 GENERATOR ACTION[2],[3]

In the previous section it was shown that if a conductor in and perpendicular to a magnetic field ϕ is moved with a velocity v perpendicularly to both the field and the conductor, an electromotive force is induced in the conductor. Because it would not be practical to move conductors in this way, coils are placed in armature slots and the coils are rotated. In practically all ac generators, usually called *alternators*, the field is rotated and the conductors are stationary. An examination of Figures 1-5(a), 1-5(b), and 1-5(c) shows a coil cross section rotating in a clockwise direction in a uniform magnetic field. The coil is rotated uniformly by a force F. In position (2) the distortion of the field around the conductors is shown as similar to that in Figure 1-3(d). The left conductor has an induced EMF toward the reader,

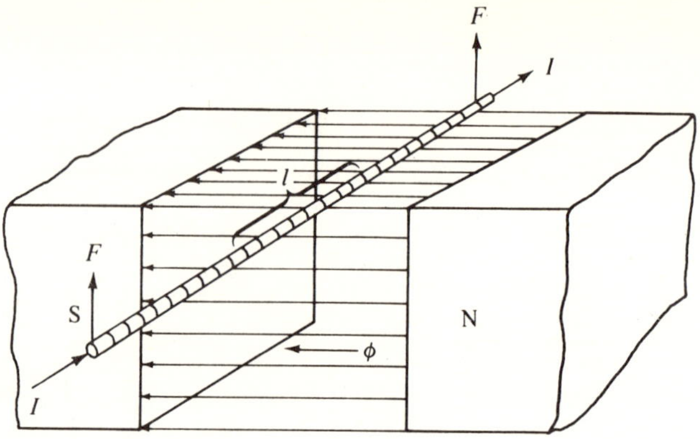

(a)

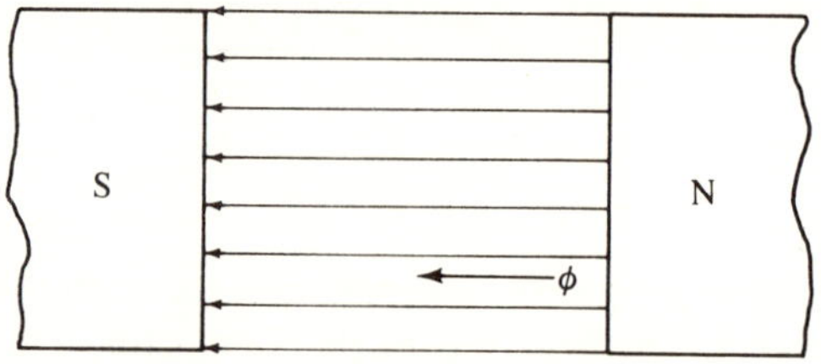

(b)

FIGURE 1-4. **(a) Conductor of length l with current I flowing through conductor causing electromagnetic force F. (b) Cross-sectional view of magnetic field ϕ. (c) Cross-sectional view of field around current carrying conductor. (d) Cross-sectional view showing interaction of field ϕ and the field around current carrying conductor.**

while the right conductor has an induced EMF away from the reader. The EMF of both conductors add together. By Faraday's law,

$$e = \frac{\phi}{t} \quad \text{or} \quad E_{\text{ave}} = \frac{\phi}{t}\,\text{V}, \tag{1-1}$$

where

E_{ave} is the average voltage induced in one conductor,
ϕ is the flux in webers (10^8 lines or maxwells),
t is the time in seconds in which the conductor is cut.

Example 1-1. Calculate the average induced EMF in a single conductor if 1.5×10^{-2} Wb are cut in 1/100 (0.01) sec.

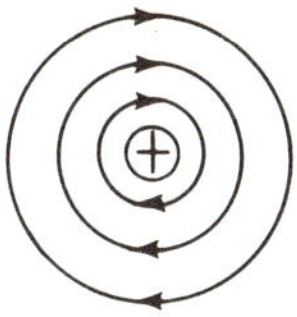

(c)

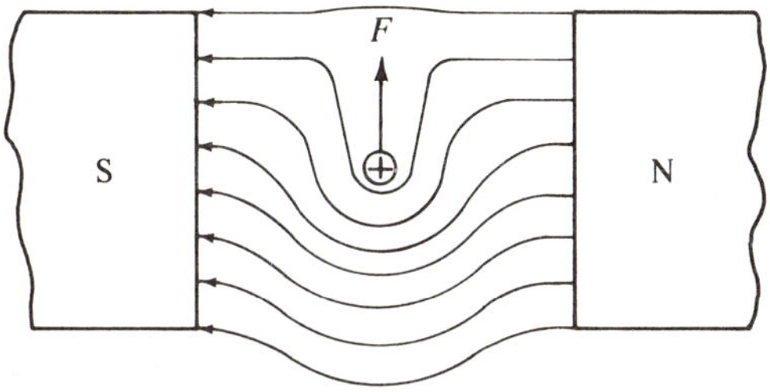

(d)

FIGURE 1-4. *Continued.*

Solution:

$$E_{ave} = \frac{1.5 \times 10^{-2}}{0.01} = 1.5 \text{ V}.$$

Equation (1-1) can be translated into linear velocity, flux density, and conductor length. It can be seen in Figure 1-5 that flux density and conductor length are constant, but the velocity component perpendicular to the flux and conductor changes as the coil rotates. In position 1 the component of velocity perpendicular to the field is zero and at this instant the EMF generated is zero. Formula (1-1) becomes

$$e = Blv \text{ V} \quad \text{(for a velocity perpendicular to the field),} \qquad \textbf{(1-2)}$$
$$e = Blv \sin \theta \quad \text{(for coil as shown in Figure 1-5),} \qquad \textbf{(1-2a)}$$

where

e is the instantaneous voltage induced,
B is the flux density in webers per square meter,
l is the length of conductor in meters,
v is the linear velocity of the conductor in meters per second,
θ is the instantaneous angle between velocity and flux in degrees.

Example 1-2. Calculate the instantaneous generated voltage of a conductor

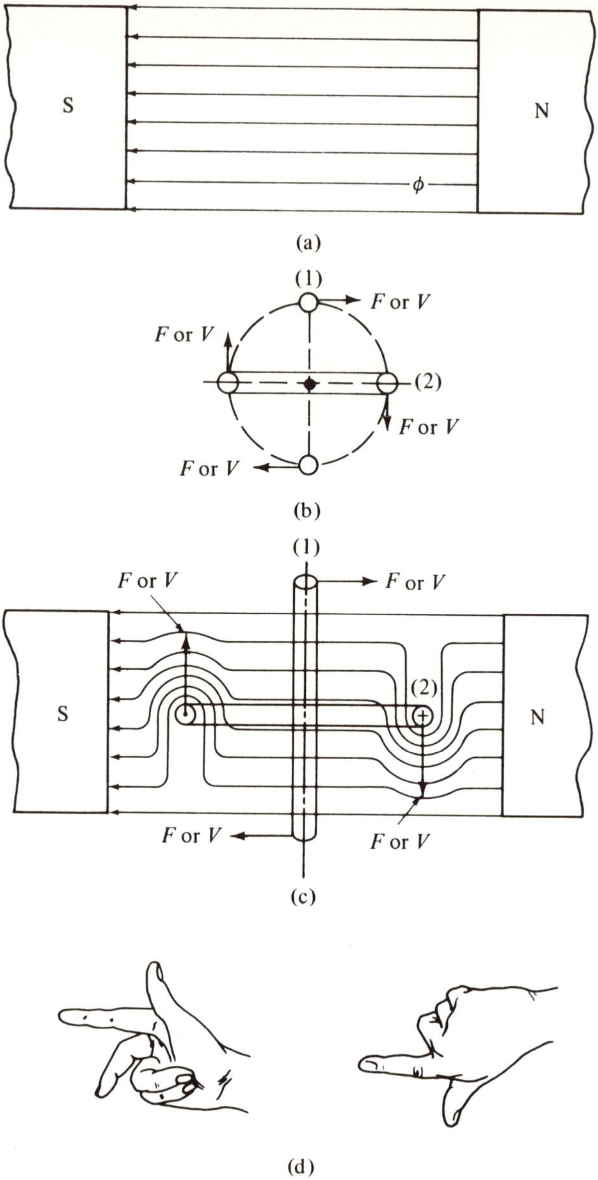

FIGURE 1-5. **(a) Cross-sectional view of magnetic field. (b) Cross-sectional view of coil turning clockwise. (c) Cross-sectional view showing distortion of flux when rotating coil is placed in field of (a). (d) Fleming's right-hand rule illustrations.**

0.30 m long in a uniform field of 1 Wb/m^2 and with conductor velocity of 5 m/sec perpendicular to the field.

Solution:

$$e = Blv \sin\theta = (1.0)(0.30)(5.0)(\sin 90°) = 1.5 \text{ V}.$$

Equation (1-2) in practical units becomes

$$e = \frac{B'l'v'}{5} \times 10^{-8} \text{ V},$$

where

B' is flux density in lines per square inch,
l' is the conductor length in inches,
v' is linear velocity of conductor in feet per minute.

Example 1-3. Calculate the instantaneous generated voltage of a conductor 15 in. long in a uniform magnetic field of 65,000 lines/in.^{2} and moving with a linear velocity of 640 ft/min perpendicular to the field.

Solution:

$$e = \frac{B'l'v'}{5} \times 10^{-8},$$

$$e = (6.5 \times 10^4)(1.5 \times 10)(6.40 \times 10^2) \times \frac{10^{-8}}{5} = 1.25 \text{ V}.$$

Generators have many conductors (coils) spaced around the armature and an even number of poles. Poles must be in pairs; for each north pole there must be a south pole to provide a proper magnetic circuit. For a dc generator, the following is the formula for the average generated EMF E_g

$$E_g = \frac{\phi PZS}{60a} \text{ V}, \quad \textbf{(1-3)}$$

where

ϕ is the flux per pole in webers,
P is the number of poles,
Z is the number of armature conductors,
S is the angular velocity (speed) in revolutions per minute (rpm),
a is the number of parallel current paths in the armature.

Example 1-4. A six-pole dc generator has an armature winding made up of 486 turns or 972 conductors with a simplex wave winding ($a = 2$). The flux per pole is 0.428×10^6 lines and the speed of armature rotation is 1200 rpm. Calculate the average generated voltage (E_g).

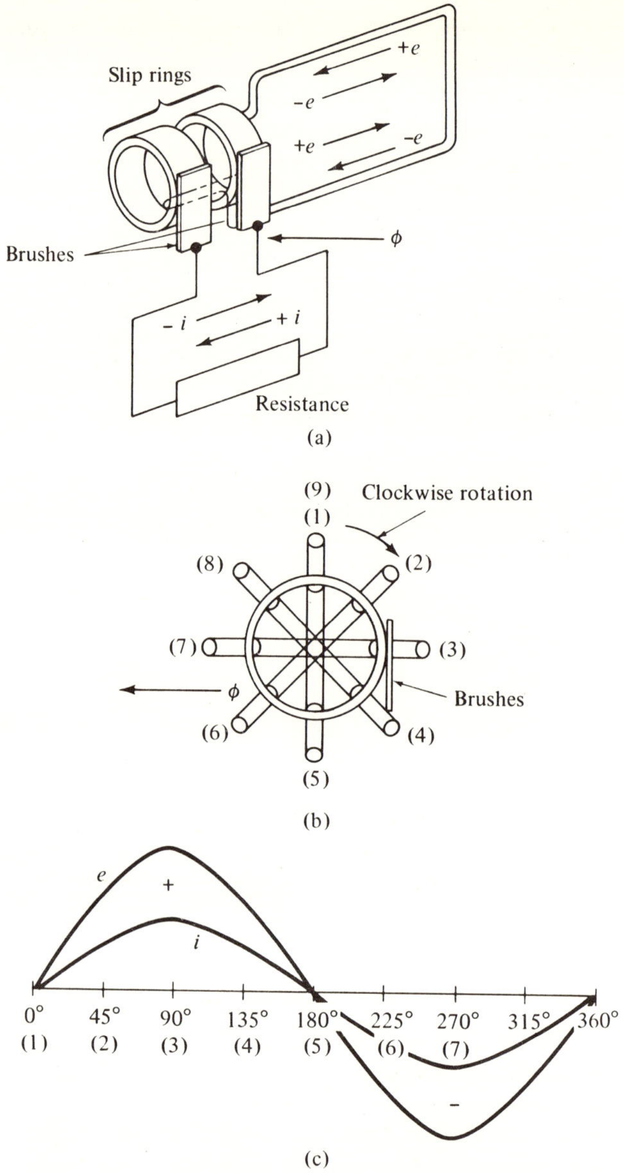

FIGURE 1-6. **(a) Coil with ends connected to slip rings. Position (1) of (b) and (c). (b) Cross-sectional view of (a). (c) Curve showing instantaneous values of induced voltage and current flow through a resistance for positions of coil rotating in uniform field.**

Solution:

$$E_g = \frac{\phi PZS}{60a}\ \text{V} \qquad (10^8 \text{ lines} = 1 \text{ Wb}),$$

$$E_g = \frac{(0.428 \times 10^6)(6)(972)(1200)}{(10^8)(60)(2)} = 250 \text{ V}.$$

The reason a voltage is induced in a rotating coil can be shown by *Lenz's law* which states

Whenever flux changes in a turn of wire, a voltage will be induced in the turn of wire in such a direction as to cause a current to flow that sets up a magnetic field opposing the change of flux.

Referring to Figure 1-5(c), the coil in position (1) links the entire magnetic field. When the coil is turned to position (2), no flux links the coil and an EMF is induced in the coil as shown. The coil in position (1) has flux through it from right to left. When the coil is turned clockwise to position (2), the flux by Lenz's law will want to be upward, so that the current in the right conductor by the right-hand rule will be away and in the left conductor will be toward the observer.

The direction of the induced EMF can readily be determined by *Fleming's right-hand rule.* This rule can be stated as follows:

Place the right hand with the thumb, index finger, and the middle finger extended with each perpendicular to the other. The index finger points in the direction of the magnetic field, the thumb points in the direction of the conductor motion, and the middle finger points in the direction of the induced voltage.

This is illustrated in Figure 1-5(d).

Basically an alternating voltage will be generated in the armature coils of generators. In Figure 1-6(a) two solid rings are shown with each end of the coil connected to a coil end. Brushes slide on the rings and a resistance is connected to the brushes. Figure 1-6(b) shows the end view of the coil in various positions as it rotates in a clockwise direction in a uniform magnetic field. Figure 1-6(c) shows the instantaneous values for each position as the coil is turned through 360 deg. Figure 1-7(a) shows the same coil connected to a split ring with each end connected to a half of the ring. Brushes are placed 180 deg from each other and connected to a resistance. In position (1) of Figure 1-7(b) the brushes are across the split sections, because at this instant the generated voltage is zero as is the instantaneous current. Figures 1-7(b) and 1-7(c) show that the direction of current is always the same but

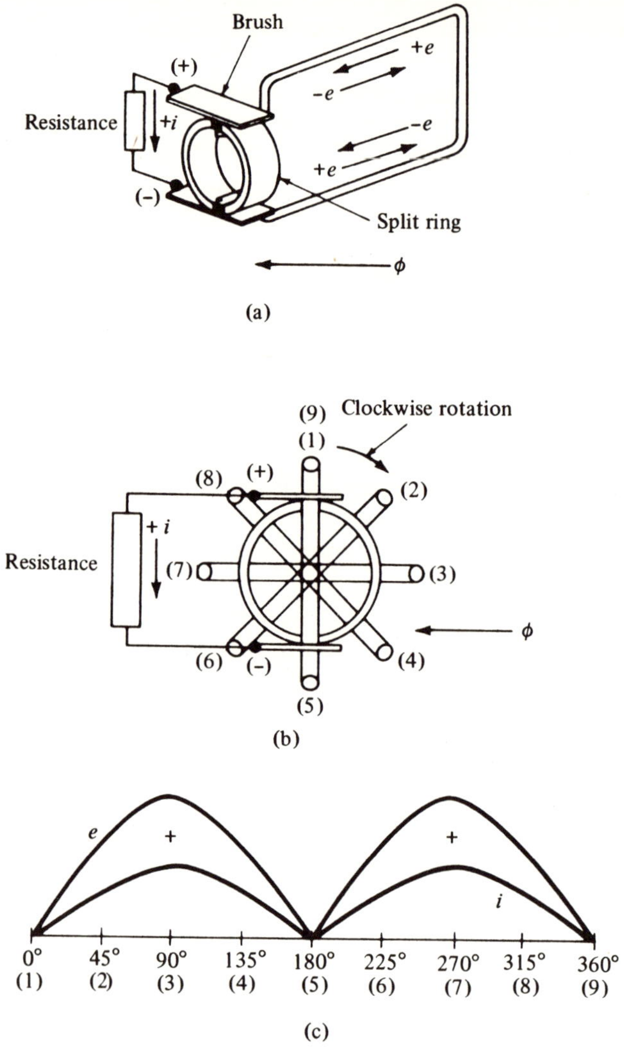

FIGURE 1-7. **(a) Coil with ends connected to halves of split ring. (b) Cross-sectional view of (a). (c) Curve showing instantaneous values of induced voltage and current flow through a resistance for positions of coil rotating in uniform field.**

varying in magnitude. In Figure 1-6(c) the trace of the induced EMF and the current are both sine waves with the current in phase with the induced voltage. In Figure 1-7(c) the loops of the current and induced voltage are also sine loops but the negative loops have been reversed. This is full wave rectification of the ac output. In the design of dc dynamos the shape of the wave is flat or square, which is the result of placing the armature coils in armature slots of magnetic material. Alternating-current generators are designed so that the generated voltage is a sine wave that can be seen on the trace of its output on an oscilloscope.

1-3 MOTOR ACTION[2],[3]

In Section 1-1 it was shown how a force is produced on a conductor placed perpendicularly in a magnetic field. In Figure 1-8(a) a magnetic field is

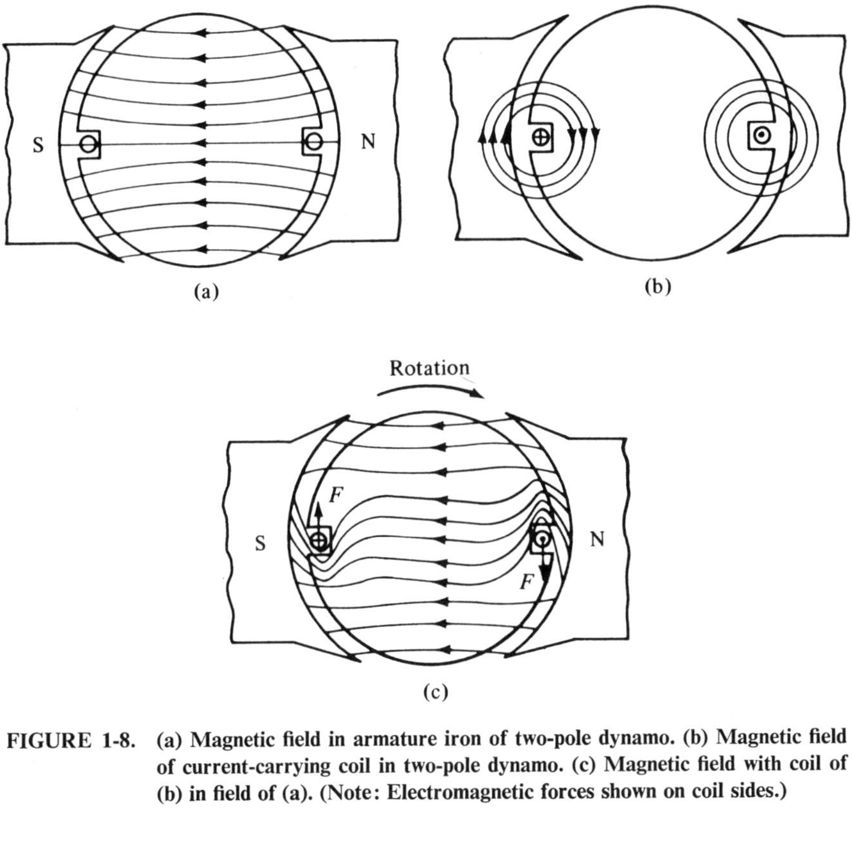

FIGURE 1-8. (a) Magnetic field in armature iron of two-pole dynamo. (b) Magnetic field of current-carrying coil in two-pole dynamo. (c) Magnetic field with coil of (b) in field of (a). (Note: Electromagnetic forces shown on coil sides.)

shown. The armature iron is shown with a coil in opposite slots. For clarity in illustrating the magnetic lines, it is assumed that the reluctance of the air-gap space is the same as the iron. Figure 1-8(b) shows the field around the conductors with the current in on the left conductor and out on the right conductor. In Figure 1-8(c) the interaction of the flux is shown. On the right conductor the flux is crowded above the conductor and thin below it. Because lines of force exert sidewise force tending to repel each other, the concentrated flux above the wire exerts a downward force on the right conductor. By a similar analysis an upward force is exerted on the left conductor. The torque exerted on the coil tends to cause the coil to rotate in a clockwise direction.

The *force* that is exerted on each conductor is proportional to the current through the conductor, the flux density B, and the length of the conductor in the uniform field. Therefore

$$F = KBIl. \tag{1-4}$$

If MKS units are used, the value of K is unity; B is in webers per square meter, I is in amperes; and l is in meters, the formula becomes

$$F = BIl \text{ N}. \tag{1-4a}$$

In CGS units the formula becomes

$$F = \frac{BIl}{10} \text{ dyn}, \tag{1-4b}$$

where

B is in gauss (maxwells per square centimeter),
I is in amperes,
l is in centimeters.

In practical units or the English system of units, the formula becomes

$$F = \frac{BIl \times 10^{-7}}{1.13} \text{ lb}, \tag{1-4c}$$

where

B is lines per square inch,
I is in amperes,
l is in inches.

Example 1-5. A single conductor is 15 in. long (active length) and placed in a uniform magnetic field whose flux density is 60,000 lines/in.^{2} and the current through the conductor is 15 A. Find the force on the conductor in pounds.

Solution:

$$F = \frac{BIl \times 10^{-7}}{1.13} = \frac{(6.0 \times 10^4)(1.5 \times 10)(1.5 \times 10) \times 10^{-7}}{1.13} = 1.2 \text{ lb}.$$

Example 1-6. If 144 conductors are placed around the circumference of the armature and 75 percent are under magnetic field influence, (a) what is the total force exerted on the armature? (b) If the diameter of the armature (mean to conductors) is 14 in., what is the torque developed? (c) If the speed is 1750 rpm, what is the horsepower developed? B, I, and l are given in Example 1-5.

Solution:

(a) Force on one conductor from Example 1-5 is 1.2 lb.

$$\text{Total force} = (1.2)(Z) \qquad \text{(percent active condition)}$$
$$= (1.2)(144)(0.75) = 129.5 \text{ lb.}$$

(b) $$\text{Torque} = Fr = 129.5\left(\frac{14}{(2)(12)}\right) = 75.4 \text{ lb-ft.}$$

(c) $$\text{Horsepower} = 2\pi\frac{TS}{33000} = \frac{TS}{5250} = 0.00019TS$$

$$\text{Horsepower} = 1.9 \times 10^{-4}(7.54 \times 10)(1.750 \times 10^{3}) = 25.$$

In the previous section *Fleming's right-hand rule* was given to tell the direction of induced voltage. The direction of the force on a conductor or the direction of rotation of the motor armature can be found by the interaction of the flux of the conductor with the main field, but this is a laborious procedure. The left-hand motor rule provides a quick way to tell the direction of the force or motion. The *left-hand rule* is as follows:

Extend the thumb, index finger, and the middle finger of the left hand at right angles to each other. Point the index finger in the direction of the flux, point the middle finger in the direction of the current flow, and the thumb will point in the direction of the force or motion.

Figure 1-9(a) shows the elements of Figure 1-4(a). Figure 1-9(b) shows the illustration of the left-hand motor rule.

1-4 THE MAGNETIC CIRCUIT[4],[5]

Because rotating machines have magnetic fields, magnetic circuits must be provided. A major part of these machines is made up of iron or magnetic steels. Figure 1-10 shows the iron and steel parts that make up the magnetic circuit of a four-pole dc dynamo. The rotating part, the *rotor*, contains the *armature core*. This part is made up of magnetic steel laminations that are about 0.025 in. thick. The laminations are often dipped in varnish or they are oxidized so that the surfaces are insulated to reduce eddy current losses.

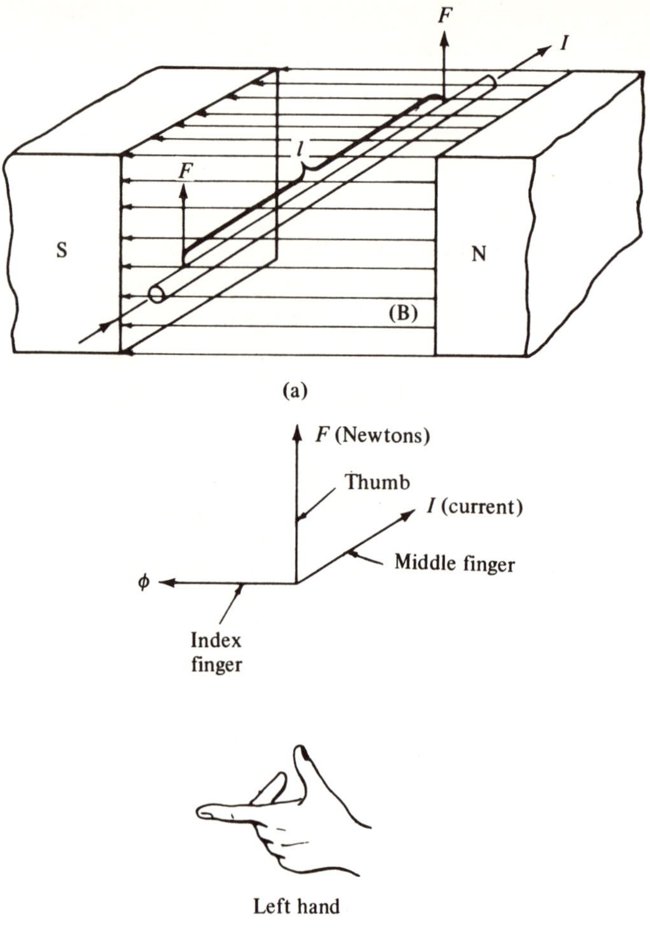

FIGURE 1-9. **(a) Conductor of length *l* with current *I* flowing through conductor with electromagnetic force *F*. (b) Illustration of the left-hand motor rule.**

Around the circumference are the slots into which the armature conductors are placed.

The field poles are made up of laminated magnet steel sheets. The poles are usually fastened to the yoke or frame with bolts. The yoke is either cast iron or steel. The new machines, especially the larger ones, have yokes of fabricated steel plate. The yoke also serves as the frame of the machine. The

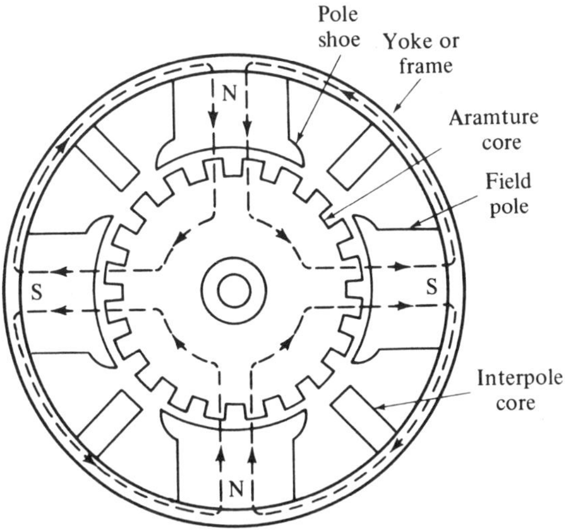

FIGURE 1-10. Magnetic circuit for dc dynamo.

field poles have an extended section called the *pole shoe*. The field coil is made to fit onto the pole and is held in place by the pole shoe. The interpole core is not a part of the basic magnetic circuit. It is placed between the main poles for the commutation function, which is discussed in the next chapter.

The basic magnetic circuit for an ac synchronous dynamo is shown in Figure 1-11. The rotor in this case shows a four-pole nonsalient pole core.

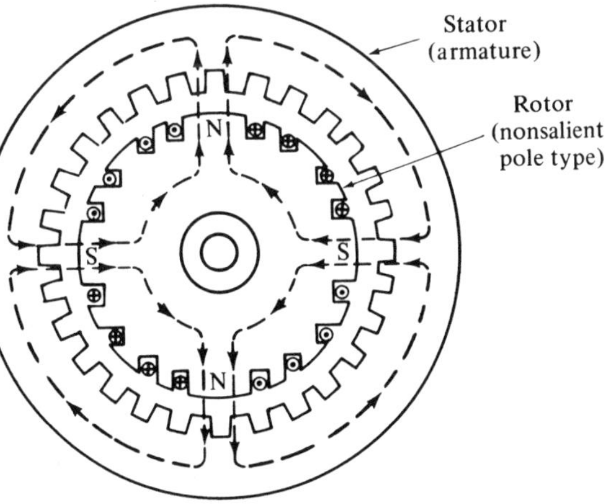

FIGURE 1-11. Magnetic circuit for ac dynamo.

Nonsalient poles have the poles induced with windings that are laid in slots. Salient poles protrude and have the coil windings around the smaller part of the core. The poles in Figure 1-10 are the salient-pole type. For synchronous machines in which the field structure rotates, salient poles usually are used for speeds below 1800 rpm. For higher speeds (3600 rpm) nonsalient poles are used. Induction motors have similar magnetic circuit construction, except that the slots are radial to accommodate the rotor winding. The parts of Figure 1-11 are made of magnetic steel laminations.

1-5 ARMATURE CONDUCTORS[1]–[3]

The armature conductors are those conductors in which the function of the machine is performed. In generators the voltages are induced in these conductors, and in motors electromagnetic forces are produced that act on these conductors. The armature conductors are the armature coils, and the conductors or coil sides are placed in the slots of the armature core. Most armature windings are whole-coiled windings. Two coil sides are placed one on top of the other in the slots as shown in Figure 1-14. The half-coil winding has but one coil side in a slot. This type is seldom used except occasionally in small ac motors, and it would not be practical for any dynamo with a commutator. The whole-coil winding will be the only type considered here.

FIGURE 1-12. **Wave coils and partially-wound armature. (Westinghouse Electric Corp.)**

The way the coils are connected makes the winding either a lap winding or a wave winding. Figure 1-12 shows lap-winding coils and Figure 1-13(a) shows a lap coil with its ends connected to adjacent commutator segments for a dc dynamo. In Figure 1-13(b) is shown a wave-winding coil with its ends con-

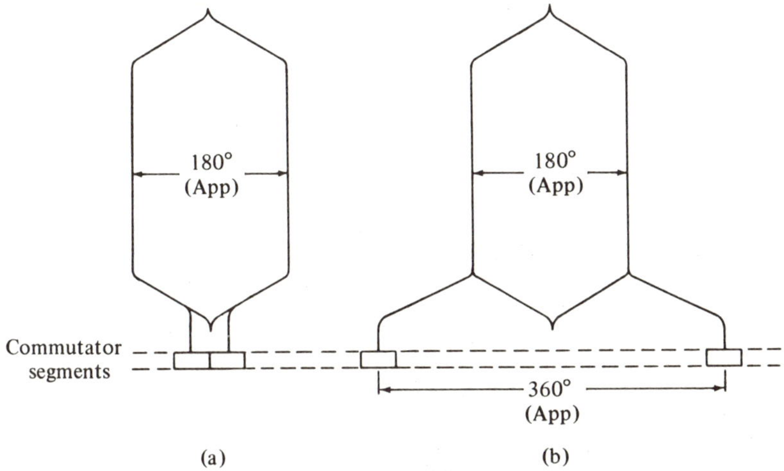

FIGURE 1-13. (a) Lap coil showing coil span and end connections. (b) Wave coil showing coil span and end connections.

nected to commutator segments approximately 360 electrical deg apart. In both cases the coil span is approximately 180 deg. Electrical degrees are as shown in Figures 1-6(c) and 1-7(c). In a two-pole dynamo when the armature turns through 360 deg or one revolution, the induced voltage in a coil also goes through 360 electrical deg. In a four-pole dynamo when the armature turns 360 deg or one revolution, the induced voltage would go through two cycles or 720 electrical deg. The span from the center of a north pole to the center of a south pole would be 180 deg. The span from the center of a north pole to the center of the next north pole is 360 electrical deg.

The coils are made with insulated wire. For larger machines insulated rectangular wire is used. The coils are wound and then shaped to the form as illustrated. After the coil is taped, it is dipped into insulating varnish or enamel and then baked. The finished coils are then placed in the slots of the armature core, first placing the bottom side into the slot. After a coil span is filled, the top sides can be placed in the top of the slot but the tops of the first slots cannot be filled until the placing of coils nears the finish. Fiber wedges are placed under the slot lip. The *slots* have a small lip on the top so

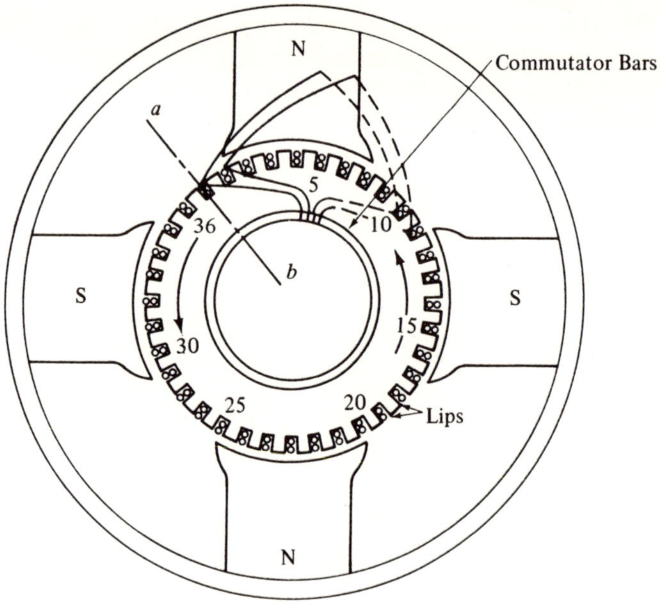

FIGURE 1-14. Four pole dc dynamo showing placing of coils in slots.

that these wedges are held in place. In Figure 1-14 these lips are shown. In this same figure two *lap coils* are shown in place and the coil ends connected to commutator segments. Cardboard strips and strips of insulating cloth are laid in the slots so that when the wedge is in place, the slot is filled snugly. In smaller machines the coils are usually of round wire of the required size coil loop; the wire is then fed directly into the slots.

It was said that the coil sides are approximately 180 electrical deg apart. Although this is the optimum condition, it is not always possible. A wave winding cannot have 180 electrical deg for the span unless "dummy" coils are used. The coil span is called *slot pitch.* If the slot pitch (Y_s) is less than 180 electrical deg, the pitch is called a *fractional pitch.* The slot pitch may be expressed as follows,

$$Y_s = \frac{S}{P} - k, \qquad \textbf{(1-5)}$$

where

Y_s is the slot pitch,
S is the number of slots in the armature core,
P is the number of poles,
K is a fraction subtracted to make Y_s an integer.

The following table shows how this formula is applied.

Number of Slots	Number of Poles	(Y_s) Slot Pitch	Span in Electrical Degrees
36	4	9	180
45	4	11	176
57	6	9	170.5
78	10	7	161.5
108	12	9	180

A simplex lap winding is depicted for a 36-slot, 4-pole, dc dynamo in Figures 1-14 and 1-15. In the later figure the armature is unfolded from line *a-b* so that the winding lies out flat showing the surface of the armature. The dotted lines represent the coil sides in the bottom of the slot. In agreement with the above table, Y_s is 9. The top of the first coil is in slot 1 and the bottom of the coil is in slot (1 + 9) or (10). The span is 180 electrical deg. The sides of the coil in slot (3) and (12) are under identical spots under alternate poles (north and south). The winding could be traced as follows: top (1), bottom (10), top (2), bottom (11), and so on to top (36), bottom (9), and then closed in commutator segment (5). The coils lie in adjacent slots and the winding does not close until all the coils have been used.

Figure 1-16 is a wave winding for a 33-slot, 4-pole, dc dynamo. The slot pitch is

$$Y_s = \frac{S}{P} - k = \frac{33}{4} - \frac{1}{4} = 8.$$

This is fractional pitch and the coil span is equal to (2 × 360/33) × 8 or 174.5 deg. The commutator pitch is 16 or 349 deg. Starting at commutator segment (1) to top slot (5), bottom slot (13) to commutator segment (17), and then to top slot (21), bottom slot (29), commutator segment (33), top slot (4), and so on, finally to bottom of slot (30), the winding closes on commutator segment (1). In Figure 1-16 the winding goes from north pole to south pole to next north pole to south pole and then to the adjacent slot under first north pole. It also is evident that slots (5), (13), (21), and (29) are not under the same relative positions under their respective poles.

The lap-winding scheme is used for ac machines and a winding for a three-phase, 36-slot armature for 4 poles is shown in Figures 1-17 and 1-18. The slot pitch is the same as in Figure 1-15, but since there are generally no commutators in ac machines, the connections are made differently. Each phase consists of 36/3 or 12 coils; 4 poles mean that there are 3 coils per

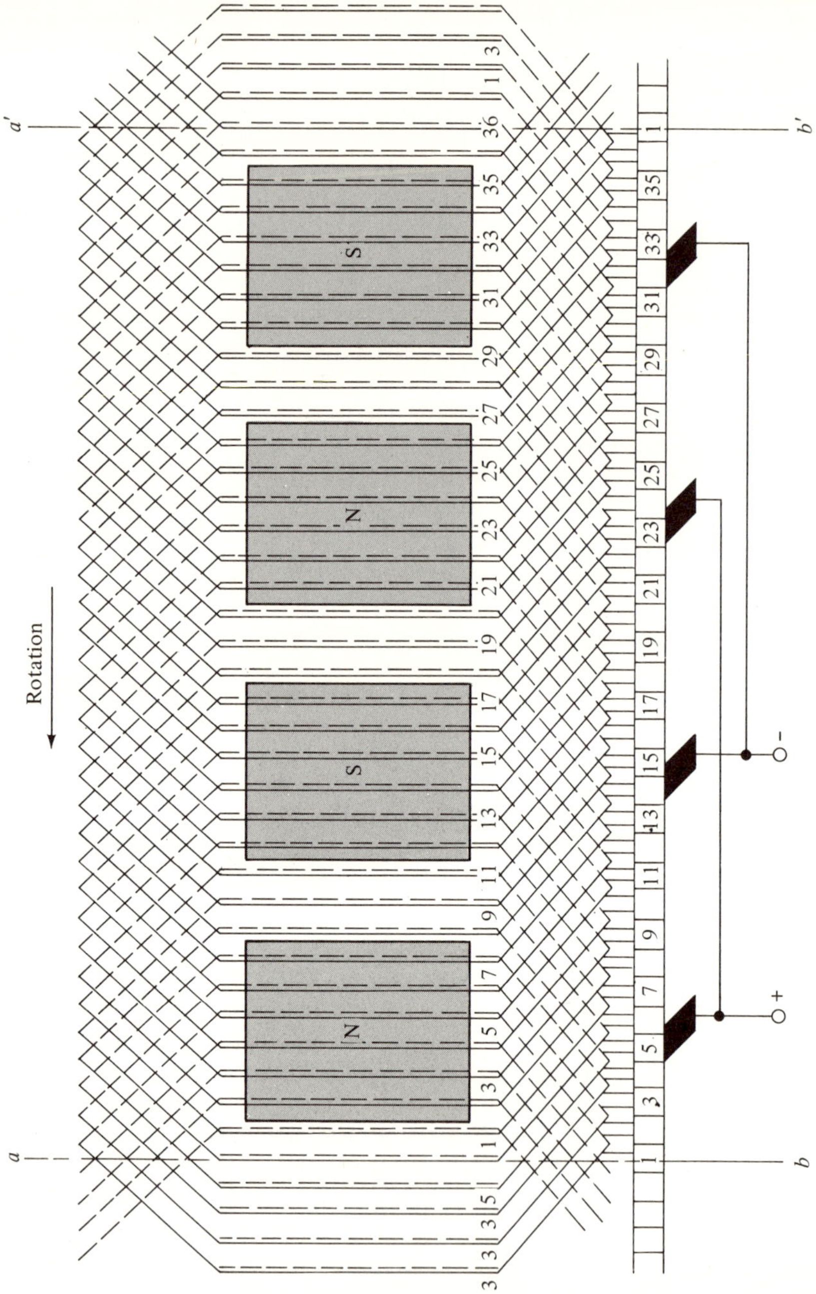

FIGURE 1-15. Flat view of armature lap winding for four-pole, 36-slot, whole-coiled winding.

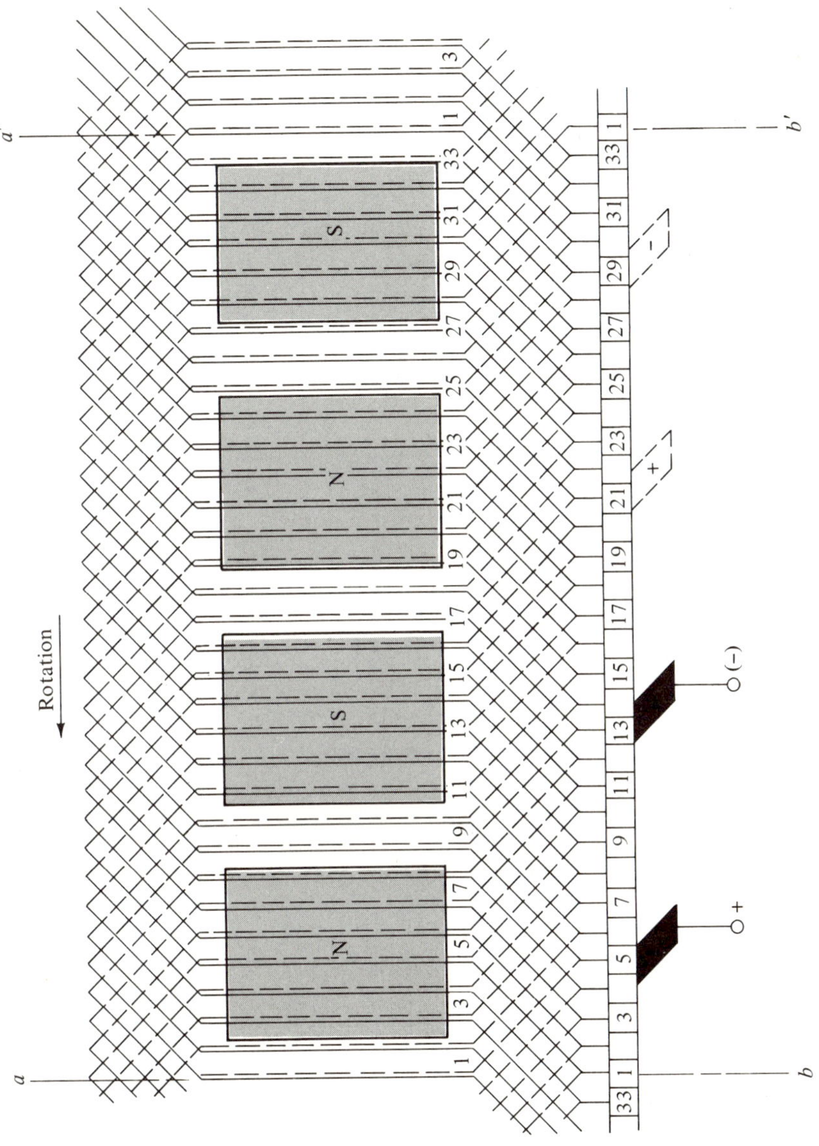

FIGURE 1-16. Flat view of armature wave winding for four-pole, 33-slot, whole-coiled winding.

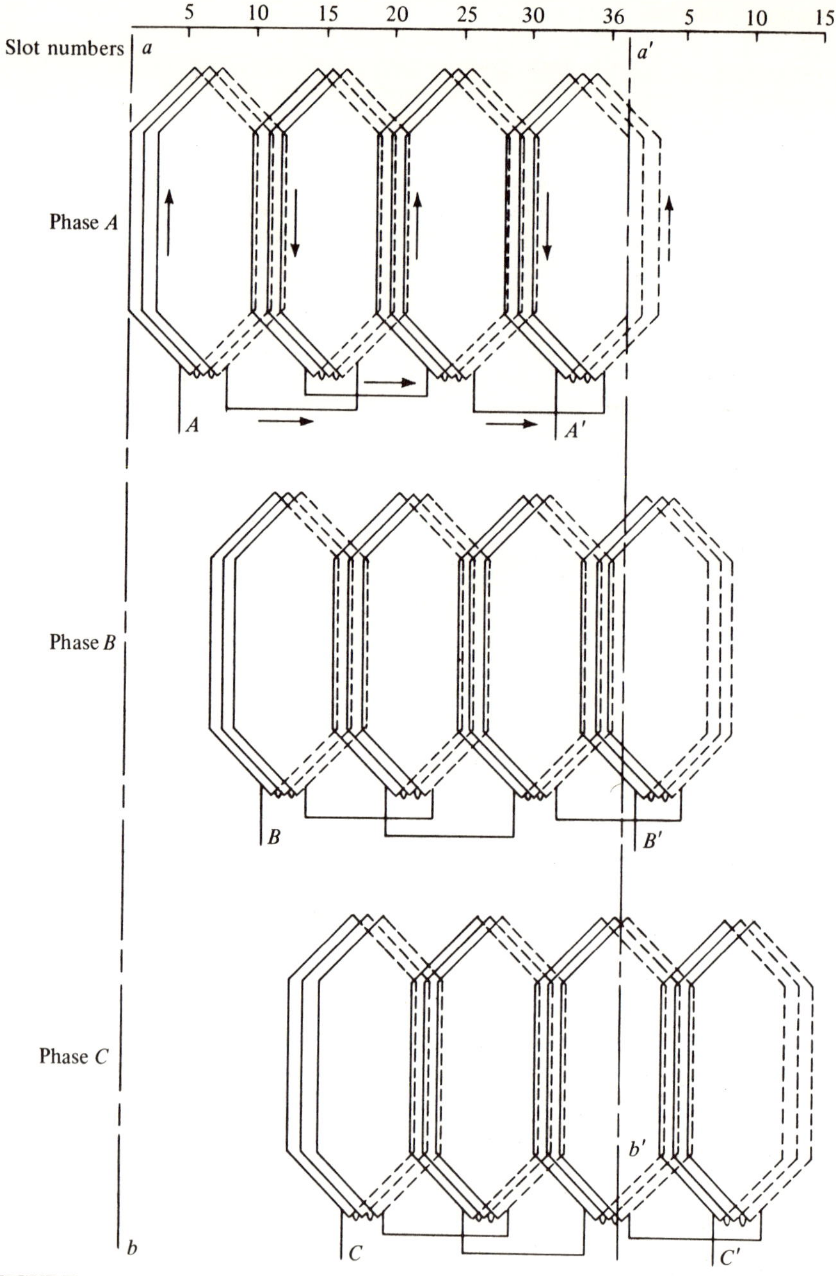

FIGURE 1-17. Lap winding showing the three phases for four poles, 36 slots. For ac dynamo armature.

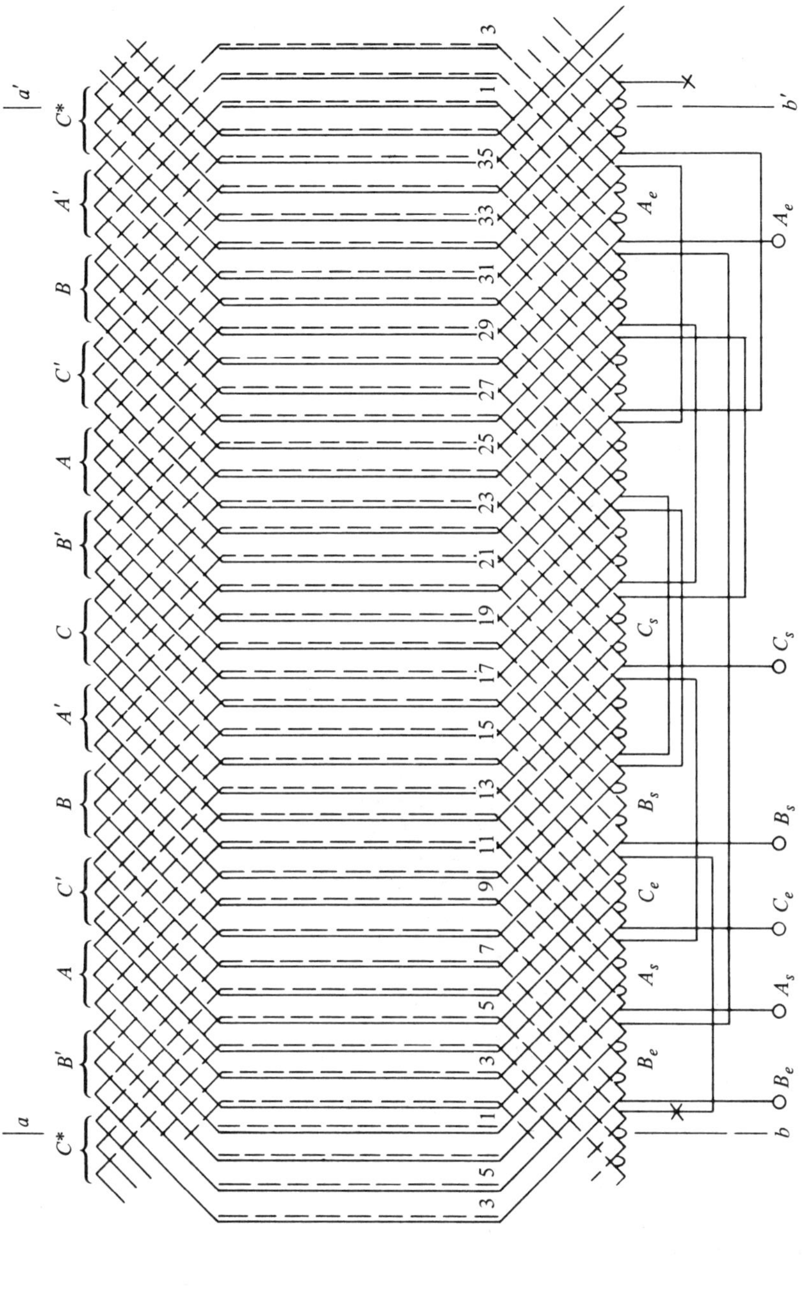

FIGURE 1-18. **Lap winding of Figure 1-17 with the three phases in place.**

phase per pole. Figure 1-17 shows each phase individually and each displaced 120 deg. Figure 1-18 shows the entire winding with the connections for each phase. This winding will be considered further in Chapters 4 and 5. Synchronous machines use this winding as the stationary armature, while in polyphase induction motors, a similar winding becomes the field winding.

All of the preceding windings were *simplex;* windings may also be *duplex, triplex,* and so on. A duplex winding has two identical simplex windings side by side while a triplex winding has three identical simplex windings side by side. There is no difference in the placement of the coils into the slots, but there is a difference in the connections of the coil ends. For dc machine windings, the ends would be connected to alternate commutator segments in the duplex winding, and every third segment in triplex winding.

Figures 1-15 and 1-16 illustrate simplex windings, each of which closes after all the coils are connected. The start of the first coil and the end of the last coil are connected to the same commutator segment. In the case of multiplex windings the closures (*re-entrancies* as they are often called) are equal to the multiplicity (plex). Closed windings are always used in dc dynamos. Figure 1-18 is an open winding because the terminals are open circuited. The open winding is used in ac machines.

1-6 INSULATION[1],[2],[4],[6]

Field coils are usually made up of wire insulated with class-A insulation. The classes of insulation are shown in Table 1-1. As seen in the table the class of insulation is determined by the maximum allowable temperature. Enough insulation must be provided to prevent short circuits between turns and to prevent grounding of conductors. The field coils include the *shunt winding* and a *series winding* for compound dynamos. The shunt coil is shaped with cardboard and then wrapped with tape (cotton) after which a series winding of large wire is placed over the shunt winding. The entire coil is then dipped in varnish or enamel and baked. Figure 1-19 shows a field coil and pole assembly which is ready to be attached to the yoke.

Armature coils for the larger dc dynamos are wound with rectangular wire. An insulating cell is placed around the long sides of the coil, after which it is wrapped with treated tape. A final layer of tape is then wrapped around the entire coil. Figure 1-12 shows wave coils for a smaller motor. In this case the insulating cells are placed in the slots and the coil sides are placed inside the insulating cells. In this figure, the armature is partially wound.

TABLE 1-1. Classes of Insulating Material and Limiting Temperature

Class of Insulation	Material	Hot-Spot Temperature °C
90 Formerly O	Cotton, silk, paper, and similar organic materials without impregnation	90
105 Formerly A	(a) The above materials immersed or impregnated in liquid dielectric; (b) molded and laminated materials with cellulose filler, phenolic resins, or other resins of similar properties; (c) films and sheets of cellulose acetate or other cellulose properties; (d) varnishes or enamels applied to conductors	105
130 Formerly B	Mica, asbestos, Fiberglas, and similar inorganic materials with a small proportion of class 105 materials as binders and fillers	130
155 Formerly F	Material of class 130 if by experience or accepted tests can operate at the given temperature	155
180 Formerly H	(a) Mica, asbestos, Fiberglas, and similar inorganic substances with binding substances of silicone compounds; (b) silicone compounds in rubbery or resinous form, or materials with equivalent properties	180
220	Materials or combinations of materials that by experience or tests can be shown to have the required thermal life at 220°C	220
Over 220 Formerly C	Mica, porcelain, glass, quartz, and similar inorganic materials	Over 220

1-7 RATING OF ELECTRICAL MACHINERY[1],[7]

The rating of a rotating machine gives the limits within which it must operate and when it is operated within these limits, it will function safely both mechanically and electrically and without excessive deterioration in its electrical or mechanical parts.

The nameplate includes the output rating, which is in horsepower for motors and kilowatts for dc generators or kilovoltamperes (kVA) for alternators. The terminal voltage, current, and speed are also included on the nameplate. For a motor the proper voltage within tolerances must be supplied

FIGURE 1-19. Main field pole assembly for a 300-hp, 600-rpm, 600-V motor. Insulation is class F, 60°C temperature rise. (Large DC Motor Business Section, General Electric Co.)

if the output horsepower and speed are to be attained. The current that the motor will draw will be within tolerances as specified on the nameplate. Tables A-1, A-2, and A-3 show the current values for motors as specified by the *National Electrical Code.*[1] These values are used to provide the proper electrical facilities of motors. For a generator the prime mover must provide the speed called for on the nameplate. For an ac generator the speed must be exactly as specified to provide the proper output frequency. It must also be provided by a mechanical input that is about 120 percent of the equivalent horsepower output.

The *temperature-rise rating* is also specified on the nameplate. The common temperature rise is 40°C for class-A insulations. If the machines are operated at continued overload conditions, the insulating materials will overheat and in time the materials will disintegrate. General-purpose machines that are self-ventilating when operating in normal surroundings have a rating of 40°C rise for continuous operation. This means that the temperature

[1] See Section 1-9.

as measured at the surface of coils will not rise to more than 40°C (104°F) above the ambient temperature. The *ambient temperature* is the temperature of the cooling medium, which in the case of self-ventilating machines is the surrounding air. It is estimated that the temperature in any part of the United States is no higher than 40°C. Adding the 40°C rise to this brings the maximum temperature to 80°C at the surface of the coils. For the average temperature of conductors as calculated by resistance measurements, 10°C is added to the surface temperature. For the *hot spot* within the conductor, 5°C is added to average temperature and this brings the hot-spot temperature to 95°C. This is 10°C below the 105°C value specified for class-A insulation. Because of this a service factor of 1.15 is specified for most machines. This is no fixed value and the NEMA code should be consulted for the service factor to be used. The hot-spot temperature is difficult to measure, especially on small machines; it can be measured by embedded thermocouples. For the temperature rise, as specified, liquid-in-glass thermometers are used when the coils are accessible. Thermocouples are used if the parts are inaccessible to liquid-in-glass thermometers.

If the duty cycle is not continuous, the machine may be operated at overload to the extent that the temperature rise will not exceed the specified temperature. For a duty cycle of one-half hour—which means the machine is operated for one-half hour and idle for one-half hour—if the machine has a 60°C rise, it will have the same life expectancy as it will have with a 40°C rise for continuous service.

A higher temperature rise for the commutator is allowed. The allowable hot-spot temperature is 65°C as compared to 55°C for the armature winding for a specified 40°C rise. Mica is used for the insulation in the commutator and therefore a higher allowable temperature is permitted.

1-8 EFFICIENCY[3]

The efficiency of a machine can be calculated by the following equation:

$$\text{percent efficiency} = \frac{\text{output}}{\text{input}} \times 100 \qquad \textbf{(1-6)}$$

$$= \frac{\text{output}}{\text{output} + \text{losses}} \times 100 \qquad \textbf{(1-6a)}$$

$$= \frac{\text{input} - \text{losses}}{\text{input}} \times 100 \qquad \textbf{(1-6b)}$$

$$= \left(1 - \frac{\text{losses}}{\text{output} + \text{losses}}\right) \times 100. \qquad \textbf{(1-6c)}$$

In the general equations, output, input, and losses are energy units, such as watthours or kilowatthours. Because energy is the product of power and time and if average energy output is divided by average input energy, time will cancel out of the expression. For this reason power is used in the equations. If a machine operates at various loads over a given period of time, such as one day (24 hr), the energy relationship must be used. Such efficiency is called *all-day efficiency*.

Efficiencies may be determined by measuring input and output power values. For a motor, the electrical input can be measured with meters and the output power can be measured by mechanical means, such as a *prony brake* or *dynamometer*. Horsepower is calculated by the following formula from mechanics,

$$\text{horsepower} = \frac{2\pi TS}{33{,}000} = \frac{TS}{5250} = 0.00019TS, \qquad \textbf{(1-7)}$$

where T is the torque in pound-feet and S is the speed in rpm.

To use the efficiency equations 1 hp is equivalent to 746 W. The method of measuring input and output for determining efficiency is impractical for large machines because of the amount of energy required and the cost of the energy. For large machines the losses are determined from tests in which only the losses need to be supplied and the efficiency is calculated with the formulas of Equations (1-6a), (1-6b), or (1-6c).

The losses are of two types:

(1) There are the rotational losses or stray power losses that are related to the rotation of the machine. These are the iron losses made up of the hysteresis loss and the eddy current loss, and the mechanical losses made up of bearing friction, brush friction, and windage (wind friction) losses. These losses are determined by a no-load test.

(2) There are the electrical losses or copper losses that are due to current flow through the various electric circuits of the machine. These losses are determined by resistance measurements using I^2R to calculate the losses in the various circuits.

In addition to the above, stray load loss which is caused by flux distortion and so forth is taken as one percent of the output for machines larger than 150 kW or 200 hp. In smaller machines this loss is omitted. Methods for finding efficiencies are prescribed by AIEE (now IEEE) standards.[2]

[2] See Section 1-9.

Example 1-7. A 10-kW generator has a stray power loss of 550 W. The electrical losses are 1400 W. What is the efficiency?

Solution:

$$\text{percent efficiency} = \frac{\text{output}}{\text{output} + \text{losses}} \times 100$$

$$= \frac{(10)(1000)}{(10)(1000) + (500) + (1400)} \times 100$$

$$= \frac{10{,}000}{11{,}900} = 83.8.$$

Example 1-8. The input to a three-phase induction motor at 440 V (line to line) is 63 A and operates at a power factor of 0.90 lagging. The output is measured with a dynamometer and the torque is 155 lb-ft at a speed of 1700 rpm. Determine the motor losses and the efficiency.

Solution:

$$\text{Horsepower output} = 0.00019TS = (0.00019)(155)(1700) = 50.2,$$

$$\text{watts output} = (50.2)(746) = 37{,}500,$$

$$\text{watts input} = \sqrt{3}\, E_L I_L \cos\theta = \sqrt{3}\,(440)(63)(0.90) = 43{,}900,$$

$$\text{losses} = \text{input} - \text{output} = 43{,}900 - 37{,}500 = 6400 \text{ W},$$

$$\text{efficiency} = \left(1 - \frac{6{,}400}{43{,}900}\right) \times 100 = 85.4 \text{ percent.}$$

Example 1-9. The input to a 220-V three-phase induction motor is 40 A at 88 percent power-factor lagging. The motor is coupled to a 250-V dc generator whose load current is 38 A. If the generator efficiency is 85 percent, calculate (a) the losses in the generator, (b) the losses of the motor, (c) the efficiency of the motor, (d) the efficiency of the motor-generator set.

Solution:

$$\text{Motor input} = \sqrt{3}\, E_L I_L \cos\theta = \sqrt{3}\,(220)(40)(0.88)$$

$$= 13{,}400 \text{ W},$$

$$\text{generator output} = EI = (250)(38) = 9500 \text{ W},$$

$$\text{generator input} = \text{motor output} = \frac{\text{generator output} \times 100}{\text{efficiency}_{\text{generator}}}$$

$$= \frac{9500 \times 100}{85} = 11{,}200 \text{ W}.$$

(a) $$\text{Generator losses} = \text{generator input} - \text{generator output}$$

$$= 11{,}200 - 9500 = 1700 \text{ W}.$$

(b) Motor losses = motor input − motor output
= 13,400 − 11,200 = 2200 W.

(c) Motor efficiency $= \frac{\text{output}}{\text{input}} \times 100 = \frac{11{,}200}{13{,}400} \times 100 = 83.6$ percent.

(d) Overall efficiency $= \frac{\text{generator output}}{\text{motor input}} \times 100 = \frac{9500}{13{,}400} \times 100$
= 71.0 percent.

(Note: Overall efficiency = (motor efficiency)(generator efficiency)
= 83.6 × 85 = 71.0 percent.)

1-9 STANDARDS AND CODES[7],[8]

Standards and codes have been established on a nationwide basis by manufacturers, distributors, consumers, engineering societies, and others. Motors and generators must meet the requirements of these standards.

(1) *NEMA standards* are the standards of the National Electrical Manufacturers Association. They define a product, process, or procedure with reference to nomenclature, composition, construction, dimensions, tolerances, operating characteristics, performance, quality, rating, and testing. They specifically cover such matters as frame sizes, torque classifications, and basis of rating. The standard for motors and generators is MG1-1967. At intervals of about six to seven months, revisions and changes to specific sections are printed so that the standard is kept up to date.

The following are voltage ratings from the standard:

Dc motors (fractional hp)	115 and 230 V
(industrial to 200 hp)	120 and 240 V
(industrial over 200 hp)	250, 500, and 700 V
Ac motors (60 Hz single phase)	115 and 230 V
(60 Hz polyphase)	110, 208, 220, 440, 550, 2300, 4000, 4600, 6600, and 13,200 V
Dc generators	125, 250, 500 and 700 V
Ac generators (single phase)	125 and 250 V
(three phase)	125, 230, 460, 575, (2400; 2500) (2400/4160Y; 2500/4330Y), (4800; 5000), 6900, 11,500, 12,500, 13,800, and 14,400 V

(2) *AIEE standards* (*IEEE*) are standards of the American Institute of Electrical Engineers, now the Institute of Electrical and Electronics Engineers. These standards concern fundamentals such as basic standards of temperature rise, rating methods, classification of insulating materials, and test codes.

(3) *ASA* (*ANSI*) *standards* are national standards established by the American Standards Association, now the American National Standards Institute, which represents manufacturers, distributors, consumers, and others concerned. AINSI (ASA) standards may be sponsored by any responsible body and may become national standards only if a consensus of those having substantial interests are reached. Standards may cover a wide variety of subjects such as methods of test, performance, definition of terms, specifications of materials, and so forth. The standards frequently are those previously adopted by NEMA and AIEE (IEEE). The chief motor and generator standard of ASA is *Rotating Machinery*, C-50.

(4) *The National Electrical Code* (*NEC*) is an AINSI (ASA) standard sponsored by the National Fire Protection Association for the purpose of safeguarding persons or buildings from electrical hazards arising from the use of electricity. It covers wiring methods and materials, protection of circuits, motors and generators, grounding, and so on.

(5) *Underwriters Laboratories, Inc.* is an independent testing organization that examines and tests devices, systems, and materials with particular reference to life, fire, and casualty hazards. It develops standards for motors and control for hazardous locations through cooperation with manufacturers. In the case of explosion-proof and dust-ignition-proof motors the Underwriters label indicates to code-enforcing authorities that motors have been inspected and adhere to Underwriters Laboratories standards for motors for hazardous locations.

(6) *Federal Specification* CC-M-641 for integral horsepower ac motors by the Federal Government covers standard motors for general government use.

(7) *IEC* (*International Electrotechnical Commission*) *standard* 72-1 Part 1; *German standard* DIN 42673; and *British standard* BSI-2960 Part 2 are standards of other countries. They are similar to NEMA standards and specify dimensions, classes of insulation, and in some cases horsepower rating.

1-10 SUMMARY

Rotating machines or motors and generators operate on the principle of electromechanical energy conversion through the medium of magnetic flux. The conversion from mechanical energy to electric energy in generators is based on the establishment of an induced electromotive force (EMF) in a conductor being made to move in a magnetic field, each perpendicular to the other. This is electromagnetic induction and by Faraday's discovery, $E_{ave} = \phi/t$ V (Equation (1-1)). For a conductor in terms of velocity in MKS units $e = BlV$ V (Equation (1-2)). In practical units $e = B'l'V' \times 10^{-8}/5$ V. The direction of the induced EMF can be determined by Fleming's right-hand rule.

When the energy conversion process is reversed, an electromagnetic force is experienced by the current-carrying conductor for motor action. The formula for this force in MKS units is $F = BlI$ N (Equation (1-4a)). In practical units $F = (BlI \times 10^{-7})/1.13$ lb. The direction of the force (motion) can be determined by the left-hand rule for motors.

The magnetic circuits of dynamos, a term for either a motor or generator in which an electromechanical energy conversion takes place, are essentially of two types. The first type, in which the armature rotates, is the type for dc dynamos; and, the second, in which the armature core is stationary, is the type used in most ac dynamos. The magnetic parts are iron, or magnetic sheet steel, or both.

The armature conductors are placed in slots in the armature core. These conductors are made into coils with straight sides and the ends of the coils are connected to form the armature winding. Half-coiled windings have one coil side per slot and whole-coiled windings have two coil sides per slot. The whole coiled is used except in some small machines where the half coil is used. In dc dynamos, the coil ends are connected to commutator segments to form either a lap or a wave winding. In lap windings the coil ends are connected to adjacent segments while in a wave winding the ends are connected to commutator segments approximately 360 deg apart. The slot pitch Y_s, the distance between the two slots in which the two coil sides are placed, is given by the formula, $Y_s = S/P - k$ (Equation (1-5)). For either type of winding this formula is used. Direct-current windings close and are called *closed windings*. Windings that do not close or open windings are used for ac machines. The ac winding is basically a lap winding with proper connection of the coil ends. Multiplex windings have two or more identical windings side by side.

All electric conductors must be insulated to prevent short circuits or grounds. Insulation materials include cotton, silk, Fiberglas, plastic materials, varnish, enamel. The classes of insulation as per the AIEE 1 standard are O, A, B, F, H, and C. Each has a maximum temperature above which it should not be operated so that the insulation will not deteriorate. Field and armature coils use class-A insulation in general-purpose machines.

Rating of electrical machinery is given for the output, terminal voltage, current, speed values, the type machine, code letter, and the temperature rise for continuous operation. The need for the nameplate is of prime importance if the machine is to be operated safely within limits and with minimum maintenance problems. The heat rise for general-purpose machines, using class-A insulation is 40°C above ambient temperature. This is for continuous operation and the temperature is measured at the surfaces of coils with liquid-in-glass thermometers.

Efficiency of a machine is expressed in percent of the useful output to the input; power values are used. Efficiency is closely related to temperature rise; for a machine that heats to a high value has a high value of loss and consequently a low efficiency. Efficiency of a motor or generator can be determined by measurement of the input and the output values. This is only done for smaller machines. The usual method is to find the losses by test and evaluate the efficiency from these losses by calculation. In such tests the losses need only be supplied. The losses are stray power or rotational losses; the electrical losses; and a stray load loss of 1 percent of the output for machines over 150 kW or 200 hp.

Standards and codes have been established nationally for the manufacture, test, application, and safety in use of motors and generators. NEMA Standard MG1-1967 is the code developed by manufacturers and used by them for ratings, nomenclature, torque classification, dimensions, and so forth. AIEE (now IEEE) standards of the Institute of Electrical and Electronic Engineers concern fundamentals such as temperature rise, classification of insulation, and test codes. ANSI (ASA) standards of the American National Standards Institute (American Standards Association) concern all such standards as NEMA and IEEE and the association acts as a control organization for standards. C-50 *Rotating Machinery* is the chief motor and generator standard. The *National Electrical Code* is concerned with proper installations and protection, and safeguarding persons and buildings. Underwriters Laboratories is an independent testing organization concerned with life, fire, and casualty hazards; its label is evidence to enforcers of codes that the product has been tested and meets the requirements for machines in hazardous locations. Federal Specification CC-M-641 is a specification for ac motors for government use. Some foreign standards are listed such as the German standard DIN 42673 which is similar to the NEMA standards.

Problems

1. Calculate the average voltage induced in a single conductor if 2.25×10^{-2} Wb are cut in 1/75 sec.
2. Calculate the average induced voltage in a conductor if 1.75×10^{6} lines are cut in 1/60 sec.
3. What is the value of flux cut if 1.5 V are induced in a conductor in 1/75 sec? (a) In webers? (b) In maxwells?
4. Derive Equation (1-2) from Equation (1-1).
5. A conductor 25 cm long is moving perpendicular to a magnetic field whose flux density is 1.25 Wb/m^2 at a velocity of 300 m/min. What is the value of the induced voltage in the conductor?
6. A certain conductor is moving at a uniform velocity in a uniform magnetic field with an induced EMF in the conductor of 5 V. (a) What is the induced EMF if the velocity is reduced 15 percent? (b) What is the induced EMF if the flux is increased 20 percent? (c) What is the induced EMF if the velocity is reduced 20 percent and the flux is increased 15 percent?
7. A conductor 15 in. long is moving at a 45-deg angle with the field and the conductor with a velocity of 1000 ft/min. If the flux density is 60,000 lines/in.^{2}, what is the voltage induced in the conductor?
8. In Figures 1-6(a) and 1-6(b) if the coil is 20 in. long and the width is 12 in. and if the uniform field has a flux density of 50,000 lines/in.^{2}, the angular velocity is 1200 rpm. (a) What is the instantaneous voltage when the coil is in position (1)? (b) In position (3)? (c) In position (4)? (d) In position (7)? (e) In position (8)?
9. Derive Equation (1-3) from preceding formulas.
10. An ideal dc eight-pole generator has an armature winding with eight current paths and 1068 armature conductors. The flux per pole is 1.13×10^{6} Mx or lines and the speed is 1150 rpm. What is the induced voltage?
11. A six-pole generator with 2.35×10^{6} lines of flux per pole and 12 current

paths and 880 armature conductors generates an EMF of 155 V. At what speed must it operate?

12. In Problem 11 if the flux is increased to 5.00×10^6 lines of flux per pole, the current paths are increased to 24 and the armature conductors are reduced to 720. What must be the speed to produce the same induced EMF?
13. A single conductor is 18 in. long and placed in a uniform field of 75,000 lines/in.^{2} and the current is 25 A. (a) What is the force on the conductor if it is perpendicular to the field? (b) What is the force on the conductor if its angle with the field is 30 deg?
14. What is the force on a conductor 45 cm long perpendicular to a field of flux density of 10,000 G if the current is 20 A? (a) If the force is in dynes? (b) If the force is in newtons?
15. A motor has 256 conductors on the armature with 78 percent under magnetic field influence. The flux density is 60,000 lines/in.^{2}, the current per conductor is 10 A, the active length of each conductor is 15 in., and the armature diameter is 10 in. (a) What is the total force on the conductors? (b) What is the torque? (c) What is the horsepower developed if the speed is 1050 rpm?
16. In Problem 15 if the flux density is increased 15 percent, the current per conductor is increased 12 percent and the speed decreased 4 percent. (a) What is the value of torque? (b) What is the horsepower?
17. An armature has 64 slots with eight conductors per slot. The poles cover 75 percent of the armature circumference and the flux density is 65,000 lines/in.^{2}; the armature length is 10 in. and the diameter of the core is 18 in.; there are eight poles and eight current paths in the armature. The developed torque is 750 lb-ft. What is the armature current?
18. In Problem 17 if the torque increases 12 percent, the flux density decreases 3 percent. What is the value of armature current?
19. A 5-kW, 120-V, dc shunt generator was operated at rated conditions for 3 hr. The thermometer on the field coil loosened and had to be removed before termination of the test. The voltage and current readings of the shunt field were taken during the test. The interpole temperature rise was 36°C and the armature coils temperature rise was 38°C. The ambient temperature was 30°C. The field resistance was 68.2 ohms at start and 82.0 ohms at the finish. All readings had leveled off. Did this generator meet the 40°C temperature rise indicated on the nameplate?
20. A 25-kW, dc generator rated at 240 V and 104 A was tested with the following data observed: Rotational losses at rated conditions were 1050

W and the electrical losses were 3150 W. What is the efficiency at rated load conditions?

21. A 5-hp, 115-V motor was tested with a dynamometer with these results: terminal voltage was 115 V; input current was 39.6 A; output force was 35.5 lb; arm was $10\frac{1}{2}$ in.; speed was 850 rpm. What is the horsepower output and what is the efficiency?
22. The input to a three-phase, 220-V, 25-hp squirrel-cage induction motor as measured with the two-wattmeter method was 22,100 W with a current of 64 A at rated voltage. The output was measured with a dynamometer with a torque of 76 lb-ft at a speed of 1730 rpm. What are the motor losses and the efficiency?
23. A motor is coupled to a generator whose efficiency is 83 percent with an output of 2 kW at 120 V. The motor has an efficiency of 84 percent. (a) What is the input to the generator? (b) What is the input to the motor? (c) What is the overall efficiency?
24. A motor-generator set has an overall efficiency of 72.5 percent. The efficiencies of the motor and the generator are equal. The input to the motor is 13.8 kW. (a) What is the efficiency of the motor or the generator? (b) What are the losses of each machine?
25. If the motor in Problem 24 is a three-phase squirrel-cage induction motor connected to a 220-V ac supply, what is (a) the input current at 85 percent lagging power factor? (b) The horsepower output of the motor? (c) If the generator is a 240-V dc generator, what is its output? (d) What is the generator current?

References

[1] Carr, C. C., *Electric Machinery*. New York: Wiley, 1958.

[2] Kosow, I. L., *Electric Machinery and Control*. Englewood Cliffs, N.J.: Prentice-Hall, 1964.

[3] Siskind, C. S., *Electrical Machines—Direct and Alternating Current*, 2nd ed. New York: McGraw-Hill, 1959.

[4] Fitzgerald, A. E. and D. E. Higginbotham, *Basic Electrical Engineering*, 2nd ed. New York: McGraw-Hill, 1957.

[5] Robertson, B. L. and L. J. Black, *Electric Circuits and Machines*, 2nd ed. Princeton, N.J.: Van Nostrand, 1957.

[6] *General Principles upon Which Temperature Limits Are Based in the Rating of Electric Equipment*, IEEE No. 1. New York: Institute of Electrical and Electronics Engineers, December, 1962.

[7] *Motor and Generator Standards*, MG1-1967, New York: National Electrical Manufacturers Association.
[8] Bulletin HBK-304, "Induction Motors—Frames 143–149, Selection and Application." Schenectady, N.Y.: General Electric Co., July 6, 1964.

The Construction and Theory of Direct-Current Machines

2

The construction of dc generators and dc motors is almost identical. Such small differences as do exist would occur, for example, in compound machines in which the series field connections must be reversed if a generator is to be used as a motor and in generators in which pole cores should have more flux retentivity. It was pointed out in the previous chapter that the armature conductors are the part of the dynamo in which the function is accomplished. In a generator a *voltage* EMF is induced in the armature conductors, whereas in a motor a *torque* (force × distance) is produced in the armature conductors. At the same time *counter torque* is acting on the armature conductors of a generator and a *counter voltage* is produced in the armature conductors of a motor. The theory of commutation, armature reaction, interpoles, and efficiency for generators and motors will be considered in this chapter.

2-1 THE MAGNETIC CIRCUIT OF THE DIRECT-CURRENT MACHINE[1],[2]

In Section 1-4 the magnetic circuits of rotating machines were discussed. Figures 1-10 and 1-14 show the magnetic circuit for a four-pole dynamo. Figure 2-1 shows the magnetic circuit and the electrical connections for a two-pole dynamo with the interpoles. Considering the magnetic circuit, starting at the north pole the flux passes across the air gap and slots to the armature core, through the armature core, through the slots and air gap to the south pole, through the south pole, through the upper and lower portions of the yoke; the two paths then combine and pass through the north pole to the point of start. There are two parallel paths through the upper and lower parts of the yoke.

In dc machines (dynamos) the poles are always stationary and fastened to the yoke. The armature core, also made of laminations that are fastened on a keyed shaft, is always the rotor. The yoke is either a cast steel cylinder

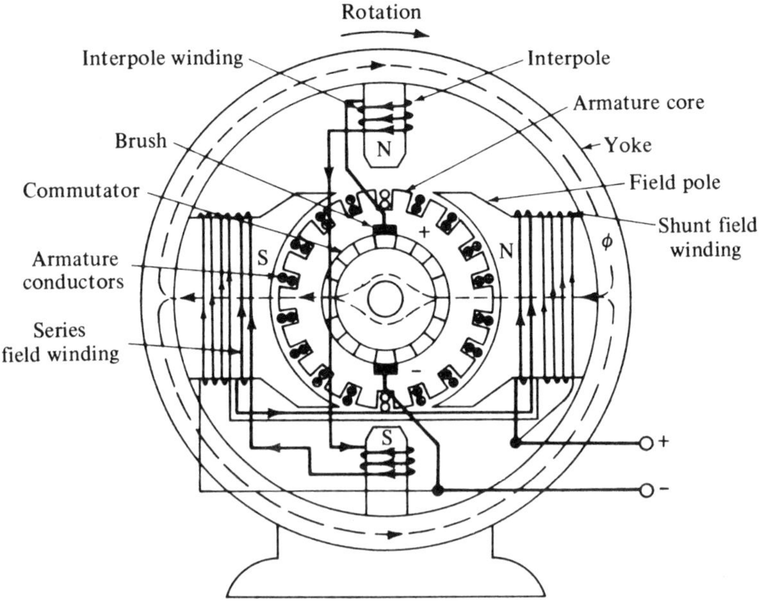

FIGURE 2-1. Two-pole dynamo, polarities shown for generator.

or a fabricated steel-plate cylinder. The poles are always of the salient-pole type. *Salient pole* means the type about which field coil winding is placed on the structure of the pole core. The structure is a projection with pole shoes to hold the field coil. These projections covering the armature circumference are the north and south poles. The ampere turns *NI* of the field coils provide the magnetomotive force to set up the flux in the magnetic circuit. In other words the salient poles are *electromagnets*.

The distribution of flux, or the flux density, in the air gap is shown in Figure 2-2. The distribution is practically a square wave, which is the desired flux distribution for dc machines. As each armature coil rotates clockwise, or in this figure moves from left to right, at a uniform speed, the induced voltage of each coil will trace an EMF of similar pattern.

2-2 THE ELECTRIC CIRCUIT OF A DIRECT-CURRENT MACHINE AND FIELD COIL CONNECTIONS[1],[3]

The type of machine depends on the way the field coils are connected with respect to the armature winding. A *separately excited* dynamo is one in which the field coils are not connected to the armature circuit. The shunt field is

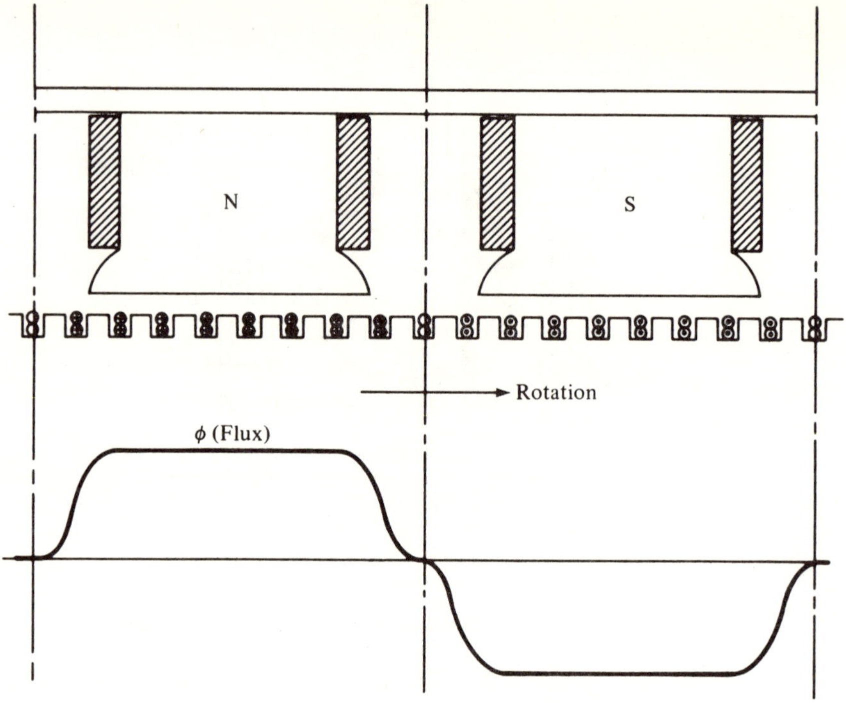

FIGURE 2-2. Air gap flux distribution for dynamo of Figure 2-1.

connected to a separate source. This connection is more common for generators than for motors. Some speed-control schemes (*Ward Leonard*) use the separate excitation for controlling the speed of a motor. Figure 2-3(a) shows the connection for a separately excited dynamo.

The *shunt* connection is shown by Figure 2-3(b). The shunt field is connected in parallel with the field winding. In Figure 2-3, a voltage source for a motor is connected to the terminals and a load for a generator is connected to the terminals.

The *series* connection is shown in Figure 2-3(c). The series field carries the full current and consequently it has relatively few turns of wire large enough in cross section to carry the full load current. This connection is more common for motors than for generators.

Figures 2-3(d) and 2-3(e) are *compound* dynamos. This type has both a shunt field and a series field. In Figure 2-3(d) the shunt field is in parallel with the armature and the series field is in series with the terminal and

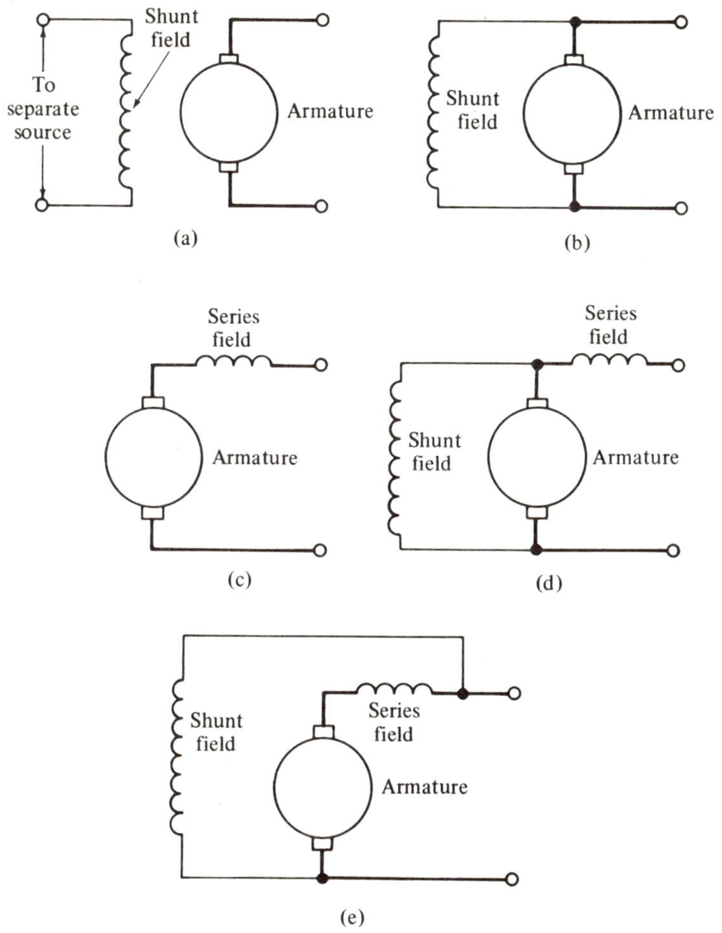

FIGURE 2-3. **(a) Electrical connections for separately excited shunt dynamo. (b) Electrical connections for self-excited shunt dynamo. (c) Electrical connections for series dynamo. (d) Electrical connections for compound, short-shunt dynamo. (e) Electrical connections for compound, long-shunt dynamo.**

carries the full load current. This is the *short-shunt* connection. In Figure 2-3(e) the series field is in series with the armature and the shunt field is in parallel with the terminals. This is the *long-shunt* connection. In this connection the series field current is the same as the armature current. The electrical connections of Figure 2-1 are for a long-shunt cumulative compound generator. If the magnetic motive force (MMF) of the ampere twins (NI) of the series field adds to the MMF of the shunt field winding, the compounding is *cumulative*. If the series field MMF subtracts from the shunt field winding MMF, the compounding is *differential.*

In ordinary practice a rheostat is connected in series with the shunt field. In a generator this rheostat controls the value of terminal voltage and in a motor it is used as a means to control speed.

The interpole or commutating pole winding and the compensating winding, if used, are not shown in the diagrams of Figure 2-2. Figure 2-4(a) shows the

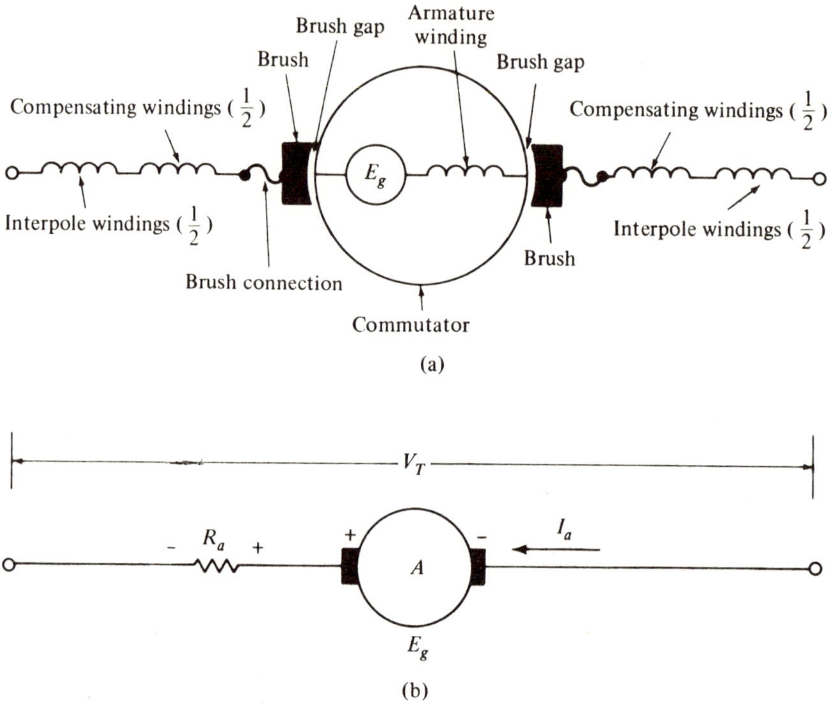

FIGURE 2-4. (a) Armature circuit of a dc dynamo. (b) Equivalent armature circuit of (a).

armature circuit of a dynamo. The interpole winding (in two parts), the compensating winding (in two parts), the brush connections or "pigtails," the brushes, and brush gaps are shown. An induced voltage E_g and the armature winding are shown inside the commutator circle. All the copper windings have a positive temperature coefficient and therefore their resistances are lumped as R_a in Figure 2-4(b). The brushes, usually carbon, and the brush gaps together have a peculiar resistance characteristic and a value of 2 to 5 V is used as voltage drop for the brush and brush gap. The induced voltage is shown as E_g. Figure 2-4(b) illustrates the armature circuit as shown on diagrams with R_a being the resistance of the windings in the circuit. In the equivalent circuit for a shunt generator,

$$V_T = E_g - V_{BD} - I_a R_a \text{ V} \tag{2-1}$$

or

$$E_g = V_T + V_{BD} + I_a R_a \text{ V}; \tag{2-1a}$$

for a motor,

$$V_T = E_g + V_{BD} + I_a R_a \text{ V} \tag{2-2}$$

or

$$E_g = V_T - V_{BD} - I_a R_a \text{ V}. \tag{2-2a}$$

Example 2-1. A shunt generator has an armature current of 78 A. The armature winding resistance is 0.065 ohm. The commutating pole winding has a resistance of 0.035 ohm. There are no compensating windings and the brush voltage drop is 3 V. If the generated voltage is 130.8 V, what is the terminal voltage (V_T)?

Solution:

$$\begin{aligned} V_T &= E_g - V_{BD} - I_a R_a \\ &= 130.8 - 3.0 - 78(0.065 + 0.035) \\ &= 130.8 - 3.0 - 7.8 = 120 \text{ V}. \end{aligned}$$

Example 2-2. A 20-hp shunt motor is rated at 72 A and 240 V. The shunt field takes 2 A. The armature winding resistance is 0.120 ohm and the interpole winding resistance is 0.03 ohm. The brush voltage drop is 4 V. What is the generated voltage (no compensating winding)?

Solution:

$$\begin{aligned} I_a &= I_L - I_f = 72 - 2 = 70 \text{ A}, \\ E_g &= V_T - V_{BD} - I_a R_a, \\ E_g &= 240 - 4 - 70(0.120 + 0.030) \\ &= 240 - 4 - 10.5 = 224.5 \text{ V}. \end{aligned}$$

2-3 VOLTAGE GENERATION IN A LAP WINDING[1],[3]

In Figure 1-6 (Section 1-2) the voltage generated in a coil that is placed in a magnetic field and rotated is shown. The trace of the generated voltage is a sine curve. If the coil were placed in slots of magnetic material as the coils are shown in Figure 2-1, the voltage generated would be similar to the flux density as shown in Figure 2-2. It would be flat or a square-wave curve. This is the ideal wave for dc machines.

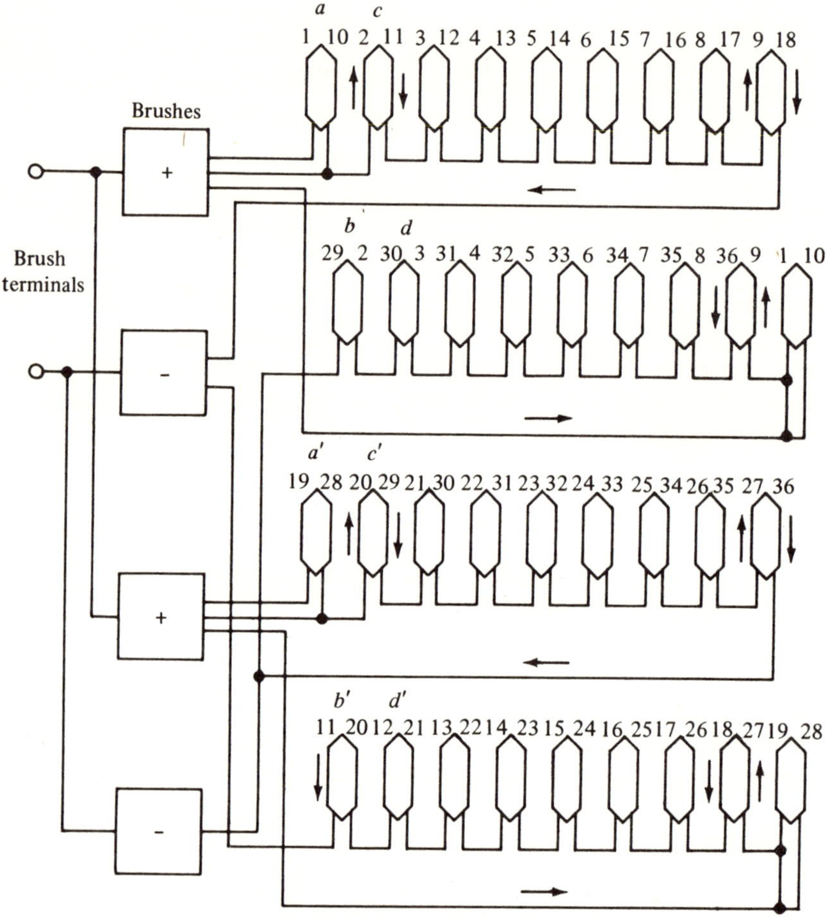

FIGURE 2-5. Schematic diagram of simplex lap winding for armature winding of Figure 1-15.

A simplex lap winding for a 36-slot, 36-coil, four-pole dynamo is shown in Figure 1-15. Figure 2-5 shows a schematic of this winding and shows the four current paths in the 36 coils. In this diagram, the current directions are those shown for a generator. A rule for the number of parallel current paths a is

$$a = mP \tag{2-3}$$

where m is the multiplicity of the winding, one for simplex, two for duplex, and so on, and P is the number of poles. Note also in Figure 2-5 and Figure 1-15 that four sets of brushes are needed. In lap windings as many sets of brushes are needed as there are poles. This is for simplex or multiplex lap windings. In multiplex windings the brushes would have to cover the bars of the parallel paths. Lap windings are used where large currents and low terminal voltages are needed.

In Section 1-2 the formula for E_g (Equation (1-3)) gave the induced voltage of a generator as

$$E_g = \frac{\phi PZS}{60a} \times 10^{-8}\ \text{V}.$$

This equation does not take into consideration the fact that not all the armature conductors are under a pole and not cutting flux. Therefore the number of conductors Z must be multiplied by a factor that represents the ratio of the number of active conductors under the poles. The following example will illustrate this.

Example 2-3. If the flux per pole of a four-pole generator is 1.93×10^6 lines and the armature is duplex lap wound with 36 coils and slots. Each coil has eight turns. The speed is 1800 rpm and 81 percent of the conductors are under poles. What is the generated voltage of the generator?

Solution:

$$Z = \text{(coils)(turns)(2 sides)} = 36 \times 8 \times 2 = 576,$$

$$E_g = \frac{\phi P(Zx\%)S}{60a} \times 10^{-8} = \frac{(1.93 \times 10^6)(4)(576 \times 0.81)(1800)}{60 \times 8} \times 10^{-8}$$

$$= 135\ \text{V}.$$

Example 2-4. (a) Design a simplex lap winding for a six-pole generator whose armature is to have 47 slots and 47 coils. Each coil has six turns.

(b) What is the coil pitch in electrical degrees? (c) If the armature rotates at a uniform speed of 1200 rpm and the flux per pole is 1.5×10^6 lines, what is the generated voltage if 80 percent of the conductors are active?

Solution:

$$\text{(a)}\ Y_s = \frac{S}{P} - k,$$

$$Y_s = \frac{47}{6} - \frac{5}{6} = 7.$$

The winding will go as follows:

Slot, top side:	1	2	3	40	41	42	43	44	45	46	47		
Commutator segment: start	1	2	3		41	42	43	44	45	46	47	1 close.	
Slot, bottom:	8	9	10	47	1	2	3	4	5	6	7		

$$\text{(b) Coil pitch} = \left(\frac{360 \times P/2}{S}\right)(Y_s)$$

$$= \left(\frac{360 \times 6/2}{47}\right)(7) = 160.8 \text{ electrical deg.}$$

$$\text{(c)}\ Z = \text{(coils)(turns/coil)(2 sides)} = (47)(6)(2) = 564,$$

$$E_g = \frac{\phi P(Z\%)S}{60a} \times 10^{-8} = \frac{(1.5 \times 10^6)(6)(564 \times 0.80)(1200)}{60 \times 6} \times 10^{-8}$$

$$= 135.4 \text{ V}.$$

In lap-wound machines of four or more poles unless the poles and air gaps are exactly the same, the voltages of the parallel paths may not be the same. Also, as the bearings wear the air gaps of the upper poles increase while those of the bottom poles become smaller. This will cause the generated voltages of the paths to be unequal with the result that heavy circulating currents flow in the armature windings and brushes which may cause heating of the windings and severe sparking at the brushes. This is more serious in larger machines. To eliminate the circulating currents through the brushes, *equalizer connections* are used. These are low-resistance connections made on the back of the armature coils between coils 360 electrical deg apart; these points should be at the same potential. If there is unbalance, the circulating current will flow through the equalizer connections. In Figure 2-5 this would mean connections between coils a and b, a' and b', c and d, c' and d', and so on.

2-4 SERIES OR WAVE WINDING[1],[3]

In Section 1-5 and Figure 1-13(b), a wave-winding coil is shown. In Figure 1-16 a wave winding is shown for a 33-slot, 33-commutator segment, 33-coil, four-pole winding. With this type winding for a four-pole dynamo, the winding progresses from a north pole to south pole, to next north pole to next south pole, and to the adjacent slot of the original north pole. The winding finally closes after all coils are included in a simplex winding. A duplex winding would have two windings in parallel with each set in alternate slots. A triplex winding would have three sets in parallel in every third slot. The calculations for the simplex winding of Figure 1-16 are (Equation (1-5))

$$Y_s = \frac{S}{P} - k \qquad \text{or} \qquad Y_s = \frac{33}{6} - \frac{1}{4} = 8.$$

The commutator pitch Y_c is calculated by the formula,

$$Y_c = \frac{C - m}{P/2} \qquad (Y_c \text{ must be an integer}), \tag{2-4}$$

where

C is the number of commutator segments,
P is the number of poles,
m is the multiplicity of the winding (plex).

In Equation (2-4) it could also be plus (+) for the $C - m$ term, but then Equation (1-5) would be $Y_s = S/P + k$. In case the plus signs are used the winding will be *progressive*. This means that after circling the armature, the coils advance one slot. A *retrogressive* winding results from the negative signs and the coils drop back one slot every time the armature is circled.

$$Y_c = \frac{33 - 1}{4/2} = \frac{32}{2} = 16 \qquad \text{(an integer).}$$

The following is the winding scheme for this problem (Figure 1-16):

Slot, top side:		1		17		33		16		32	...	18	
Commutator segment: start	30		13		29		12		28	...	14		30 close.
Slot, bottom:		9		25		8		24		...	10	26	

Figure 2-6 shows the schematic for this wave winding. There are two current paths and two sets of brushes are all that are needed. Therefore, in a wave winding the number of current paths is equal to two times the multiplicity

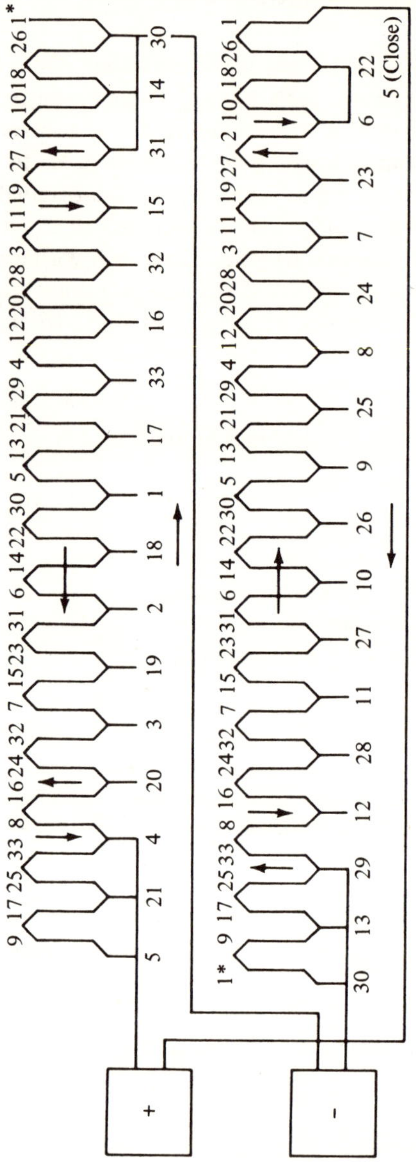

FIGURE 2-6. **Schematic diagram of simplex wave winding for armature winding of Figure 1-16.**

of the winding. A 6-pole triplex wave winding has $(2 \times 3) = 6$ paths. A 10-pole duplex wave winding would have 4 current paths. The wave-wound machine therefore has higher voltage and lower current rating than a lap-wound machine for the same number of conductors. One exception would be the 2-pole dynamo because all two-pole dynamos have wave windings.

In the series winding the number of commutator segments must be such that when multiplied by the multiplicity, it will be odd in addition to fulfilling Equation (2-4). It was for this reason that 33 slots were chosen for this simplex wave winding. An examination of the 36-slot, 4-pole lap example shows some of the problems in the calculation of wave windings. The slot pitch is 9, but Y_c comes out to $17\frac{1}{2}$. If 18 is used, the following sequence follows: segment 1-slot 8-slot 17-segment 19-slot 26-segment 1. Here the winding closes on one round. A scheme that could be used would be to employ 35 commutator segments and tape the ends of one coil. Such a coil is called a *dummy* coil. In this case $Y_s = 9$ and $Y_c = (35 - 1)/2 = 17$. The scheme is as follows: (coil in slots 1 and 10 is the "dummy")

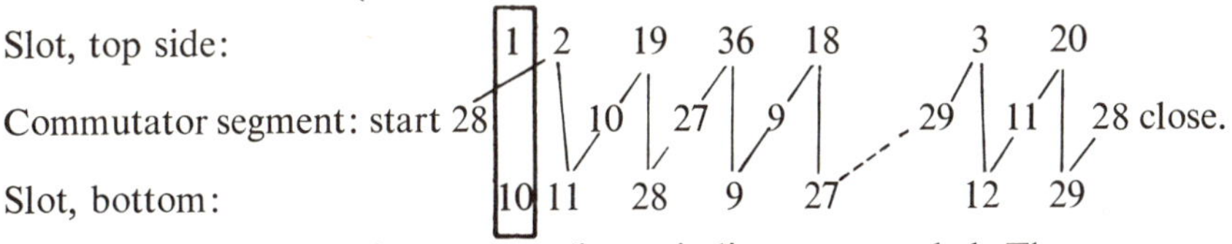

With the wave winding no equalizer windings are needed. The armature voltages add in going around the circumference of the armature and any irregularities are simply added equally for the two paths where one is clockwise and the other counterclockwise. Figure 2-6 shows the upper path as counterclockwise while the bottom path as clockwise. The generated voltage and coil pitch are similar to the lap winding in the previous section.

Example 2-5. Calculate the generated voltage of Example 2-3 if the winding is a simplex wave winding with one dummy coil.

Solution:

$$Z = (\text{coils})(\text{turns/coil})(2\text{ sides}) = 35 \times 8 \times 2 = 560,$$

$$E_g = \frac{\phi P(Z\%)S}{60a} \times 10^{-8} = \frac{(1.93 \times 10^6)(4)(560 \times 0.81)(1800)}{60 \times 2} \times 10^{-8}$$

$$= 525 \text{ V}.$$

Example 2-6. Repeat Example 2-4 for a simplex wave winding with 49 coils, 49 slots, and 49 commutator segments. The flux per pole, turns per coil, and speed are as in Example 2-4.

Solution:

$$\text{(a)}\ Y_s = \frac{S}{P} - k = \frac{49}{6} - \frac{1}{6} = 8$$

$$Y_c = \frac{C - m}{P/2} = \frac{49 - 1}{6/2} = 16$$

Slot, top side: 1 17 33 49 ------- 2 18 35

Commutator segment: start 45 12 28 44 11----- 46 13 29 45 close.

Slot, bottom: 9 35 41 8 --------- 10 26 43

$$\text{(b) Coil pitch} = \left(\frac{360 \times P/2}{S}\right)(Y_s) = \frac{360 \times 6 \times 8}{2 \times 49}$$

$$= 176.5 \text{ electrical deg.}$$

$$\text{(c)}\ \ Z = \text{(coils)(turns/coil)(2 sides)} = 49 \times 6 \times 2 = 588,$$

$$E_g = \frac{\phi P(Z\%)S}{60a} \times 10^{-8} = \frac{(1.5 \times 10^6)(6)(588 \times 0.8)(1200)}{60 \times 2} = 423 \text{ V.}$$

2-5 ARMATURE REACTION[2]–[4]

Armature reaction is the effect of the armature MMF on the flux distribution under the field poles. At no load for either a generator or a motor, the armature conductor current is small and therefore the effect of armature MMF on the field flux is slight. At load considerable current (I_a/a) will flow through the armature coils. The armature acts like a solenoid with an iron core (the armature core). This MMF sets up a magnetic field as shown in Figure 2-8(a). The flux of the field is as shown in Figures 2-7(a) and (b). The resultant flux is shown in Figures 2-9(a) and (b). In the later figure the neutral axis is shifted from its original position by an angle θ in the direction of generator rotation. (For a motor the angle θ is in the opposite direction of rotation.) (As we will see in the next section, the shift of the neutral axis has a marked effect on commutation). This means that in a generator the flux will be more dense under the trailing pole tip and that the neutral shifts by an angle θ in the direction of rotation. The greater the load, the greater will be the angle and the more dense is the flux under the trailing pole tip. For a motor the angle shifts toward the direction of rotation and the flux is more dense under the leading pole tip. In Figures 2-7, 2-8, and 2-9, the (b) figures show the flux distribution under the poles. In Figure 2-9(b) the area under

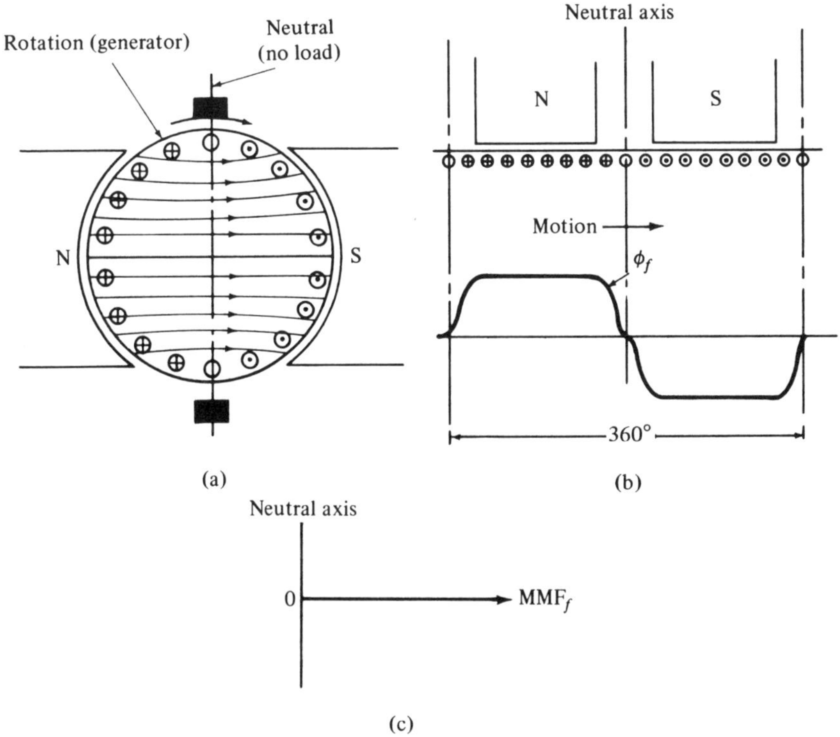

FIGURE 2-7. (a) Magnetic flux in dynamo at light load (main field). (b) Flux distribution under pole faces at light load. (c) Field MMF phasor.

the curve is less than the area under the curve for the field flux alone. This is due to saturation at higher flux densities and to the demagnetization effect of the armature reaction.

The (c) figures show the MMFs in phasor form. In Figure 2-7 the phasor is horizontal and to the right. In Figure 2-8 the phasor is vertical and in a downward position. In Figure 2-9 the phasors are added together. Because of the shift of the neutral axis, the MMF$_a$ of the armature has shifted by the angle θ. The component which is MMF$_a$ ($\sin\theta$) is opposite to MMF$_f$, and this is the demagnetizing component as represented by phasor $\overline{OB}$. The demagnetizing ampere turns NI_d are equal to

$$NI_d = \frac{2\theta Z}{360}\left(\frac{I_a}{a}\right) = \frac{\theta Z I_a}{180a}, \qquad \textbf{(2-5)}$$

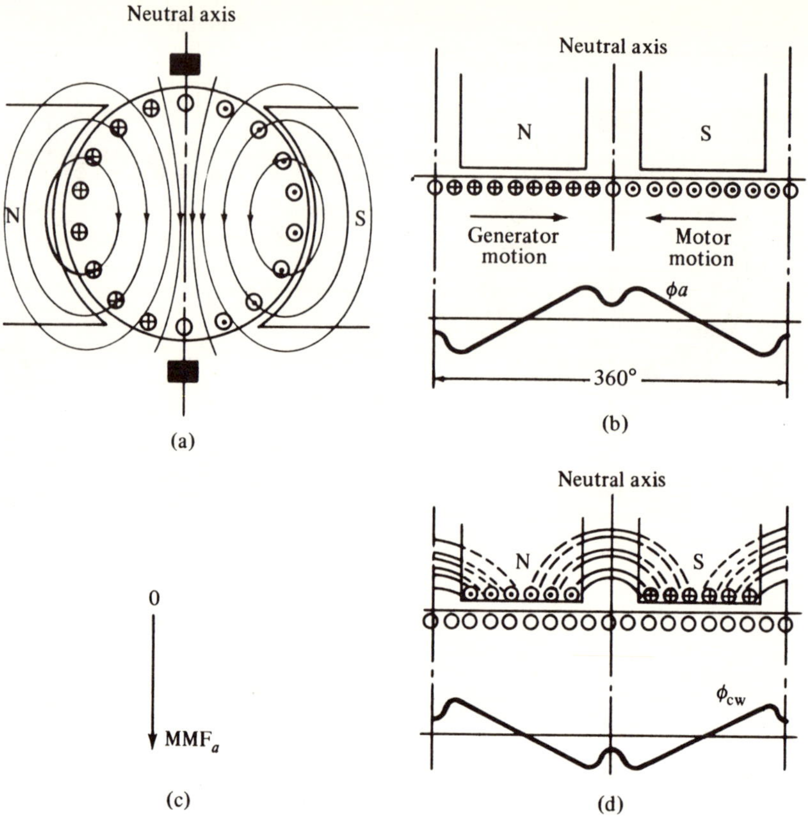

FIGURE 2-8. **(a) Armature flux in dc dynamo due to current flow in armature conductors. (b) Flux distribution under pole faces of the armature flux of (a). (c) Armature MMF phasor. (d) Compensating windings in pole faces, and flux distribution of compensating winding under poles.**

where

θ is the angle of the shift of the magnetic neutral,
Z is the number of armature conductors,
a is the number of current paths in the armature.

The component $\overline{OA}$ which is perpendicular to MMF_f is the *cross magnetization* MMF NI_c and is equal to

$$NI_c = \frac{(180 - 2\theta)ZI_a}{360a}, \tag{2-6}$$

where the constants have the same definitions as in Equation (2-5).

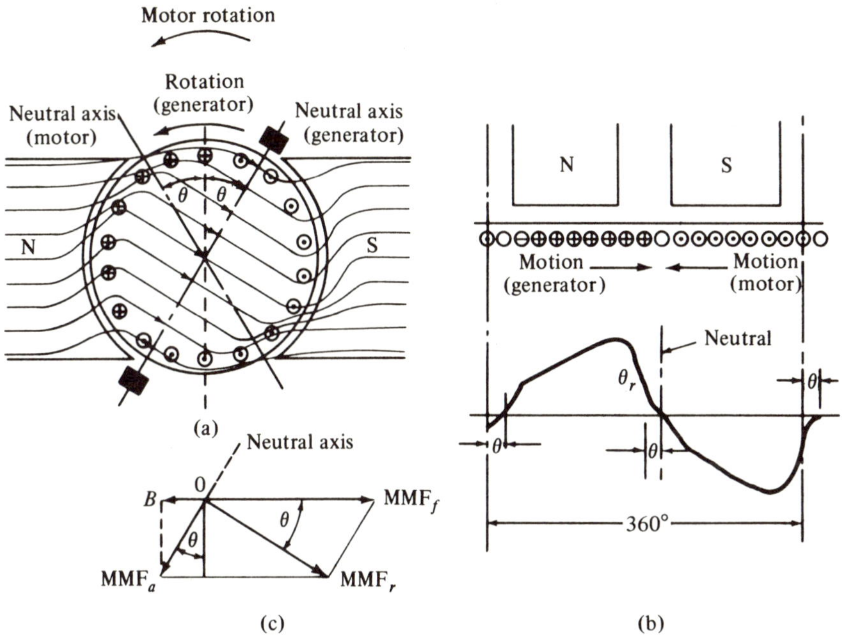

FIGURE 2-9. (a) Resultant magnetic flux in dc dynamo under load. (b) Resultant magnetic flux distribution under pole faces for flux of (a). (c) Phasor diagram of resultant MMF in dc dynamo.

Example 2-7. In Example 2-6 for a 6-pole generator with a simplex wave winding of 49 coils and 6 turns per coil, calculate the demagnetizing ampere turns NI_d of the armature if I_a is 100 A and the angle of neutral axis shift is 8 deg. Also calculate the cross-magnetizing ampere turns NI_c.

Solution:

$$NI_d = \frac{\theta Z I_a}{180a} = \frac{(8)(588)(100)}{(180)(2)} = 1310 \text{ A turns},$$

$$NI_c = \frac{(180 - 2\theta)ZI_a}{360a} = \frac{(180 - 2 \times 8)(588)(100)}{(360)(2)} = \frac{(164)(588)(100)}{720}$$

$$= 13{,}400 \text{ A turns}.$$

Compensating windings are windings that carry armature current as shown in Figure 2-8(d). These windings will set up an MMF that is practically equal but opposite to that due to the armature MMF. The flux distribution under the poles is shown in this figure.

2-6 COMMUTATION[1],[2],[4]

The commutation process is the changing of the current direction in the armature coils as the coils pass from the influence of a set of north-south poles to the influence of the following south-north poles. This happens in a very short time (0.001 sec or less) and if there were no self-inductance in the armature coils and no armature reaction, the process would be quite simple. The brushes are set on the neutral axis of the dynamo, the coil whose ends are connected to commutator segments that are shorted by the brush has zero induced voltage in it, and no current flows. In Figures 2-7(a) and (b) the coil sides of such a coil are shown. In Figures 2-10(a), (b), and (c) the sequence is shown for a lap-wound coil. In Figure 2-10(a) the coil is shown just before the coil is to undergo commutation. Current flows from commutator segment *A*, up the left side of coil, into commutator segment *B* and through the (+) brush to the external circuit. The current from this coil is joined from the preceding coil, which has just completed its commutation process.

In Figure 2-10(b) the coil has moved one-half commutator segment to the right. If this were the armature winding depicted in Figure 1-15, which has 36 segments, the coil would have moved through 5 space deg or 10 electrical deg. At 1800 rpm for the speed, this represents a time change of 0.000462 sec. The currents as shown are as follows: from the following coil to segment *A*, to (+) brush and from the preceding coil through segment *B* to (+) brush and to external circuit. The coil itself is short circuited by the (+) brush and segments *A* and *B*. If there were no self-inductance in the coil, no current would flow in the coil; but because the coil is in slots of magnetic material, there is self-inductance. In accordance with Lenz's law current will continue to flow as shown by the dotted current arrows, through the coil, segment *B*, brush and segment *A*. This current, especially at rated load, would cause sparking at the brush and heating in the coil and brush. Two things can be done to eliminate this current flow.

(a) The brushes can be set slightly ahead of the neutral axis, so that an opposing EMF is induced in the coil that would neutralize the short-circuit current due to the self-inductance. This would be quite simple except, as was mentioned in the preceding section, that the neutral axis shifts as the load changes. For a uniform load this method is all right but it is impracticable for a changing load.

(b) The second method would be with commutating poles in order that

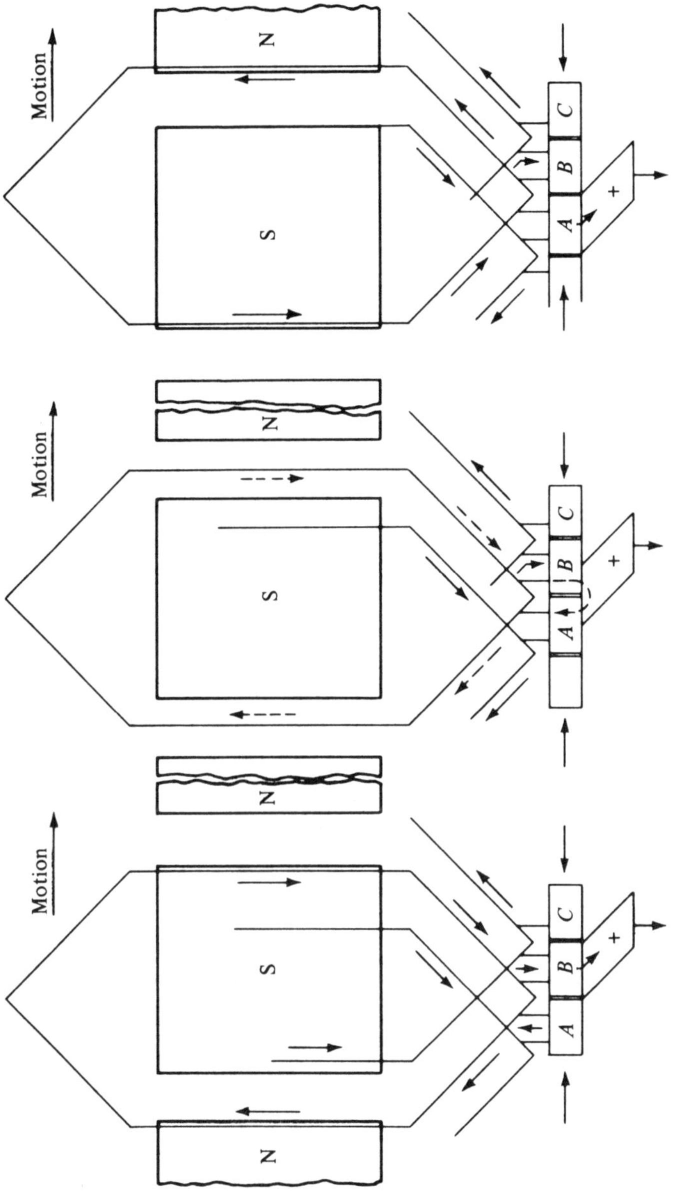

FIGURE 2-10. (a) Armature coil about to undergo commutation. (b) Armature coil undergoing commutation. Coil is short-circuited by (+) brush. Half-way point. (c) Armature coil just after commutation has been completed.

the proper flux be provided the opposing EMF in the coil. This is the subject of the next section and it is the method that is preferred to the first.

In Figure 2-10(c) the coil is shown as having moved another one-half segment or 5 space deg. This is 0.000462 sec later than that shown in (b). Now the current flow is as follows: from the following coil to segment *A* to (+) brush, and from the preceding coil to segment *B*, segment *B* to the coil under consideration, up the right side, down the left side of coil, into segment *A*, to (+) brush, and to external load. The left side of the coil is now under the south pole and the right side is under the next north pole. Commutation is now completed in this coil, the current has been reversed, and in this example the time was 0.000924 sec. Each coil does the same thing as it approaches the neutral axis. The same thing also happens at the negative brush with the current flow being in the opposite directions. Figure 2-11

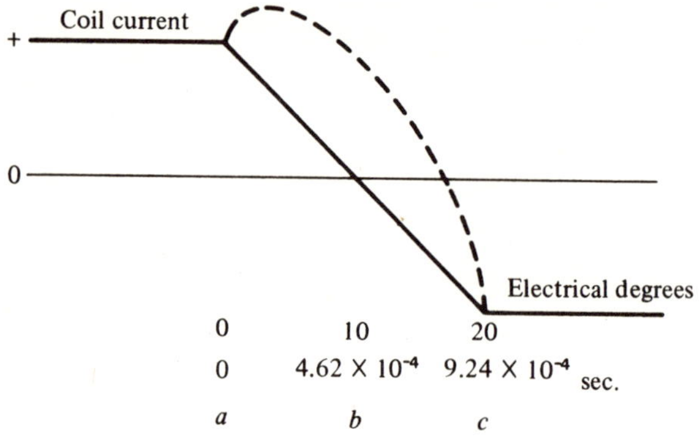

FIGURE 2-11. Ideal coil current during commutation for Figure 2-10.

shows the ideal graphical coil current flow during commutation. By proper design of commutating poles or brush setting the ideal is practically attained. The dotted line shows the coil current if no correction were made. The time and degrees on the figure are for the 36-coil, 4-pole dynamo at a speed of 1800 rpm.

2-7 COMMUTATING POLES[1],[2]

In the last section it was pointed out that due to self-inductance of the armature coils there would be heating and sparking of the commutator during commutation. It was also pointed out that the brushes should be placed on the neutral axis and that for a varying load the axis shifts. It was mentioned that to eliminate the short-circuit current the brushes should be advanced slightly beyond the neutral axis. All this is inconvenient and impracticable. *Commutating poles* (often called *interpoles*) solve these problems.

These pole cores are thin straight poles and, in some cases, are beveled; they cover about one slot pitch of the armature core. The excitation is heavy

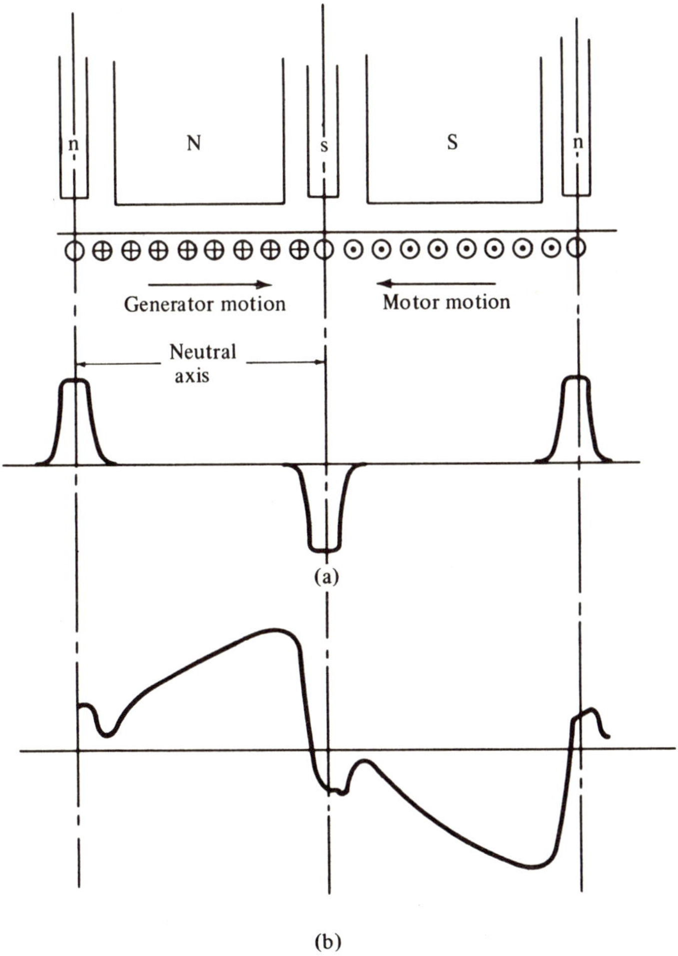

FIGURE 2-12. (a) Distribution of commutating pole flux. (b) Distribution of flux under pole faces with main flux, armature flux, and commutating pole flux.

wire of few turns. As shown in Figure 2-4(a) these windings are connected in the armature circuit. Since the armature current is practically the same as the line current, the MMF is proportional to the line current. The flux from the commutating poles is proportional to the load current if the flux density in the core is kept within the straight-line portion of the magnetization curve. The interpoles are placed between the main poles on the no-load neutral axis.

Figures 2-1 and 2-12 show the location of these poles. For a generator the sequence is main north, south commutator, main south, north commutator, and so on, in the direction of armature rotation. For a motor the sequence is N, n, S, s in the direction of rotation. Figure 2-12(a) shows the flux distribution of the commutating poles for a generator. Figure 2-9(b) shows the flux distribution of the main flux and the armature reaction effect. Figure 2-12(b) shows the effect of the commutating pole on the main flux. The no-load neutral axis is where the brushes are placed, and the coil undergoing commutation will have a small induced EMF induced in it that is opposite to the pre-commutation EMF. In addition with the proper adjustment it will overcome the lag to change to the new direction due to the inductance of the coil.

The commutating poles do not correct the wave form of the flux under the main poles. This is the function of the compensating windings. The commutating poles simply provide, at any load, the flux needed to provide for the quick current direction change in the coil undergoing commutation.

In smaller machines one commutating pole for every pair of main poles is often used. The MMF for such a pole would have to be double when two poles are used. Larger machines, above $7\frac{1}{2}$ hp, operate better if there are as many commutating poles as there are main poles.

2-8 TORQUE OF DIRECT-CURRENT MOTORS[1],[5]

Torque may be defined as a force that tends to produce rotation about a point times the radial distance from that point when the force is perpendicular to the radial distance. In a dc motor (or any motor) when current is flowing through the armature conductors, a force results on each conductor. The *force on a conductor* is given by the formula in Equation (1-4c),

$$F = \frac{BlI}{1.13} \times 10^{-7} \text{ lb.}$$

The terms were defined in Section 1-3. Figure 2-13 shows the forces acting on the armature conductors. The torque on one conductor is

$$T_Z = \frac{BlIr}{13.56} \times 10^{-7} \text{ lb-ft,} \tag{2-7}$$

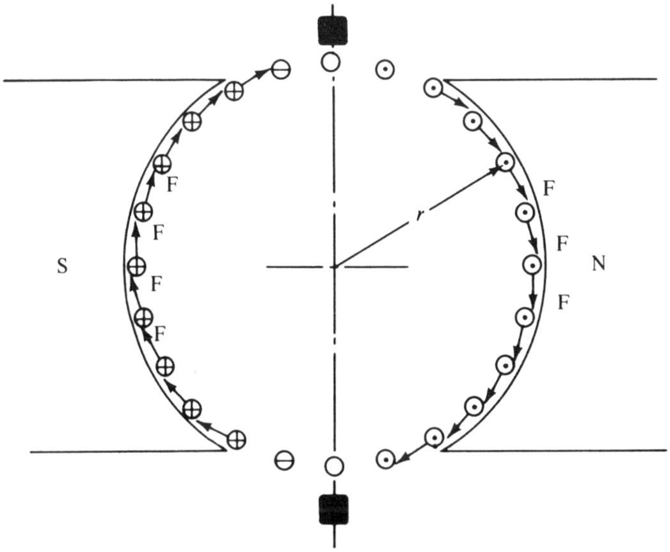

FIGURE 2-13. Torque developed by the armature conductors of dc motor.

where T_Z is the torque in pound-feet (pound-foot is the unit generally used rather than foot-pound, which is the unit for energy and work) and r is the distance from the axis of rotation (center of shaft) to the center of the conductor in inches. To find the total torque Equation (2-7) is multiplied by the number of active armature conductors. *Active conductors* are those under poles. In one revolution of the armature each conductor is acted on each time it comes under a pole; therefore it is acted on P (number of poles) times. The current in each armature conductor is I_a/a. The flux per pole is equal to BA or $B(2\pi rl/P)$ or B is equal to $\phi P/2\pi rl$. Substituting these values in Equation (2-7) times Z,

$$T = \frac{(\phi P/2\pi rl)(I_a/a)(l)(Z)}{13.56}(r) \times 10^{-7}.$$

Solving the above,

$$T = \frac{0.1173\phi P I_a Z}{a} \times 10^{-8} \text{ lb-ft} \tag{2-8}$$

where

ϕ is the flux in lines or maxwells,
P is the number of poles,
I_a is the armature current,

Z is the number of armature conductors,
a is the number of parallel current paths.

It should be noted that except for the constant, the formula for torque is similar to that for the generated voltage of a generator, Equation (1-3a),

$$E_g = \frac{\phi PZS}{60a} \times 10^{-8}.$$

For a given motor the values of P, Z, and a are constants and the formula can be written,

$$T = K_t \phi I_a. \qquad \textbf{(2-9)}$$

In this formula

$$K_t = \frac{0.1173PZ \times 10^{-8}}{a}.$$

Example 2-8. The armature of a dc motor has 550 conductors and a diameter (mean of conductors) of 20 in. and an axial length of 24 in. If 78 percent of the conductors lie under poles with a flux density of 60,000 lines/in.^{2} and each conductor carries 30 A, find (a) the total rotational force on the armature, and (b) the armature torque.

Solution:

(a) $F_t = \left(\frac{F}{\text{conductor}}\right)(Z_{\text{act}}) = \left(\frac{BlI}{1.13} \times 10^{-7}\right)(Z_{\text{act}}),$

$$F_t = \left[\frac{(60{,}000)(24)(30)}{1.13} \times 10^{-7}\right][(550)(0.78)] = 1640 \text{ lb.}$$

(b) $T = F_t r = (1640)\left(\frac{20}{2 \times 12}\right) = 1367$ lb-ft.

Formula (2-7) could have been used to find T,

$$T = \frac{BlIr}{13.56} \times 10^{-7}(Z_{\text{act}}) = \frac{(60{,}000)(24)(30)(20)}{13.56(2)} \times 10^{-7}(550 \times 0.78),$$

$$T = 1367 \text{ lb-ft.}$$

Formula (2-8) could also have been used as follows. Assume that the motor has four poles and is triplex wave wound (any assumption of poles and winding will work).

$$\frac{\phi}{\text{pole}} = BA = (60{,}000)\left(\frac{24 \times 20 \times 0.78}{4}\right) = 17.68 \times 10^6 \text{ l.}$$

$$a = 2 \times 3 = 6 \text{ paths}; \qquad I_a = 6 \times 30 = 180 \text{ A.}$$

$$T = \frac{0.1173\phi P I_a Z}{a} \times 10^{-8} = \frac{(0.1173)(17.68 \times 10^6)(4)(180)(550)}{6} \times 10^{-8},$$

$$T = 1367 \text{ lb-ft.}$$

Example 2-9. If the motor of Example 2-8 has its field flux reduced 15 percent and its armature current increased 30 percent, find the torque developed under the new conditions.

Solution:

Formula (2-9), $T = K_t \phi I_a$, will be used. In this motor the torque varies as the flux and as I_a. The new torque will be (100 − 15) or 85 percent of its former torque because of reduced flux. By the same reasoning the torque will be (100 + 30) or 130 percent of its former value due to increase of armature current. Therefore

$$T = \frac{85}{100} \times \frac{130}{100} \times 1367 = (0.85)(1.30)(1367) = 1510 \text{ lb-ft.}$$

Torque is independent of the speed of a motor or any mechanical machine in itself. The output of a motor is mechanical energy and the rate of the energy output is the power output in horsepower. The horsepower formula as given in Equation (1-7) is horsepower $= TS/5250 = 0.00019TS$. If the first form of this formula is solved for T, the following is the formula for torque T,

$$T = \frac{5250 \text{ hp}}{S}, \tag{2-10}$$

where T is in pound-feet and S is the speed in rpm. The power developed in the armature of any dynamo is equal to $E_g I_a$. In a motor E_g is the counter EMF and it is always less than the impressed supply voltage. Because 746 W is the mechanical equivalent of 1 hp, the horsepower developed by a motor is $(E_g I_a)/746$. Substituting this in Equation (2-10), the torque developed, T_d, becomes

$$T_d = \frac{5250}{S} \times \frac{E_g I_a}{746} = 7.04 \frac{E_g I_a}{S}. \tag{2-11}$$

If the value of E_g in Equation (2-3a) is substituted in Equation (2-11), the formula for torque in Equation (2-8) will result. Equations (2-7), (2-8), (2-9), and (2-11) all are for the torque developed by the armature of the motor. Equation (2-10) is a formula for the torque output. The difference between the torque output and torque developed is utilized to supply rotational losses of the motor. It was mentioned that torque was not dependent on speed. In Equations (2-10) and (2-11) torque is inversely proportional to speed. In both cases the value of torque is not in itself dependent on speed, but rather is the value when both the power and the speed are given. Torque may depend indirectly on speed just as values of E_g and I_a are dependent on speed. Speed is fully discussed in the next chapter.

Example 2-10. (a) What is the output torque of a 25-hp motor if it operates at 1200 rpm? (b) What is the torque output if the power output is increased 50 percent and the speed is decreased 10 percent?

$$\text{(a) } T = \frac{5250 \text{ hp}}{S} = \frac{5250 \times 25}{1200} = 109.5 \text{ lb-ft.}$$

$$\text{(b) } T = 109.5 \frac{150/100}{(100 - 10)/100} = 109.5 \frac{1.5}{0.9} = 182 \text{ lb-ft.}$$

2-9 EFFICIENCY OF DIRECT-CURRENT MACHINES[1],[3]

The efficiency of any machine is the ratio of the useful output to the input. It is normally taken as the ratio of power output to the power input expressed in percent. This may be done because the efficiency is taken for steady state conditions. The expressions for efficiency as given in Section 1-8 in Equation (1-6) are given here in terms of power,

$$\text{Percent efficiency} = \frac{P_o}{P_i} \times 100 \tag{2-12}$$

$$= \frac{P_o}{P_o + P_l} \times 100 \tag{2-12a}$$

$$= \frac{P_i - P_l}{P_i} \times 100 \tag{2-12b}$$

$$\text{Percent efficiency} = \left(1 - \frac{P_l}{P_o + P_l}\right) \times 100. \tag{2-12c}$$

Efficiencies can be determined by *direct measurements.* For generators the output is measured by electrical meters, while the input would be found by measuring the input to a calibrated motor. For motors the output would be determined with a *prony brake*, *calibrated generator*, or *dynamometer* (mechanical) while the input is measured with *electrical meters.* This method is good for small machines, but for large machines beyond 10 to 15 hp, the power demand becomes costly and the accuracy of the efficiency decreases. For the larger machines, the efficiency is found by tests for finding the losses and the use of Equation (2-12c).

The losses of dc dynamos are classified as (a) the rotational (or stray power) losses; (b) the electrical or copper losses; (c) the stray-load loss.

The rotational losses are those losses caused by rotation of the armature and can be subdivided into (a) friction and windage losses and (b) core (or iron) losses. The friction and windage losses consist of bearing friction, brush friction, wind friction (air or cooling gases) losses. All machines need ventilation and in small and medium machines the surrounding air is made to flow through them. Large machines, especially the large turbogenerators (ac) in power stations, are cooled by inert gas. The armature core losses are subdivided into hysteresis loss and eddy current loss. Hysteresis loss is the loss due to reversal of the MMF in the iron and eddy current loss is due to electric currents being induced in the iron.

The copper losses are those losses due to the current flow through the electric circuits. They are the VI or RI^2 losses due to resistance. For a compound wound dynamo with compensating windings the electrical losses are armature winding ($I_a^2R_a$); interpole winding ($I_a^2R_i$); series field ($I_a^2R_s$) for long shunt connection and ($I_L^2R_s$) for the short shunt connection; compensating windings ($I_a^2R_c$); brush and brush contact ($V_{BD}I_a$); shunt field ($I_f^2R_f$) or (V_fI_f). For the brush drop (V_{BD}), it is taken from 2 to 5 V.

The stray load loss includes such losses as caused by armature reaction, commutation, and so on. This is arbitrarily taken as one percent of the output for machines larger than 200 hp power or 150 kW. In smaller machines it is disregarded.

The tests that must be made are a stray power test. This is a no-load test at rated speed and rated value of E_g. For a motor $E_g = V_T - V_{BD} - I_aR_a$. For a generator $E_g = V_T + V_{BD} + I_aR_a$. The resistances of the various parts are measured by the *voltmeter-ammeter method.* The values are corrected to 75°C, which is the approximate operating temperature of the windings. Figure 2-14(a) shows the connection diagram for measuring the armature winding resistance, the interpole resistance, and the compensating winding resistance if there is a compensating winding. Note that V_a is measured across the commutator segments and if several readings are taken for an average value, the voltmeter terminals must be on the same segments for all readings. The shunt field resistance is measured with the same type of circuit. A bridge may be used because the resistance has a higher value than the armature circuit resistances. Figure 2-14(b) shows the connection diagram for measuring the stray power loss. Here a starting resistance is shown but a drum controller starter or any other type of starter may be used. One precaution is in order—never start across the line. The following example illustrates this method for determining efficiency.

Example 2-11. The following data were collected for a 5-kW, 125-V, 40 A, long-shunt, flat-compound generator. The resistance measurements at 25°C were as follows: shunt field, 104.6 ohms; armature winding (between commutator segments), 0.25 ohm; commutating pole winding, 0.021 ohm; compensating winding, 0.016 ohm; series field winding, 0.018 ohm; voltage

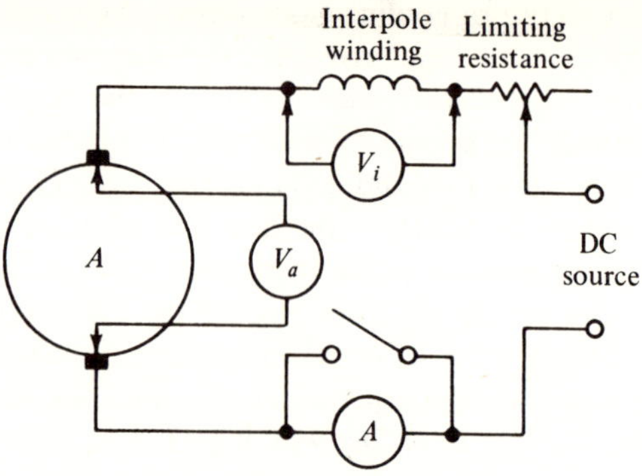

(a)

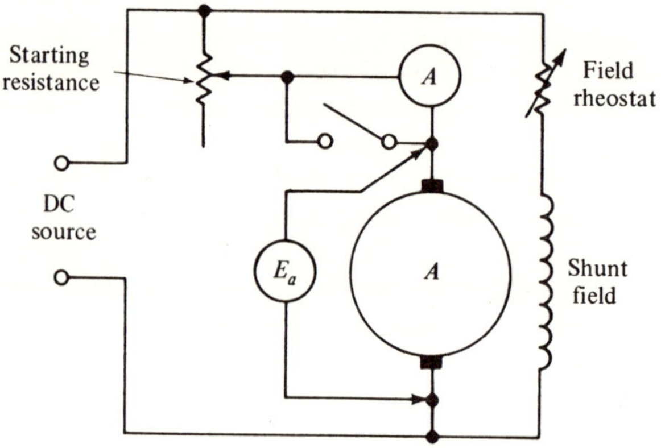

(b)

FIGURE 2-14. (a) Connection diagram for measuring armature circuit resistance values. (b) Connection diagram for measuring stray power loss.

brush drop, assume 2 V. The stray power loss was taken with V_a (Figure 2-14b) = 137.3 V and the stray power loss with corrections was 275 W. Determine the full load and one-half load efficiencies.

Solution:

Correcting the resistances to 75°C, we have the following:

Shunt field (R_f): $$104.6 \frac{234.5 + 75}{234.5 + 25} = 125 \text{ ohms}$$

Armature winding (R_a):	$0.25 \dfrac{309.5}{259.5}$	=	0.298 ohm
Commutator pole winding (R_c):	$0.021 \dfrac{309.5}{259.5}$	=	0.025 ohm
Compensating winding (R_{co}):	$0.016 \dfrac{309.5}{259.5}$	=	0.0191 ohm
Series field (R_s):	$0.018 \dfrac{309.5}{259.5}$	=	0.0215 ohm

Rated load: $I_f = \dfrac{125}{125} = 1 \text{ A}$

$I_a = I_L + I_f = 40 + 1 = 41 \text{ A}$

Stray power loss: 275 W

Armature circuit:

$$R_{ac} = R_s + R_{co} + R_c + R_a$$
$$R_{ac} = 0.298+0.0215+0.0191+0.025$$
$$= 0.3636 \text{ ohm}$$

Armature circuit loss:	$I_a{}^2R_{ac} = (41)^2(0.3636)$	=	611 W
Shunt field loss:	$V_TI_f = 125 \times 1$	=	125 W
Brush drop loss:	$V_{BD}I_a = 2 \times 41$	=	82 W
		total losses,	1093 W

Percent efficiency (Eq. 2-12a) $= \dfrac{P_o}{P_o + P_l} \times 100 = \dfrac{5 \times 1000}{5000 + 1093} = 82.1$

Percent efficiency (Eq. 2-12c) $= \left(1 - \dfrac{P_l}{P + P}\right) \times 100 = \left(1 - \dfrac{1093}{6093}\right) \times 100$

$= 82.05$

Equation (2-12c) will give a more accurate answer when using the slide rule for computations.

One-half load: (approximate)

Assuming a terminal voltage of 127 V for flat compounding, we have

$$I_L = \frac{W}{V_T} = \frac{2500}{127} = 19.7 \text{ A}$$

$$I_f = V_T/R_f = 127/125 = 1.015 \text{ A}$$
$$I_a = I_L + I_f = 19.7 + 1.015 = 20.72 \text{ A}$$

$$E_g\ (\tfrac{1}{2}\ \text{load}) = V_T + V_{BD} + I_a(R_{ac}) = 127 + 2 + 20.72(0.3636)$$
$$= 127 + 2 + 7.53 = 136.5\ \text{V}$$
$$E_g\ (\text{full load}) = 125 + 2 + 41(0.3636) = 141.9\ \text{V}$$

Stray power: $SP_{FL}\dfrac{(E_g\ \text{at}\ \frac{1}{2}\ \text{load})}{(E_g\ \text{at full load})}$

$$275\,\frac{136.5}{141.9} = 265\ \text{W}$$

Armature circuit:	$I_a{}^2R_{ac} = (20.72)^2(0.3636)$	= 156 W
Shunt field loss:	$V_TI_f = 1.015 \times 127$	= 129 W
Brush drop:	$V_{BD}I_a = 2 \times 20.7$	= 41.4 W
	total losses,	591.4 W

$$\text{Percent efficiency} = \left(1 - \frac{P_l}{P_o + P_l}\right) \times 100 = \left(1 - \frac{591}{2500 + 591}\right) \times 100$$
$$= 80.86$$

In the last example the efficiency at full load was higher than the efficiency at half load. Maximum efficiency should be near rated load. The *maximum efficiency* occurs when the fixed losses (stray power and shunt field) are equal to the varying losses (those losses associated with the armature electrical losses). In the last example, at rated conditions, the fixed power losses were 400 W and the varying losses were 693 W. At half load the stray power and shunt field losses were 394 W and the varying losses 170 W. A rough check shows a maximum efficiency of about 82.4 percent at three fourths the rated load.

A motor efficiency problem is solved in a similar manner, with the following changes. The input is the basis for the calculation and the losses are subtracted from the input power for the output. The output watts are converted to the equivalent horsepower. For example, a 5-hp motor at 120 V is rated at 40 A. Rated armature current would be about 38.5 A. At half load 19.25 or 19 A might be used for the approximate one-half load. The generated voltage is less than the terminal voltage as well, so in the stray power test the value of E_g is impressed across armature brushes.

2-10 CURRENT, TORQUE, AND VOLTAGE RELATIONS IN A DIRECT-CURRENT MACHINE[5]

Current Relations. The current relations of the different parts of dc machines can easily be determined from the schematic diagrams of the electrical connections of the machine. *Kirchhoff's current law*—the algebraic sum of the currents flowing into a junction is equal to zero—may be applied. In Section 2-2 and Figure 2-3 the various connections for dynamos were given. Figure 2-15(a) and 2-15(b) are shown for a *long-shunt compound* connection for a generator and a motor. The dotted lines will change the connection to a short-shunt connection. In the generator the EMF is E_g in series with R_a (the armature-winding resistance) within the larger circle

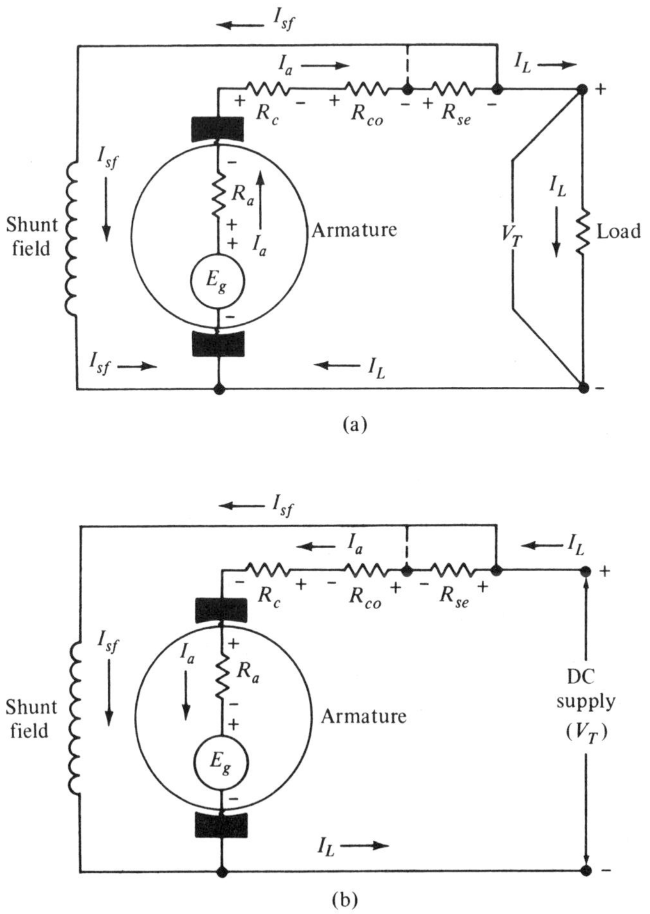

FIGURE 2-15. **(a) Long-shunt schematic diagram for a dc generator, showing currents and voltage polarities. (b) Long-shunt schematic diagram for a dc motor, showing currents and voltage polarities.**

representing the armature winding. Starting at E_g to the (+) terminal, the same current I_a, the armature current passes through R_a, brushes, R_c, R_{co}, and R_{se}. By Kirchhoff's current law,

$$I_a = I_{sf} + I_L,$$
$$I_a = I_c = I_{co} = I_{se}.$$

For the short-shunt connection,

$$I_a = I_{sf} + I_L,$$
$$I_a = I_c = I_{co},$$
$$I_{se} = I_L.$$

For the motor the dc supply is the external voltage source. For the long-shunt connection, Figure 2-15(b),

$$I_L = I_{sf} + I_a,$$
$$I_a = I_{se} = I_{co} = I_c.$$

For the short-shunt connection (dotted lines),

$$I_L = I_{se},$$
$$I_L = I_{sf} + I_a,$$
$$I_a = I_{co} = I_c.$$

We have not numbered these equations because in solving machine problems the circuit should be sketched and circuit analysis used with voltages and resistances in their proper places. For this reason the above formulas should not be memorized.

Voltage Relations. The voltage relations make use of *Kirchhoff's voltage law*, which may be stated, in going around any closed traverse, the algebraic sum of the EMF and the potential drops is equal to zero. In a generator, again referring to Figure 2-15(a), the only EMF is the generated voltage, E_g. V_T is the terminal voltage at the terminals to produce current flow through the load. For the long-shunt generator,

$$E_g - I_aR_a - I_aR_c - I_aR_{co} - I_aR_{se} - V_T = 0$$

or

$$V_T = E_g - I_a(R_a + R_c + R_{co} + R_{se}).$$

For the short-shunt generator (dotted lines),

$$E_g - I_aR_a - I_aR_c - I_aR_{co} - I_LR_{se} - V_T = 0$$

or

$$V_T = E_g - I_a(R_a + R_c + R_{co}) - I_LR_{se}.$$

For the motor-connected long shunt, Figure 2-15(b),

$$V_T - I_aR_{se} - I_aR_{co} - I_aR_c - I_aR_a - E_g = 0$$

or

$$E_g = E_c = V_T - I_a(R_a + R_{co} + R_{se} + R_c).$$

For the motor short-shunt connection (dotted lines),

$$V_T - I_LR_{se} - I_aR_{co} - I_aR_c - I_aR_a - E_g = 0$$

or

$$E_g = E_c = V_T - I_LR_{se} - I_a(R_{co} + R_c + R_a).$$

These voltage equations are not numbered. Like the current equations, they are not to be memorized. The same current circuit diagram can be used and with Kirchhoff's voltage law and Ohm's law the solution can handily be found.

Torque Relations. In the preceding discussion of voltage relations it was determined that in a generator an EMF E_g was generated in the armature and subtracting the current resistance drops, the remainder—the terminal voltage—was left to produce current flow through a load. Also, in a motor after subtracting the current resistance drops from the terminal voltage, there remained a counter EMF or generated voltage to balance the remainder of the terminal voltage. Similar torque relations exist as will be shown.

In a motor, torque is produced as can be seen in Figure 2-16(b). After the torque losses have taken care of the rotational losses, the remaining torque is available at the shaft to drive an external mechanical load. It remains to show that similar to counter EMF in a motor, there should be counter torque in a generator. In Figure 2-16(a) it is obvious that current will flow in the armature and this current will produce torque in a counterclockwise direction by applying *Fleming's left-hand motor rule.* This torque is opposite to the direction of rotation and so is opposite to the torque supplied by the prime mover. The difference is the torque loss to take care of the rotational losses, just as in the motor.

In Section 2-8, a formula (Equation (2-11)) for torque was given,

$$T = \frac{7.04E_gI_a}{S}.$$

It would appear that the developed torque depends on speed. However, by substituting the value for E_g in Equation (2-11), the formula became in Equation (2-9),

$$T = K_t\phi I_a,$$

where $K_t = 0.1173PZ \times 10^{-8}/a$. In this last formula T is not dependent on speed, but only as speed affects E_g in Equation (2-10). Also, speed is affected by ϕ and I_a and in this dependent way torque is affected by the speed. In a dc

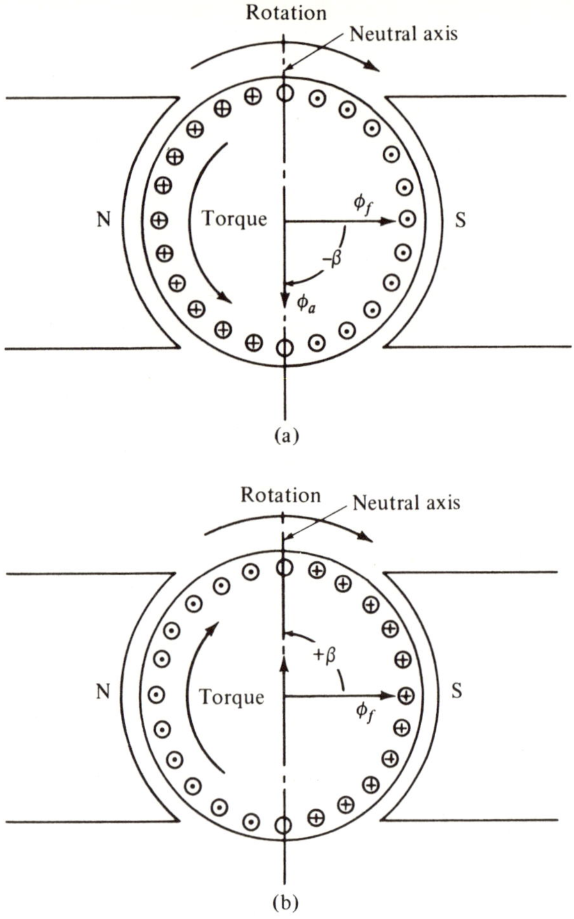

FIGURE 2-16. (a) Dc generator showing torque and torque angle. (b) Dc motor showing torque and torque angle.

motor there is a relationship between torque and speed. This will be discussed in the next chapter on characteristics of dc machines.

In Figures 2-16(a) and (b) the torque angle β is shown. In a generator the neutral axis shifts to the right which increases the $-\beta$ angle. In the motor the neutral axis shifts to the left, toward the direction of rotation, making $+\beta$ greater than 90 deg. In both cases fewer conductors are effective in producing torque. Obviously if the torque angle β becomes 0 or 180 deg there will be no torque developed. In dc machines the angle of shift does not change more than 10 to 12 deg, so there will be no point of breakdown torque as in synchronous or induction motors.

2-11 SUMMARY

In this chapter it was shown how dc machines are constructed, including the magnetic circuit and the electric circuit. The rotor is the armature core and is made of laminated steel stampings fastened to a steel shaft. The pole cores, also of laminated steel, are fastened to the yoke which supports, through end bells, the armature shaft in bearings and the brush rigging. The field pole windings can be connected in shunt (parallel) or in series, or both connections may be used to provide a shunt, series, or compound machine. In addition a separately excited machine is also possible. A complete magnetic and electrical diagram is shown in Figure 2-1 and the schematic diagrams for shunt are shown in Figure 2-2.

The lap and wave windings were considered, with a lap winding having current paths (a) equal to the number of poles times the multiplicity. A wave winding has current paths two times the multiplicity. The slot pitch is calculated in both types of windings by Equation (1-5) given in Chapter 1,

$$Y_s = \frac{S}{P} - k.$$

The commutator pitch for a lap winding is equal to the multiplicity of the winding. In a wave winding the commutator pitch is equal to Equation (2-4),

$$Y_c = \frac{C - m}{P/2}.$$

For a given generator the generated voltage is

$$E_g = K\phi S$$

where $K = (PZ \times 10^{-8})/60a$.

Armature reaction is the reaction of the armature flux with the main flux from the field poles. In a generator this causes the flux to be more dense under the trailing pole tip of the pole face, while in a motor the flux is most dense under the leading pole tip. The neutral axis shifts by a small angle θ, the angle increasing as the load increases; for a generator it shifts in the direction of rotation and for a motor it shifts toward the direction of rotation. The demagnetizing MMF is given by Equation (2-5),

$$NI_d = \frac{\theta Z I_a}{180a}.$$

The cross-magnetizing MMF is given by Equation (2-6),

$$NI_c = \frac{(180 - 2\theta) Z I_a}{360a}.$$

Commutation is the process whereby the current is reversed as it passes from the influence of a north-south set of poles to the influence of the next south-north set of poles, the south pole being the same pole in both cases. This must be done quickly and can be done by advancing the brushes slightly beyond the magnetic neutral axis or by commutating poles on the no-load neutral. In a generator the polarity is main north, commutating south, main south, commutating north. For a motor the polarity of the commutating poles is opposite that of a generator. The windings of commutating poles are connected in series with the armature.

Torque is developed by a motor in the armature conductors with respect to the shaft of the motor. Torque is a force times a distance (both perpendicular), tending to cause rotation about a point from which the distance is measured. The torque for a motor is given by Equations (2-8)–(2-11),

$$T = \frac{0.1173 \phi P I_a Z}{a} \times 10^{-8} \text{ lb-ft},$$

$$T = K_t \phi I_a,$$

$$T = \frac{5250 \text{ hp}}{S} \text{ (mechanical)},$$

$$T = \frac{7.04 E_g I_a}{S}.$$

Efficiency of machines may be calculated by measuring the power input and power output. Percent efficiency is given by Equation (2-12),

$$\text{percent efficiency} = \frac{P_o}{P_i} \times 100.$$

This is the method for small machines, but for large machines the electrical losses are calculated by I^2R or VI, the stray power loss measured by test. From the losses and output (for a generator) or input (for a motor) the efficiency is calculated. In the stray power test it is important that the proper value of E_g is impressed across the brush terminals. For a motor

$$E_g = V_T - V_{BD} - I_a R_a$$

and for a generator

$$E_g = V_T + V_{BD} + I_a R_a.$$

The stray power or rotational loss varies directly as the induced voltage, E_g. The other efficiency formulas are given in Equations (2-12a, b, c),

$$\text{percent efficiency} = \frac{P_o}{P_o - P_l} \times 100 = \frac{P_i - P_l}{P_i} \times 100$$

$$= \left(1 - \frac{P_l}{P_o + P_l}\right) \times 100.$$

Equation (2-12a) is used in the case of generators and (2-12b) is used for motors. Equation (2-12c) can be used for generators as written and for motors using P_i for $P_o + P_l$. The latter formula yields more accurate slide-rule results.

The current relations are shown for both a long-shunt and a short-shunt connection compound machine. These relations are based on Kirchhoff's current law and are not repeated here. For the solution of dc machinery problems the circuit should be used and previous experience in circuit solution should be applied.

The voltage relations are also given for compound machines and for the same reasons as for current relations are not repeated here. Kirchhoff's voltage law and Ohm's law are used in the voltage relations.

The torque relations are shown by the four torque equations previously given. It might be stated that the three electrical formulas are for the developed torque. The torque to provide for the rotational losses is used from the developed torque and the balance is available at the shaft as load torque. In a generator there is a counter torque that is equal to the torque provided by the prime mover less that used to supply the rotational losses. Torque is, itself, not dependent on speed but is dependent on flux ϕ and armature current I_a. Because these values are dependent on speed, there is a torque-speed relationship in dc motors.

Problems

1. A 25-kW, 250-V shunt generator has a shunt field resistance of 62.5 ohms, a commutating pole-winding resistance of 0.025 ohm, an armature resistance of 0.160 ohm; use 4 V for brush drop. Calculate the generated voltage.
2. A shunt generator has a generated voltage of 276 V. The armature resistance is 0.087 ohm and the commutating pole winding resistance is 0.02 ohm. The shunt field resistance is 50 ohms; use 4 V for brush drop. Calculate the generator rating if the rated armature current is 205 A.

3. A 120-V, 10-hp shunt motor rated at 76 A has a shunt field resistance of 60 ohms, a compensating winding resistance of 0.04 ohm, a commutating pole-winding resistance of 0.05 ohm, and an armature resistance of 0.14 ohm; use 2 V for brush drop. Calculate the generated voltage.
4. A 25-hp long-shunt compound motor has a generated voltage of 214.6 V; an armature current of 86 A; an armature resistance of 0.16 ohm; a series field resistance of 0.06 ohm; a commutating pole resistance of 0.04 ohm; use 3 V for brush drop. (a) Calculate the value of terminal voltage; (b) if the field current is 3 A, calculate the power input.
5. (a) Design a simplex lap winding for a four-pole generator with an armature of 39 slots and 39 coils. Each coil has six turns. (b) If the armature rotates at 1150 rpm, and the flux per pole is 2.1×10^6 lines with 75 percent of the conductors under pole faces, calculate the value of generated voltage. (c) Calculate the coil pitch.
6. (a) Design a simplex wave or series winding for a four-pole dynamo whose armature has 29 coils, 29 slots, and 29 commutator segments. (b) Calculate the generated voltage if each coil has eight turns. The flux per pole is 1.75×10^6 lines and the speed is 1750 rpm; 75 percent of the conductors are under pole faces.
7. Design a simplex wave winding with 47 coils, 47 slots, and 47 commutator segments. The dynamo has six poles. Also calculate the value of coil pitch.
8. The armature current in Problem 5 is 86 A and the neutral axis shifts 7.5 deg. Calculate the demagnetizing and cross-magnetizing ampere turns.
9. In Problem 6, if the armature current is 50 A and the angle the neutral axis shifts is 9 deg, calculate the demagnetizing and cross-magnetizing ampere turns.
10. A motor has 464 armature conductors, each carrying a current of 25 A. The dimensions of each coil is 15×12 in. wide; 75 percent of the conductors are under poles. The flux density is 70,000 line/in.^{2} (a) Calculate the force on each conductor, and (b) calculate the developed torque.
11. The motor of the preceding problem has four poles and has a duplex lap winding with an armature current of 200 A. Calculate the developed torque with the use of Equation (2-8). Also calculate the output torque if the motor is rated at 60 hp, 240 V, and 850 rpm.
12. (a) In the preceding problem the flux is reduced 5 percent and the armature current is reduced 50 percent; calculate the new value of torque. (b) If the torque in the preceding problem is to remain unchanged and

the flux is decreased 10 percent of its original value, calculate the value of the necessary armature current.

13. A 250-V, 100-kW, 1150-rpm generator has a generated voltage of 275 V and an armature current of 410 A. (a) Calculate the value of the developed counter torque at rated output. (b) If the machine is to be used as a motor at 240 V (terminal) and an armature current of 390 A, a 4-percent reduction in the flux per pole from part (a), calculate the developed armature torque. (c) If the generated voltage is 215 V in part (b), what will be the speed of the motor?
14. An eight-pole, 850-rpm, shunt dynamo with eight current paths has an armature whose length is 15 in. and whose diameter is 24 in. It has 576 armature conductors and an armature resistance of 0.05 ohm. The brush drop is 2 V; 78 percent of the armature circumference is covered by pole faces. The current per path is 22 A. The air gap is $\frac{1}{8}$ in. The generated voltage is 230 V at rated speed and flux. Calculate (a) the developed torque, (b) the flux per pole, (c) the flux density, (d) the force on one conductor, and (e) the terminal voltage.
15. The dynamo of the preceding problem is driven by a prime mover at 935 rpm and furnishing a load so that the armature current per path is 22 A. (a) Calculate the generated voltage and the terminal voltage; (b) calculate the counter torque developed if the flux is increased 3 percent from previous problem; (c) calculate the power output of the generator at these conditions and if the shunt field current is 5 A.
16. A 100-hp motor is rated at 240 V, 341 A, 650 rpm, and is cumulative compound with a developed torque of 875 lb-ft at rated conditions. The armature winding has 14 current paths and 956 armature conductors. The machine has 14 poles. The rated armature current is 330 A. The flux per pole is 2.37×10^6 lines. If the flux per pole is increased 12 percent and the armature current is increased 15 percent, calculate the new value of developed torque.
17. A 25-hp motor operates at 86 percent efficiency at rated load; 30 percent of the losses are stray power loss and 70 percent are copper losses. Calculate the stray power loss and the copper losses at rated load.
18. A 250-V, 50-kW, long-shunt compound generator was tested and the following data were found:

Stray power losses:	3550 W
Shunt field resistance at 75°C:	50 ohms
Series field resistance at 75°C:	0.015 ohm
Interpole resistance at 75°C:	0.010 ohm

Armature winding resistance at 75°C:	0.055 ohm
Brush drop:	assume to be 3 V

Calculate the efficiency at rated load.

19. In the preceding problem maximum efficiency occurs when the fixed losses (stray power and shunt field) equal the variable losses. Calculate (a) the armature current at maximum efficiency, (b) the output at which maximum efficiency occurs, and (c) the maximum efficiency, assuming flat compounding and the same terminal voltage throughout. Also the same voltage brush drop.
20. In the preceding two problems calculate the stray power losses at $\frac{1}{4}$, $\frac{1}{2}$, $\frac{3}{4}$, and $1\frac{1}{4}$ loads.
21. Calculate the electrical losses for the loads of the preceding problem.
22. Calculate the efficiencies at the loads of Problems 20 and 21.
23. The output torque of a 10-hp, 240-V, 38-A motor is 81.0 lb-ft at rated speed of 650 rpm. Calculate the efficiency and the losses if 35 percent of the total losses are rotational losses.
24. A 75-hp, 240-V long-shunt motor having a rated input current of 263 A was tested and the following were the results:

Stray power loss, corrected:	3150 W
Armature winding resistance at 75°C:	0.037 ohm
Series field resistance at 75°C:	0.008 ohm
Commutating pole winding at 75°C:	0.005 ohm
Assume brush drop of 3 V:	
Shunt field resistance at 75°C:	34.3 ohms

For armature currents of $\frac{1}{4}$, $\frac{1}{2}$, $\frac{3}{4}$, rated and $1\frac{1}{4}$ rated calculate the stray power losses.

25. Calculate the electrical losses for the five values of armature current in the preceding problem.
26. Calculate the efficiencies and power outputs for each of the preceding currents.
27. Plot efficiency versus horsepower output cure from the preceding problem. Determine the efficiency at rated output and the value of efficiency at maximum output and the value of output at maximum efficiency.

References

[1] Kosow, I. L., *Electrical Machinery and Control*. Englewood Cliffs, N.J.: Prentice-Hall, 1964.

[2] ROBERTSON, B. L. and L. J. BLACK, *Electric Circuits and Machines*, 2nd ed. Princeton, N.J.: Van Nostrand, 1957.
[3] SISKIND, C. S., *Electrical Machines—Direct and Alternating Current*, 2nd ed. New York: McGraw-Hill, 1959.
[4] KU, Y. H., *Electric Energy Conversion*. New York: Ronald Press, 1959.
[5] CARR, C. C., *Electrical Machinery*. New York: Wiley, 1958.

Operating Characteristics of Direct-Current Machines

3

In the preceding chapter the theory of the voltage generated, torque relations, armature reaction, commutation, armature windings, and efficiency of dc machines were considered. In this chapter the operating characteristics of dc machines will be discussed. The characteristics important to generators are the terminal voltage—load characteristics, magnetization curves, and effects of speed on these characteristics. The characteristics important to motors are the speed, torque, current, and efficiency load. The methods of starting motors, speed control, and control such as direction of rotation, stopping, will also be considered in this chapter. The subject of starting and control, even for dc motors, is too broad a subject to be covered completely, but the basic principles and some basic circuits will be considered.

3-1 LOAD VOLTAGE CHARACTERISTICS OF DIRECT-CURRENT MACHINES[1],[2]

The load voltage characteristic of a generator is the *terminal voltage*, or the voltage across the terminals to which the load is connected, plotted against the load in watts or kilowatts, with the speed held at a consant value. An ideal generator would have the same terminal voltage at any load value, but this is not the case for any actual generator. The expression for the generated voltage of a generator is $E_g = K\phi S$, where $K = (PZ \times 10^{-8})/60a$. The speed of the prime mover can be kept constant, but because of armature reaction the flux ϕ will not be constant. The terminal voltage V_T is dependent on the generated voltage by the expression, $V_T = E_g - I_aR_a$, where R_a is the sum of the resistances in the armature circuit including the brush and brush-gap resistance. The load voltage characteristics for the separately

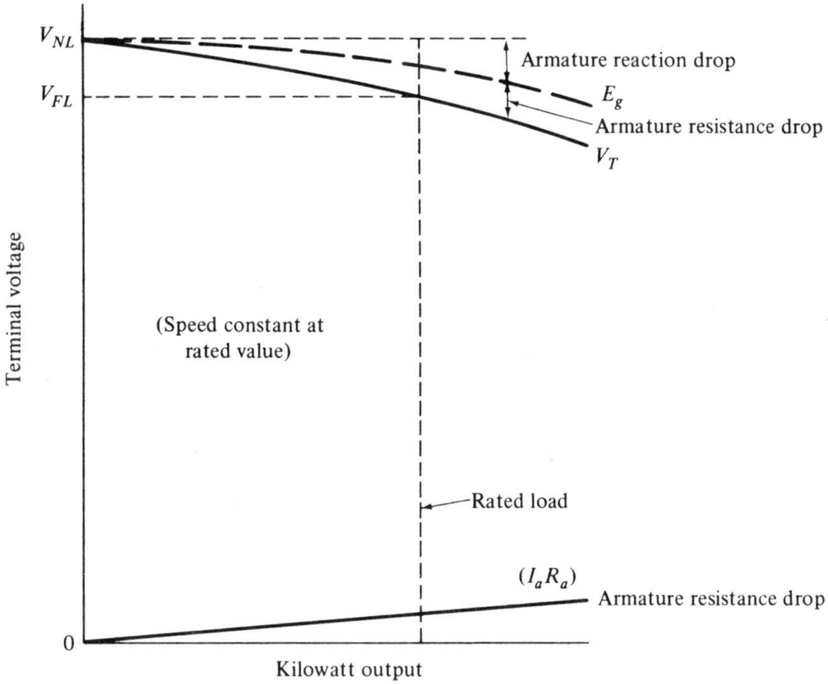

FIGURE 3-1. Terminal voltage—load characteristic of separately excited shunt generator.

excited shunt and self-excited shunt, compound, and series generators are examined in this section.

The *separately excited* load voltage characteristic is shown in Figure 3-1. The flux supplied by a separate source is a constant value but as the load increases, the flux is decreased by armature reaction MMF. The curve (broken line) marked E_g shows the effect of armature reaction. It drops slightly from no load to full load. This line is the generated voltage curve. The armature resistance drop I_aR_a is subtracted from the generated voltage E_g and the difference is the terminal voltage,

$$V_T = E_g - I_aR_a.$$

The solid line V_T is the external characteristic of terminal voltage versus load. The two causes for a drop from the no-load value are (a) the drop due to armature reaction and (b) the drop due to the armature circuit resistance, I_aR_a.

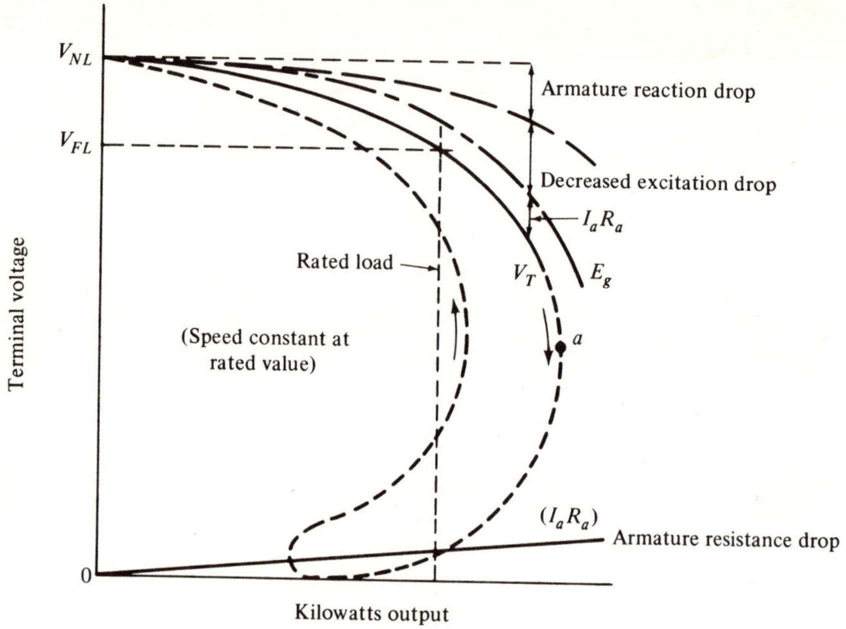

FIGURE 3-2. External shunt generator terminal voltage—load characteristic.

The *self-excited shunt-generator* load voltage characteristic is shown in Figure 3-2. Like the separately excited generator, there is an armature reaction drop and the drop due to I_aR_a armature circuit resistance. In this case because the generator is self-excited, there is a drop of potential across the shunt field that gives the additional drop of decreased excitation. In the figure after the drops caused by the armature reaction and decreased excitation are subtracted from the no-load voltage, the generated voltage line E_g remains. Subtracting the armature resistance drop I_aR_a from the generated voltage, $V_T = E_g - I_aR_a$, yields the external characteristic of terminal voltage versus load. This characteristic drops off slightly more than the characteristic of the separately excited shunt generator. The three drops are (a) armature reaction, (b) reduced excitation because of armature reaction, and (c) armature resistance voltage, I_aR_a. These drops all occur simultaneously and each is dependent on the other, and the only curve obtained by test is the load characteristic curve itself. If the curve is continued as shown by the dotted line by adding load, a point *a* is reached that is called the *breakdown point.* This is the maximum load that the shunt generator can carry. If at this point the load is reduced and then increased again, the dotted line would go back to the no-load value. This part of the curve is impractical and a generator should not be loaded to the breakdown point.

The *compound generator* may have a variety of characteristics. In Figure 3-3 the load characteristics of an *undercompound*, a *flat compound*, an *over-*

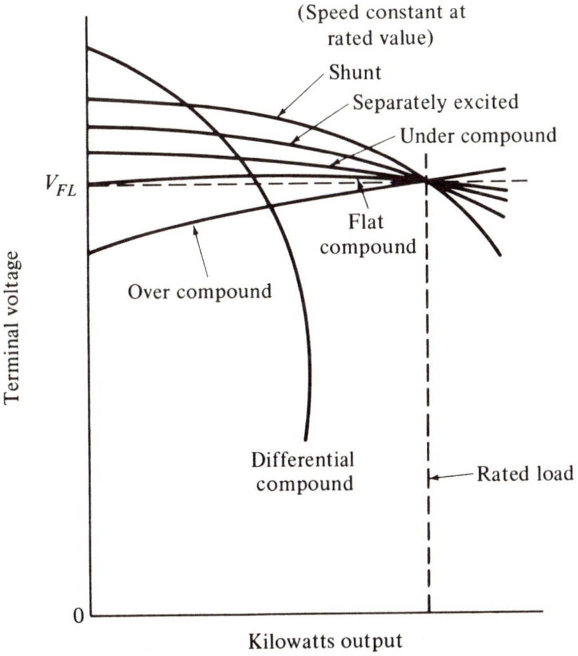

FIGURE 3-3. **External terminal voltage—load characteristics of shunt, separately excited, under compound, flat compound, over compound, and differential compound generators.**

compound, and a *differential compound* generator are shown together with the self-excited shunt and separately excited shunt generator. They all have the same full-load voltage at rated load, except the differential compound generator that cannot attain rated output.

The undercompound generator has a more flat load voltage characteristic than the shunt or separatedly excited generator. The series field adds some MMF to the shunt field MMF as the load increases, but not enough to make the full-load voltage higher than the no-load voltage. The next curve is the flat compound generator in which the no-load voltage is equal to the full-load voltage. This type is close to the ideal, but at other loads there is a small variation from the rated voltage. The next curve is the overcompound generator in which the full-load voltage is higher than the no-load voltage. Here as the load increases the series field adds more MMF to the main field than necessary to overcome the armature reaction. This type of generator can be made flat compound by placing a low resistance across the series field winding of just the right value to divert some current from the series field. The last curve is the differential compound generator. Here the series field is connected so that the MMF of the series field is opposite to the main flux. Care must be taken not to operate a differential compound generator at too low a value of terminal voltage, for the series field may predominate and the polarity may reverse.

The mark of how well a generator performs is how little the terminal voltage drops from no load to full load. *Voltage regulation* is a term that defines the performance of a generator. Voltage regulation is the voltage change in percent as the load is changed from a specified load to zero with a constant-field rheostat setting and at a constant speed. The specified load is usually taken as rated load and the expression may be stated as

$$\text{percent voltage regulation} = \frac{V_{NL} - V_{FL}}{V_{FL}} \times 100. \qquad \textbf{(3-1)}$$

A flat compound generator has zero percent regulation, while an overcompound has a negative regulation. The lower the percentage, the better the regulation.

Example 3-1. A compound generator has a rated load voltage of 125 V. When the load is removed, the voltage is 130 V. Calculate the percent voltage regulation.

Solution:

$$VR = \frac{V_{NL} - V_{FL}}{V_{FL}} \times 100 = \frac{130 - 125}{125} \times 100 = 4.4 \text{ percent.}$$

Example 3-2. An overcompound generator has a rated load voltage of 250 V and a percent regulation of −4.0. (a) Calculate the no-load terminal voltage. (b) If the rating of the generator is 50 kW, calculate the load if the load voltage is 247.5 V (assume straight-line characteristic).

Solution:

(a) Solving Equation (3-1) for V_{NL},

$$V_{NL} = \frac{(V_{FL})(\%VR)}{100} + V_{FL} = \frac{(250)(-4.0)}{100} + 250,$$

$$V_{NL} = -10 + 250 = 240 \text{ V}.$$

(b) $$\frac{\text{rated load}}{V_{FL} - V_{NL}} = \frac{\text{load?}}{V_T - V_{NL}} \quad \text{(by similar triangles)},$$

$$\text{load?} = (\text{rated load})\frac{V_T - V_{NL}}{V_{FL} - V_{NL}} = 50\,\frac{247.5 - 240}{250 - 240} = 37.5 \text{ kW}.$$

A *series generator* does not have many practical applications. An examination of Figure 2-3(c) indicates that the current could build up indefinitely and were it not for limitations of the voltage drops (I_aR_a and I_aR_{se}) and armature reaction, it would do so. The armature current and load current are the same and load current is used for the abscissa for the load characteristic instead of load. A maximum voltage V_m is reached and beyond this point the voltage drops off rapidly due to saturation of the iron. Figure 3-4 shows a typical load voltage characteristic. The series generator finds some use on dc transmission systems to boost the voltage to compensate for the drops at varying current values. In Europe where dc is used for high-voltage transmission systems it is used in the *Thury system*. In this respect the curve from e_r to near the V_m point is the characteristic used. Beyond the maximum voltage point the load current is practically constant for a wide variation of terminal voltages. This feature makes the generator useful as a dc welder in which the heat of the arc I^2R should be uniform for all voltage values.

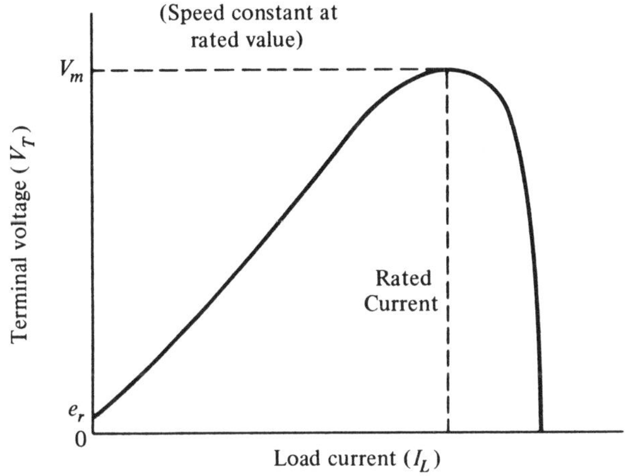

FIGURE 3-4. External terminal voltage—load current characteristic of a series generator.

3-2 PERFORMANCE FROM MAGNETIZATION CURVES[3]–[5]

The magnetization curves are often called *saturation curves* because of the fact that when magnetic materials are magnetized by magnetomotive force (MMF), they become saturated with magnetic flux and no matter how much the MMF is increased the flux lines do not increase. The unit for MMF is the ampere-turn and in a dc machine the field turn is a constant. Therefore, in a dc shunt generator the field current is proportional to the field MMF. The abscissa of the magnetization curve is field current. Since $E_g = K\phi S$ and if the speed is kept constant, the generated voltage is a function of the flux ϕ and for a given generator, a function of flux density B. The ordinate of the magnetization curve is the generated voltage E_g, which is the open-circuit terminal voltage. In Figure 3-5 the magnetization curve for a shunt generator is shown. This curve is obtained by separately exciting the shunt field and running the generator at constant rated speed. The field current is

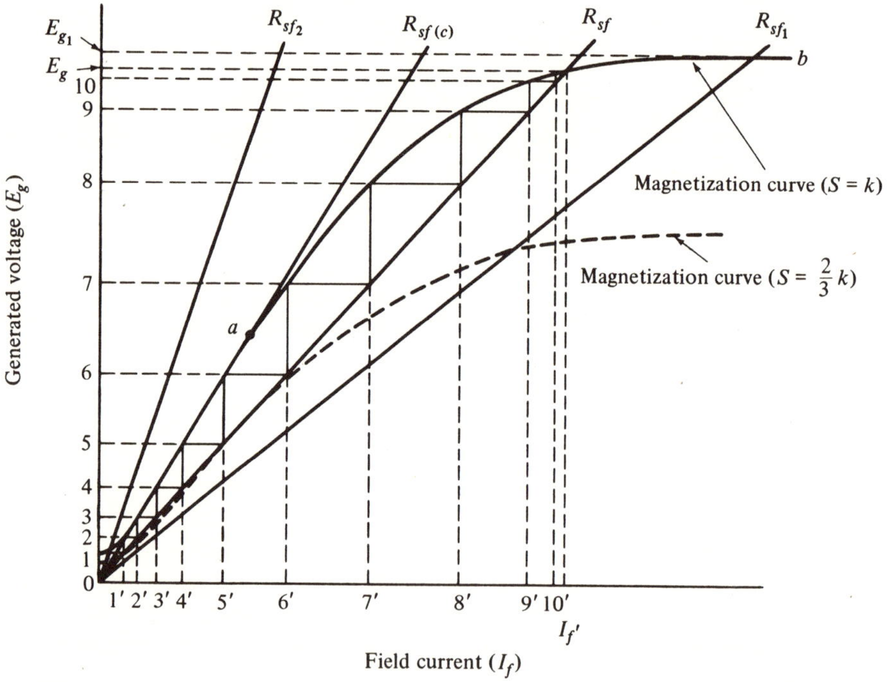

FIGURE 3-5. Magnetization curves (F_g — I_f) of shunt generator showing build-up at rated speed.

increased from zero to a value for maximum voltage. Care must be taken that in increasing the field current it not be increased beyond the desired value and then reduced to the desired value because of the hysteresis of the iron. In fact, if the values of field current are decreased in steps to zero, a curve above the first curve is obtained. With I_f at zero, there is a small voltage. As the field current is increased, the curve bends into a straight line and as I_f increases still more, it bends and finally becomes a straight horizontal line. The end of the straight-line portion is a and at this point the iron starts to saturate slowly, at first, but increases as I_f increases. Finally at point b the magnetic circuit is saturated and no more flux can be put into the magnetic circuit.

The *build-up process* for a self-excited shunt generator for a value of shunt field resistance represented by the slope of the line (R_{sf}) is as follows: The residual magnetism provides a voltage marked 1. This voltage gives a value of field current 1′. Field current 1′ produces a voltage 2. Voltage 2 increases the field current to 2′, which produces a voltage 3; and so on until the point of E_g is reached for a value of field current $I_{f'}$. This is at the point where the resistance line crosses the magnetization curve.

If the slope of the resistance line were made less, which means less resistance (such as a smaller value represented by the line R_{sf1}), the value reached by the generated voltage would be E_{g1}. Also, if the resistance were increased to a value represented by the line $R_{sf(c)}$, which coincides with the straight-line portion of the magnetization curve, the ordinate at a would be the value of voltage the generator would generate. This last line is often called the *air-gap line*. It is the *critical-resistance line* for the shunt field circuit resistance; for if the circuit resistance is higher than the value represented by this line, the generator will not build up an appreciable voltage. Line R_{sf2} represents a higher value of resistance; the voltage it will produce is only slightly above the residual voltage value.

The dotted magnetization curve is shown for a constant speed of two thirds rated speed value. For each value of I_f, the generated voltage is two thirds the value at rated speed.

The following are the four conditions for buildup of a generator

(1) There must be residual magnetism. This is obvious from the magnetization curve; for if there is no residual voltage, there will be no field current to give the larger value of voltage. In this respect a motor does not need residual magnetism; if a motor is used as a generator, it may have to be *flashed* before it will build up. *Flashing* is the applying of a small voltage across the field to start the buildup

process. This may have to be done to a generator if it has not been used for some time.

(2) The resistance of the shunt field circuit must be less than the critical value. This was explained previously. Often a shunt field rheostat simply needs to be cut out. Open shunt field circuit will act the same way.

(3) The armature circuit and field circuit must be connected in the proper relationship. If there is residual voltage, the current will flow through the field in a negative direction and the residual voltage will drop off a slight bit.

(4) The speed of the prime mover may be too low. A check of the speed of the prime mover may indicate that the speed is less than that specified on the nameplate.

The external load characteristic of a generator can be obtained graphically from the magnetization curve. The number of field turns and armature turns must be known. By use of Equation (2-5) for finding the demagnetizing

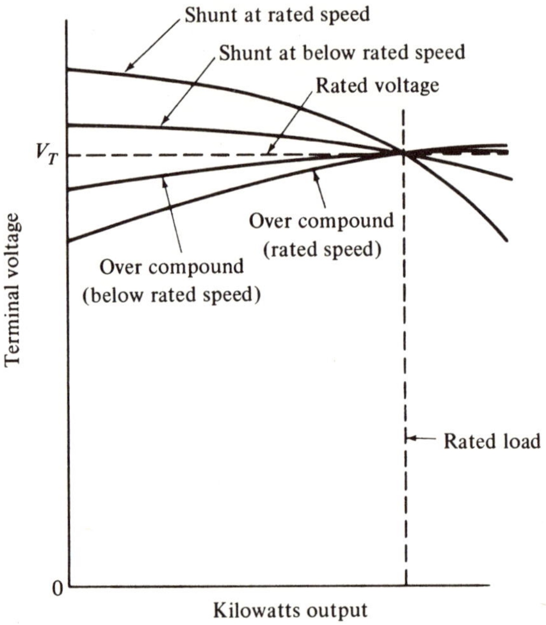

FIGURE 3-6. External terminal voltage—load characteristic for rated and below-rated speed for shunt and over-compound generator.

MMF of the armature and translating this to a reduced value of field current, a new value of E_g can be found on the curve. Subtracting the I_aR_a drop will yield the value of terminal voltage. The process is approximate since it is difficult to determine accurately the demagnetizing reduction of field current. The process is beyond the scope of this book but a complete discussion may be found in the references.

In Figure 3-5 a magnetization curve for reduced speed was shown. In Figure 3-6 are shown the load voltage characteristics for a shunt and an overcompound generator operating at reduced speed. The voltage regulation is better for the below speed curves. However, to attain rated voltage would mean considerable overexcitation and high current values, which in turn would mean heating. In the case of Figure 3-5 rated voltage could not be attained for the two-third speed even at no load. However, shunt and compound generators always have a field rheostat to adjust the voltage. If the speed is too low, however, the generator should not be operated for long periods of time.

3-3 OPERATING CHARACTERISTICS OF SHUNT AND COMPOUND MOTORS[2],[4]

The *torque load* characteristic of the motor has the same significance as the load voltage load characteristic of the generator. From the formula given in Equation (2-9) $T = K_t\phi I_a$, it can be seen that torque varies directly with the magnetic flux and the armature current. In a shunt motor the MMF is constant because the field is connected to the source, and therefore the flux is practically constant except for the slight drop due to armature reaction MMF. In a shunt motor $I_a = I_L - I_{sf}$ and because the field current is about 2 to 5 percent of the line current, the armature current is practically the same as the line current. Therefore, the *torque characteristic* (torque versus output) is practically a straight line as shown in Figure 3-7. For the cumulative compound motor the torque characteristic is a slightly concave curved line as shown by the dotted line on Figure 3-7. The series field turns add MMF to the shunt MMF as the load increases. As pointed out in Section 3-5 the series motor torque is a parabolic curve with the torque varying nearly as the square of the current. So the degree of compounding determines the amount of curve in the torque characteristic.

The *speed characteristic* (speed versus load) is considered next. From the generated voltage formula $E_g = K\phi S$; solving for S,

$$S = \frac{E_g}{K\phi}, \tag{3-2}$$

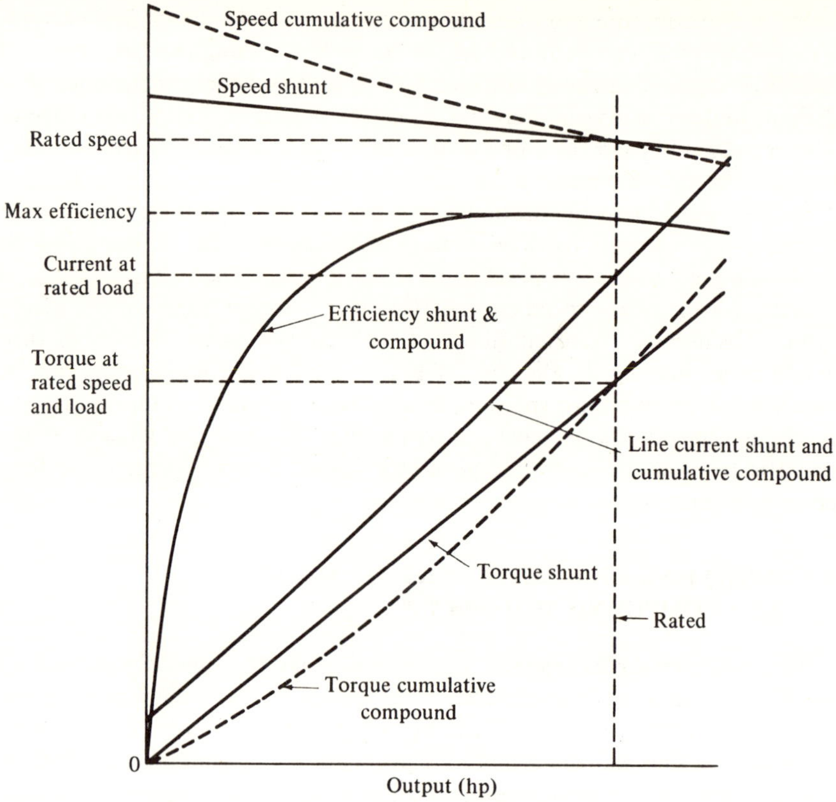

FIGURE 3-7. **Characteristics of dc Shunt and cumulative compound motors (speed, efficiency, line current and torque versus load in hp).**

where

S is the speed of the motor in rpm,
E_g is the generated voltage (counter EMF for a motor),
ϕ is the magnetic flux in line.

For a shunt motor, $E_g = V_T - I_a R_a$, and therefore

$$S = \frac{V_T - V_{BD} - I_a R_a}{K\phi}, \tag{3-2a}$$

where

S and ϕ are as previously defined,
V_T is the terminal voltage of the supply,
I_a is the armature current,

V_{BD} is the brush drop voltage,
R_a is the resistance of the armature circuit.

The shunt motor speed versus load curve is practically a straight line dropping from 2 to 10 percent. For the cumulative compound motor drops are more than for the shunt motor. For a compound long-shunt motor the speed is

$$S = \frac{V_T - V_{BD} - I_a(R_a + R_{se})}{K\phi}. \tag{3-2b}$$

The denominator increases as I_a increases with the load because of the series field MMF. If everything else remains the same, the speed decreases. The numerator also decreases because of the voltage drops of the armature circuit and the series field subtracting from the terminal voltage.

A measure of the drop in motor speed is the *percent speed regulation* given by the formula,

$$\text{percent speed regulation} = \frac{S_{NL} - S_{FL}}{S_{FL}} \times 100, \tag{3-3}$$

where S_{NL} is the no-load speed in rpm. S_{FL} is the full load or rated speed in rpm. This formula is similar to the percent voltage regulation; and the lower the value of the percent regulation, the better the motor maintains constant speed as load is applied. The term *constant-speed motor* designates a motor operating at normal speed, the speed being constant or practically constant as in the dc shunt motor. The cumulative compound motor is a *varying-speed motor.* A varying-speed motor is a motor the speed of which varies with the load. The speed regulation of a cumulative compound motor may be from 10 to 25 percent. It does have a definite no-load and full-load speed.

The efficiency characteristics are similar for shunt and compound motors. In both motors the maximum efficiencies run from 85 to 90 percent. As the efficiency curve shows, the maximum efficiency occurs at about 75 percent of rated load with the efficiency at rated load just a slight bit less.

The line current characteristic is shown and it is a line with a slight concave curve. At no load there is a value of current that is enough to supply the stray power losses. This curve is relatively unimportant since the rated current value is on the nameplate of the machine. Also, Appendix Table A-1 shows the full-load currents for dc motors as specified by the National Electrical Code.

The differential compound motor has its series turns opposing the main or shunt field turns. No characteristic is shown, but it has a definite no-load speed and unless the series field supplies a weak MMF, the speed will tend to rise; because of the counter torque of the load, however, it cannot, and so the operation at an appreciable load usually is unstable. It may be used in special cases where the speed must be constant, but then the series field is weak; ordinarily there is little use for this type of motor. In testing compound motors, however, the person testing such motors should be aware of the results of a series field connected in differential manner.

Example 3-3. A 230-V shunt motor has an armature resistance of 0.25 ohm and a line current of 40 A when it is operating at rated load and at a speed of 1200 rpm. At what speed will it run at no load if the armature current at no load is 4.0 A and the field takes 1.0 A? Assume no effect of armature reaction and a full-load brush drop of 3 V and a no-load brush drop of 2 V. Also, what is the percent speed regulation?

Solution:

$$I_a = I_L - I_{sf} = 40 - 1 = 39 \text{ A},$$

$$S = \frac{V_T - V_{BD} - I_a R_a}{K\phi};$$

at full load:

$$1200 = \frac{230 - 3 - 39(0.25)}{K\phi},$$

$$1200 = \frac{217.25}{K\phi},$$

$$I_{aNL} = 4 - 1 = 3 \text{ A},$$

$$S_{NL} = \frac{230 - 2 - 3(0.25)}{K\phi} = \frac{227.25}{K\phi}.$$

Ratio of speed at no load to the speed at full load:

$$\frac{S_{NL}}{1200} = \frac{227.25/K\phi}{217.25/K\phi}$$

or

$$S_{NL} = 1200\frac{227.25}{217.25} = 1255 \text{ rpm},$$

$$\text{percent } SR = \frac{S_{NL} - S_{FL}}{S_{FL}} \times 100$$

$$= \frac{1255 - 1200}{1200} \times 100 = 4.58.$$

Example 3-4. A 120-V long-shunt cumulative compound motor whose armature resistance is 0.30 ohm and whose series field resistance is 0.06 ohm has a rated speed of 1700 rpm when the armature current is 24 A. Calculate the no-load speed if the armature current drops to 2.5 A and the flux is reduced 85 percent of its full-load value. Brush drop at full load is 2 V and 1 V at no load. Calculate the speed regulation.

Solution:

$$S = \frac{V_T - V_{BD} - I_a(R_a + R_{se})}{K\phi},$$

$$1700 = \frac{120 - 2 - 24(0.30 + 0.06)}{K\phi} = \frac{109.4}{K\phi},$$

$$S_{NL} = \frac{120 - 1 - 2.5(0.30 + 0.06)}{K(0.90\phi)} = \frac{118.1}{0.90(K\phi)},$$

$$S_{NL} = 1700\,\frac{118.1/0.90(K\phi)}{109.4/K\phi} = 1700\,\frac{118.1}{(109.4)(0.90)} = 2040 \text{ rpm},$$

$$\text{percent } SR = \frac{2040 - 1700}{1700} \times 100 = 20.0.$$

3-4 SPEED CONTROL OF SHUNT AND COMPOUND MOTORS[1],[2]

The methods for controlling the speed of shunt and cumulative compound motors are the same and for this reason are considered together. In a compound motor the series field is added in the armature circuit. Basically there are two fundamental methods for controlling the speed of a shunt or compound motor. Looking at the speed formulas, Equations (3-2) and (3-2a),

$$S = \frac{E_g}{K\phi} = \frac{V_T - V_{BD} - I_a R_a}{K\phi},$$

either the value of counter EMF E_g can be changed or the value of flux ϕ can be changed. From the second formula it is apparent that the value of counter EMF can be changed in several ways while the flux change means a change in the field MMF. The methods of speed changes follow.

(1) *Field resistance control* is a method whereby the value of the excitation flux is changed. To decrease the flux, a variable resistance is added to the shunt field circuit. An increase in the shunt field circuit resistance will decrease the shunt field current, resulting in a decreased field MMF and finally reducing the flux. In the formulas a decrease in the denominator will increase the value of speed. In Figure 3-8(a) the field rheostat is shown by R_f. Added resistance lowers the field current and flux. This method will only increase the speed from the

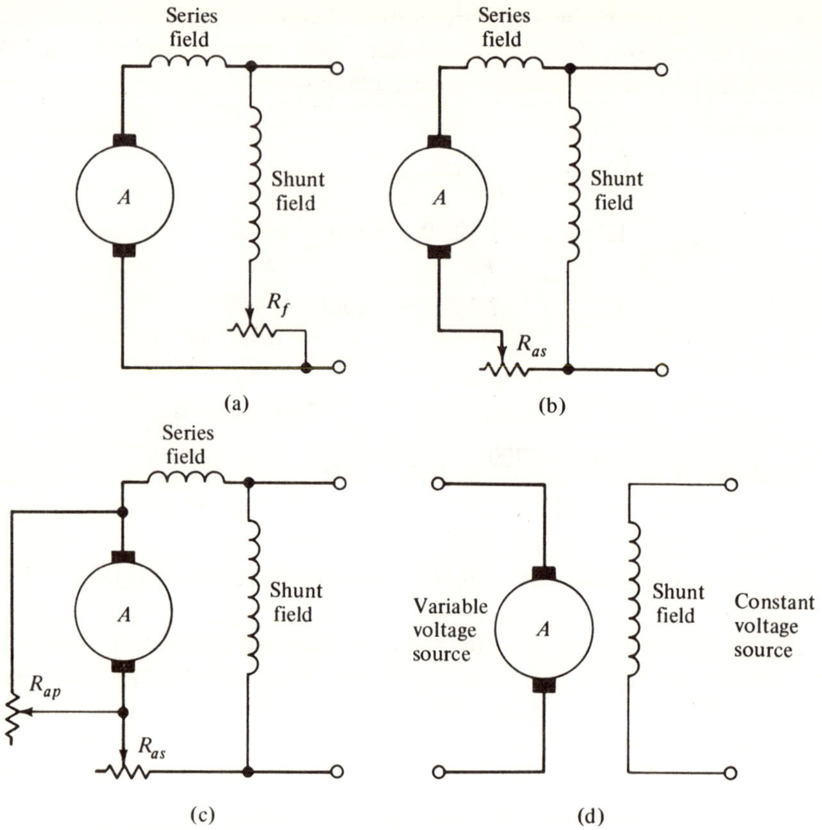

FIGURE 3-8. Schematic diagrams for speed control. (a) Shunt field resistance. (b) Series armature resistance. (c) Series and shunt armature resistance. (d) Variable armature voltage.

basic speed of the motor to a higher speed. This method has advantages. It provides smooth control of speed; it is relatively inexpensive; it is efficient since little power is lost by resistance heating; and the speed regulation is little affected. In Figure 3-9 curve *a* shows the basic speed characteristic of a shunt motor and curve *d* shows the speed characteristic for a particular value of shunt field resistance added. Care must be used not to operate motors at too high a speed. The NEMA standard MG-1 lists maximum speeds for motors. With a weakened field and high values of armature current and with the effect of armature reaction, the speed may become unstable and the motor may "run

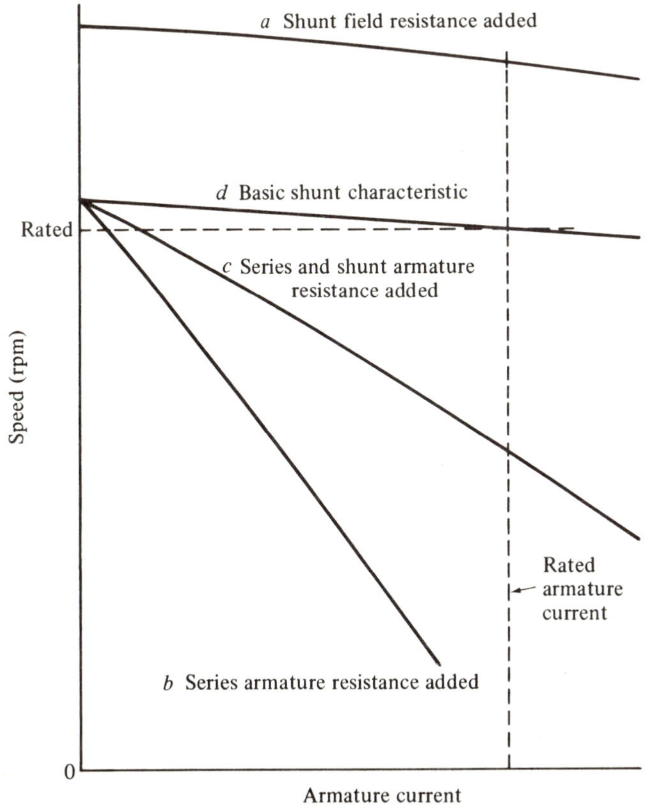

FIGURE 3-9. Speed—armature current characteristics. Curve *a*: shunt field resistance added. Curve *b*: series armature resistance added. Curve *c*: series and shunt armature resistance added. Curve *d*: shunt characteristic, no resistance added.

away." Shunt motors very often have *stabilizing windings* that are really a weak cumulative motor but with few series turns. Without these windings, the shunt speed characteristic often turns upward near rated values and beyond. If the basic speed characteristic turns up, extreme care should be taken in operating such a motor at overspeed. A further precaution in overspeed operation is called for because at high speeds commutation difficulties may arise. If excessive sparking takes place, the commutator may become scorched. A third precaution with overspeed operation is necessary in starting. A shunt motor should always be started with full field and the speed adjustment made after it has attained its basic speed.

(2) *Series armature resistance* control is the adding of resistance in the armature circuit. This method is shown in Figure 3-8(b). Examining the second speed formula, the added I_aR_{as} voltage drop will cause the value of counter EMF to lower the speed because the numerator becomes $V_T - V_{BD} - I_a(R_a + R_{as})$. In Figure 3-9 curve *b* shows the speed characteristic with a relatively large fixed value of resistance added in the armature circuit. This will provide speeds below the basic value of speed. Such resistance is simple and easy to install; it is often the same as that used for starting. Starting is taken up in Section 3-7. The cost of such resistors is relatively high because large amounts of power must be dissipated as heat. The efficiency of this system for controlling speed is extremely low. The loss is represented by $I_a^2R_{as}$. For resistance of this type, cast iron grid resistors are usually used. These grids do not offer smooth steps because they are connected into fixed steps. For smaller motors, in combination with field control, armature resistance provides an inexpensive speed-control system.

(3) The *series* and *shunt armature resistance* method as shown in Figure 3-8(c) combines an armature shunt resistance in addition to the series resistance in the armature circuit. The shunt resistance R_{ap} shunts some of the armature circuit current from the armature. The numerator of the speed equation becomes $V_T - V_{BD} - I_aR_{as} - (I_a - I_{ap})(R_a)$. The last term becomes smaller so the numerator is larger. In Figure 3-9, curve *c* shows the speed characteristic as higher than for series armature resistance alone. An example will illustrate this, too. In the torque equation, $T = K_t\phi I_a$, however, the torque also decreases. If R_{ap} is too low, the torque may become unstable. This system could be used where the load torque is variable and the speed is less likely to change. In the armature resistance alone, the torque is constant while the speed changes.

(4) *Armature voltage control* is a system whereby the impressed voltage across the armature is varied. The schematic diagram for this system is shown in Figure 3-8(d). This system is discussed in Section 3-6. When the separate source is another generator, the system is known as the *Ward Leonard system.*

Example 3-5. A 5-hp, 120-V dc motor has a rated armature current of 39 A and its rated speed is 1750 rpm. If the armature resistance is 0.25 ohm, (a) calculate the speed with 1.5 ohms added in the armature circuit; (b) calculate the value to resistance to be added to the armature circuit for a speed of 1200 rpm at rated armature current (brush drop is 2 V).

Solution:

(a) $$S = \frac{V_T - V_{BD} - I_a R_a}{K\phi} = \frac{120 - 2 - 39(0.25 + 1.5)}{K\phi},$$

$$1750 = \frac{120 - 2 - 39(0.25)}{K\phi},$$

$$S = 1750\,\frac{118 - 39(1.75)/K\phi}{118 - 9.75/K\phi} = 1750\,\frac{49.75}{108.25} = 800 \text{ rpm}.$$

(b) $$\frac{1200}{1750} = \frac{120 - 2 - 39(0.25 + R_{as})}{120 - 2 - 39(0.25)} = \frac{108.25 - 39R_{as}}{108.25},$$

$$39R_{as} = 108.25 - \frac{1200}{1750}(108.25) = 108.25 - 74.2,$$

$$R_{as} = \frac{34.05}{39} = 0.87 \text{ ohm}.$$

Example 3-6. With the same motor as that of Example 3-5, in addition to the series resistance of 1.5 ohms in series with the armature, 3.0 ohms are shunted across the armature, making the armature current 20 A. Calculate the speed.

Solution:
Referring to Figure 3-10, by the expression for V_T,

$$\begin{aligned} I_l R_{as} + (I_l - 20)(R_{ap}) &= V_T, \\ I_l(1.5) + (I_l - 20)(3.0) &= 120, \\ 1.5I_l + 3I_l - 60 &= 120, \\ 4.5I_l &= 180, \\ I_l &= 40 \text{ A}. \end{aligned}$$

The expression for the voltage across the armature V_a,

$$V_a = (I_l - 20)(3.0) = (40.0 - 20.0)(3.0) = 60 \text{ V},$$

$$S = \frac{60 - 2 - 20(0.25)}{K\phi} = \frac{53}{K\phi}, \qquad \textbf{(A)}$$

$$1750 = \frac{120 - 2 - 39(0.25)}{K\phi} = \frac{108.25}{K\phi}. \qquad \textbf{(B)}$$

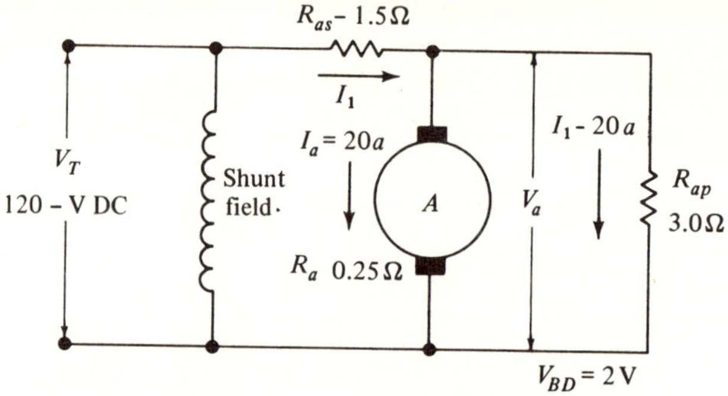

FIGURE 3-10. Circuit diagram for Example 3-6.

Ratio of (A) above to (B) is

$$\frac{S}{1750} = \frac{53/K\phi}{108.25/K\phi}$$

or

$$S = 1750 \frac{53}{108.25} = 856 \text{ rpm}.$$

3-5 SPEED-TORQUE CHARACTERISTICS OF SERIES MOTORS[2],[5]

The series motor has the field windings connected in series with the armature as shown in Figure 2-2(c). The field current and the armature current are the same. The torque expression, $T = K_t\phi I_a$, becomes $T = K_t' I_a^2$ if saturation of the field core is neglected or when the machine operates on the straight-line portion of the magnetization curve. The torque output to horsepower output is shown in Figure 3-11 by curve *b*. It is parabolic in shape to near the rated load and then due to saturation of the magnetic circuit, the line becomes straight in the overload region.

The speed output horsepower is shown in Figure 3-11 by curve *a*. The speed formula, Equation (3-2b),

$$S = \frac{V_T - I_a(R_a + R_{se})}{K\phi}$$

becomes

$$S = \frac{V_T - I_a(R_a + R_{se})}{K_l I_a},$$

because the magnetic flux is proportional to I_a as mentioned above. There-

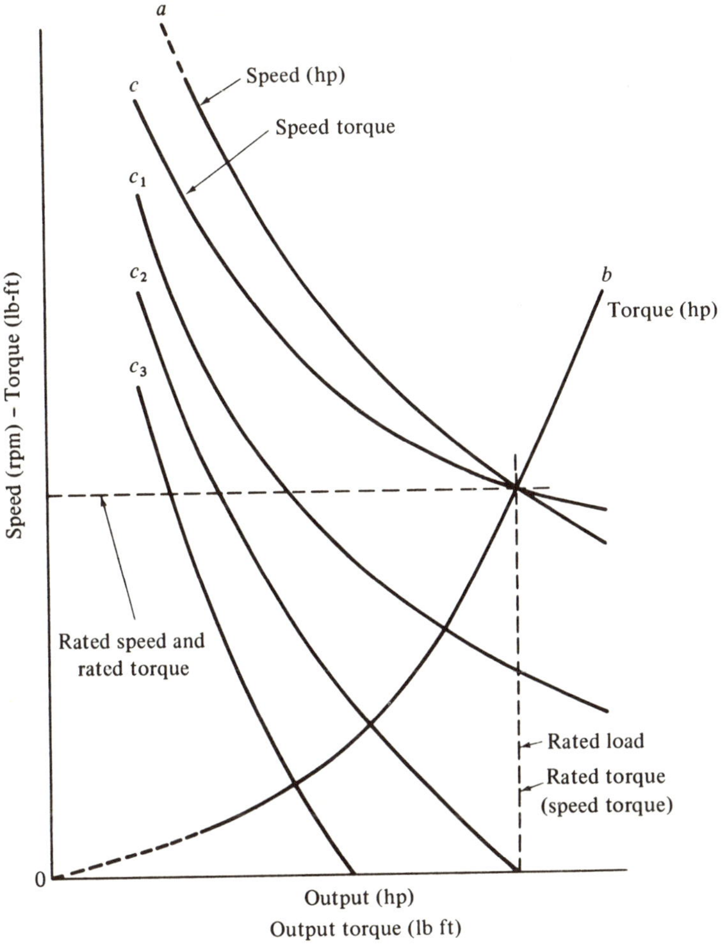

FIGURE 3-11. Series motor characteristics. Curve *a*: speed—horsepower output. Curve *b*: torque—horsepower output. Curve *c*: speed—output torque.

fore, for a light load I_a is small; the numerator becomes larger and the denominator becomes small, making the speed very high. At heavy loads I_a is large; the numerator becomes small and the denominator becomes large, which causes the speed to be low. This calls for precautionary measures; a series motor should be coupled to its load and not belt connected unless there is an overspeed centrifugal switch for protection. As with any motor, the maximum speed is limited. These overspeed values are listed in NEMA standards MG-1. There is no definite no-load speed and therefore there is no speed-regulation value for a series motor. The motor is a varying-speed motor. The speed can be adjusted by adding resistance. The series motor is an adjustable-speed motor. A diverter that has a low value of resistance can be placed in parallel with the series field to reduce the MMF of the field, lowering the flux to increase speed. The speed would be raised at the expense of lowered torque value. The series motor has a high starting torque and is used in hoists, cranes, freight elevators, and traction motors.

Figure 3-11, curve C, shows a typical speed-torque output curve for a series motor. Curve C_1 is a speed-torque curve with resistance added. Curves C_2 and C_3 are the speed-torque curves with more resistance added. From these curves the output torque can be read with a speed measurement. The torque developed is equal to the output torque plus the torque utilized to supply the stray power losses. From Equation (2-10), we find $T_d = (7.04E_gI_a)/S$, E_gI_a = power developed, and the power output is expressed by the formula,

$$P_o = \frac{T_oS}{7.04}\text{ W}, \tag{3-4}$$

$$P_o = \frac{T_oS}{5250}\text{ hp}. \tag{3-4a}$$

These formulas provide a method of determining the stray power loss of a motor (see Example 3-8 below).

Example 3-7. A 10-hp, 240-V, 850-rpm series motor has an armature resistance of 0.36 ohm and a series field resistance of 0.14 ohm. Its rated current is 38 A. Calculate the speed when the line current is 20 A assuming that the flux is proportional to the field current; assume a 3-V brush drop.

Solution:

$$S = \frac{V_T - V_{BD} - I_a(R_a + R_{se})}{K\phi}.$$

The unknown speed is

$$S = \frac{240 - 3 - 20(0.36 + 0.14)}{K(20/38)\phi} = \frac{237 - 10}{0.526K\phi} = \frac{227}{0.526K\phi},$$

at rated,

$$500 = \frac{240 - 3 - 38(0.36 + 0.14)}{K\phi} = \frac{218}{K\phi},$$

$$\frac{S}{500} = \frac{227/(0.526K\phi)}{218/K\phi} = \frac{227}{(0.526)(218)},$$

$$S = 500\frac{227}{218(0.526)} = 988\text{ rpm}.$$

Example 3-8. A 240-V, dc series motor has a load current rating of 38 A and 1150 rpm. The armature resistance is 0.15 ohm and the series field winding is 0.10 ohm. From the speed-output torque curve, the rated value of torque is 49 lb-ft at rated speed of 1150 rpm. The brush-drop resistance is included in the armature resistance. Calculate the stray power loss.

Solution:

$$E_g = V_T - I_a(R_a + R_{se}) = 240 - 38(0.150 + 0.10),$$
$$E_g = 230.5 \text{ V}.$$

Power developed is

$$P_d = E_g I_a = 230.5 \times 38 = 8750 \text{ W}.$$

Power output is

$$P_o = \frac{(T_o S)}{7.04} = \frac{49 \times 1150}{7.04} = 8000 \text{ W},$$

$$\text{stray power} = P_d - P_o = 8750 - 8000 = 750 \text{ W}.$$

3-6 ADJUSTABLE VOLTAGE CONTROL FOR DIRECT-CURRENT MOTORS[1],[6]

In Section 3-4, a fourth method of speed control, armature voltage control, was mentioned. The Ward Leonard system was also mentioned as a system whereby the variable voltage is supplied by a rotating machine —namely, a shunt generator. In the speed formula for a shunt motor, $S = (V_T - V_{BD} - I_a R_a)/K\phi$, V_T, the voltage applied to the motor armature is varied and the motor is separately excited. This variable voltage in the Ward Leonard system is supplied by a separately excited generator.

The schematic diagram is shown in Figure 3-12 for a basic Ward Leonard speed-control system. As shown three rotating machines are needed for the controlled motor as follows: (a) a *driving motor*, which usually is a constant speed, three-phase, ac induction motor; (b) a *small exciter* (generator) to provide excitation for the motor and the generator; and (c) a *separately excited generator*, which provides the variable voltage for the armature of the controlled motor. These three machines are coupled together. Where a dc source is available, however, the exciter is not needed. A field rheostat can be inserted in the motor field circuit. From the torque equation, $T = K_t \phi I_a$, if the motor flux is changed by adding resistance in the field circuit to decrease the flux, the armature current I_a will have to increase if the torque is constant. Loads usually are such that the torque requirement is constant. The diagram

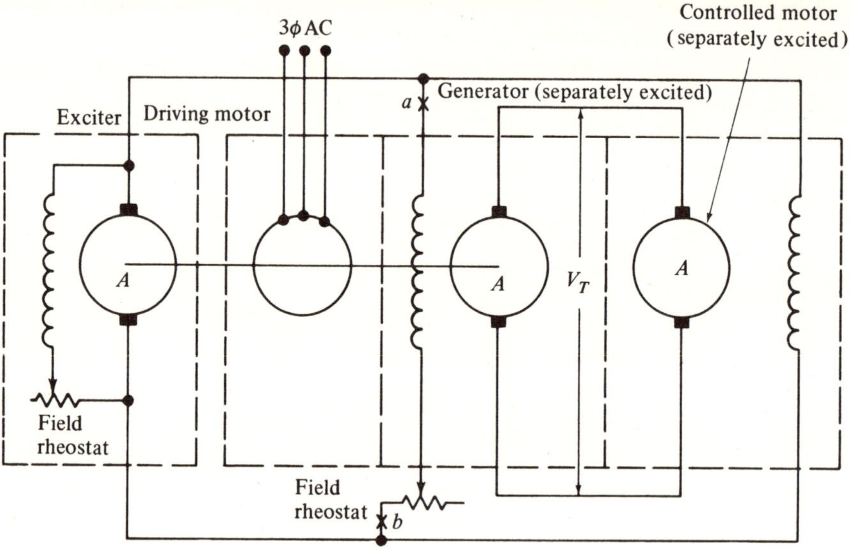

(a)

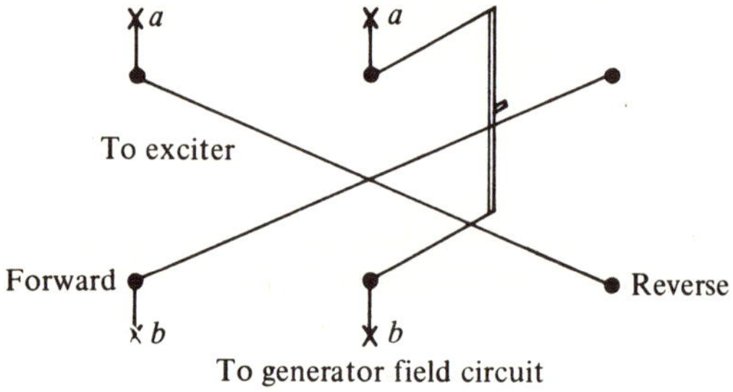

(b)

FIGURE 3-12. (a) Ward Leonard speed-control schematic diagram. (b) Exciter switch for reversing rotation of motor.

of Figure 3-12 is for unidirectional operation. In Figure 3-12(b) a reversing switch is shown. The armature field circuit could be broken at *a* and *b* with the left side of the switch connected to the exciter or source and the center connected to the generator field.

In the Ward Leonard system of control there is a smooth and stepless change in speed from standstill to maximum. Armature currents are never excessive unless the motor rheostat is cut in or the motor is overloaded. Reversal of direction of rotation is easily accomplished without excessive armature currents and without having to add series armature resistance. The speed regulation of the motor is good at any speed of operation. The motor operates at high efficiency as compared to the armature resistance control method. There are several disadvantages to this system, however, the first of which is the overall cost of the machines needed. Each motor that is controlled needs its individual system. The second disadvantage is that the overall efficiency is low. If the controlled motor is operating at rated load, the overall efficiency would be around 60 percent. If the motor operates for long periods at light load, the overall efficiency becomes quite low. However, the advantages far offset the disadvantages for such uses as operation of passenger elevators, rolling and paper mill drives, large hoists, and many other large applications that require adjustable speeds.

The basic system for large units in which speed ranges are great must be modified by the use of two exciters in opposition or by the use of a control generator such as the *Amplidyne*, *Rototrol*, or *Regulex* generators. The basic system is a rotary amplifier since a small amount of change of excitation power changes the larger power output of the load. So for large loads a greater amplification is needed to improve the response time of the basic system.

For small motors having fractional horsepower to low integral horsepower ratings, the variable dc source can be a gas or *thyratron tube rectifier*, or a *silicon-controlled rectifier* operated from an ac source. For dc motors from 1 to 15 hp, ac power is rectified by silicon-controlled rectifiers. For larger motors to several hundred horsepower, rectification of ac by *magnetic amplifiers* with silicon diodes may be used for the variable dc source. The silicon-controlled rectifier (SCR) and the magnetic amplifier with silicon diodes are replacing the gas-filled tube-type rectifiers because of maintenance and reliability factors. In the Ward Leonard system itself, the silicon-controlled rectifier is replacing the small dc rotating exciter to supply the variable dc excitation of the generator that supplies the variable voltage of the controlled-speed motor.

3-7 STARTING OF DIRECT-CURRENT MOTORS[1],[6],[7]

A motor at the instant of start is at standstill and in the starting operation must be accelerated to its running speed. It has been established that a motor has a generated voltage (counter EMF) when it is running. When the motor is running, the *armature current* is given by the formula,

$$I_a = \frac{V_T - K\phi S}{R_a}, \tag{3-5}$$

where

V_T is the applied terminal voltage in volts,
K is a constant depending on the physical properties of the motor,
ϕ is the flux per pole in lines (maxwells),
R_a as used here is the resistance of the armature circuit.

From this equation when the motor is at standstill, S (speed) is zero and the counter EMF ($K\phi S$) is zero, and I_a would be quite large because R_a is a fraction of an ohm. Therefore when starting any dc motor, resistance must be inserted in the armature circuit as shown in Figure 3-13(a) and (b). This

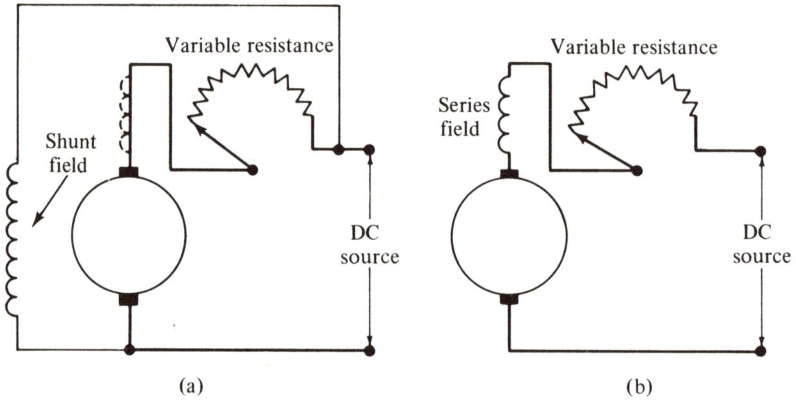

FIGURE 3-13. (a) Shunt or compound motor with starting resistance. (b) Series motor with starting resistance.

resistance must be cut out as the motor accelerates to attain the rated speed.

The starting procedure may be accomplished with manual starters or automatic starters. Some manual starters that use the starting resistance for speed control are called *controllers* because they also function to control the motor. Controllers are taken up in the next section of this chapter. Manual starters are of two types: the three-point starters and the four-point starters.

The three-point starter often is referred to as the three-point face-plate starter, because for larger motors the device is mounted on an open-face plate (usually slate). Starters for smaller motors are enclosed in a metal box. Figure 3-14 shows a wiring diagram and Figure 3-16(a) shows the schematic

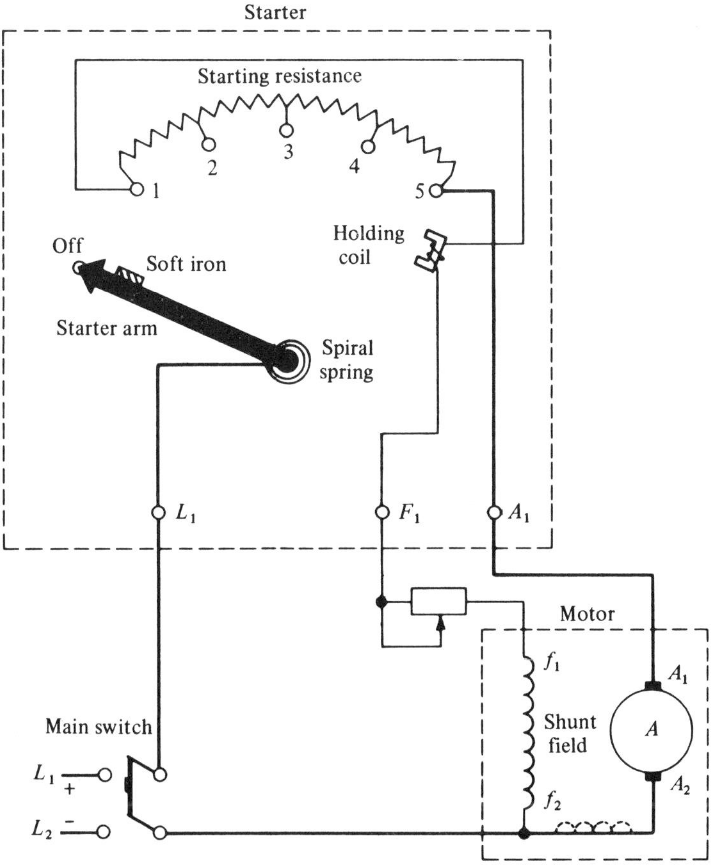

FIGURE 3-14. Manual three-point starter for shunt or compound motor.

diagram of the three-point starter. These diagrams show a four-step starter: as the arm is brought up the resistance is cut out in four steps. In series with the shunt field of the motor is a holding coil that holds the arm in place with the soft iron keeper. The main switch must be opened to stop the motor. When the main switch is opened, the arm is returned to the off position by a spiral spring and the arm is in position for the next start. If examination of the shunt field circuit shows that too much resistance is cut in to increase the speed, the spiral spring pull could return the arm to the off position and the motor would stop. Also if the voltage of the source would drop, unless the drop is large, the arm would not drop to off. Therefore it can be said that this type of starter does not provide for undervoltage protection.

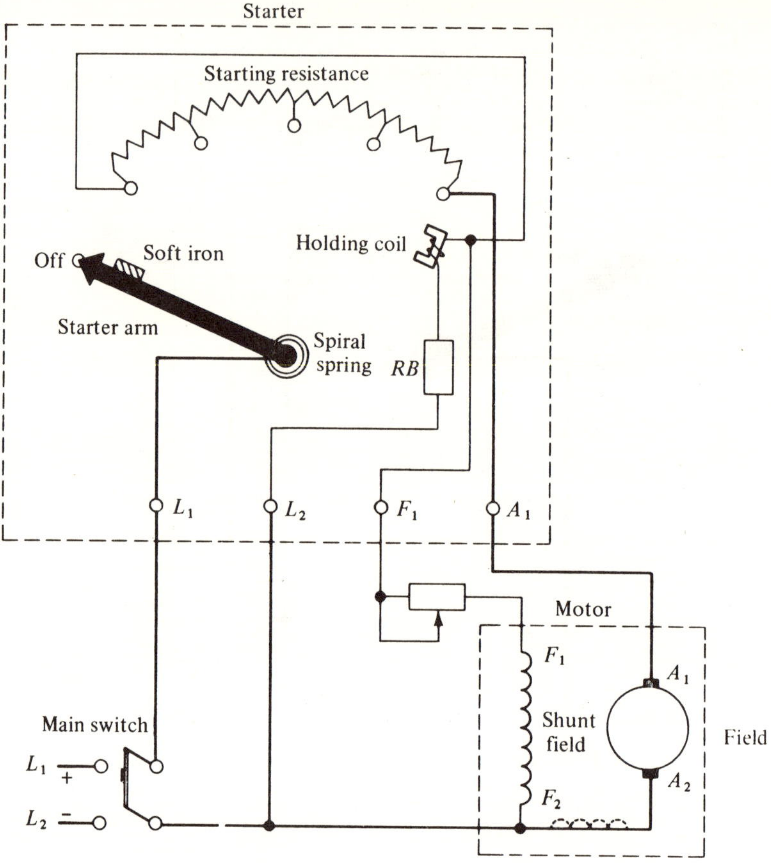

FIGURE 3-15. Manual four-point starter for shunt or compound motor.

The four-point starter does provide for undervoltage protection. Figure 3-15 shows the wiring diagram of a four-point starter while Figure 3-16(b) shows the schematic diagram. In this starter the holding coil is across the line with a protective resistance *PR* in series. There are three parallel paths and a change in the shunt field circuit will not affect the holding circuit. Also, the holding coil circuit can be adjusted with the series protective resistance to stop the motor at a preset value of undervoltage. One disadvantage that the three-point starter does not have is open-shunt field protection. To stop the motor with the four-point starter, the main switch is opened just as with the three-point starter.

Manual starters are not adaptable for large motors, especially over 15 hp

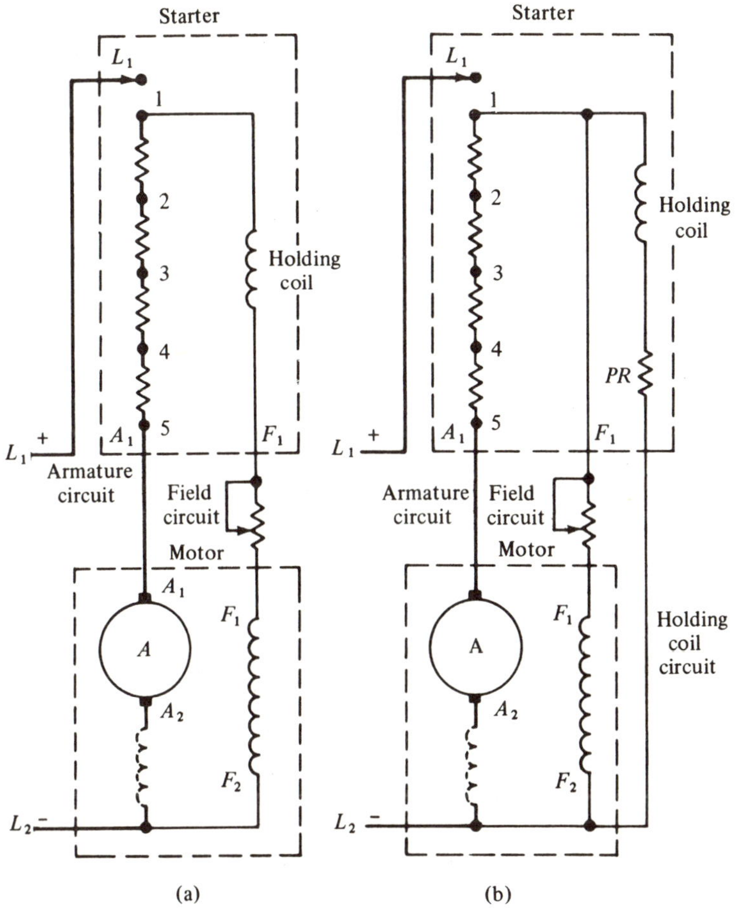

FIGURE 3-16. (a) Schematic diagram for three-point starter (Figure 3-14). (b) Schematic diagram for four-point starter (Figure 3-15).

rating. The contacts pit because of arcing at the contacts. The operator may be haphazard in using the manual starter, either pulling the arm up too quickly or too slowly. If the starting arm is pulled up too fast, the motor may be damaged; if the arm is drawn up too slowly, the starting resistors may overheat and be damaged. On the other hand manual starters are lower in cost than automatic starters. The maintenance cost is less for the manual starter than for the automatic starter. An advantage of the manual starter is that the starting time can be more easily controlled to meet the varied starting situations. Manual starters are more compact than automatic starters, but at the point of operation they take up more space than a start and stop button of the automatic starter.

Automatic starters are of two basic types: (a) the *current* or *load-limit* type and (b) the *definite-time* type. In the current or load-limit type, the speed of acceleration will depend on the load through the amount of current the motor draws. The definite-time type has a prescribed time for the acceleration irrespective of the load. In some process operations this may be an important factor.

There are numerous schemes employed for the automatic load-limit type depending on the types of relays and contactors that are used. The series-relay current limit, the counter-EMF relay, and the current-limit contactor are the most commonly used types. Figure 3-17 shows a three-step starter using *counter-EMF relays.* The relays in turn energize contactors which short out the steps of starting resistance. The operation of the relays depends on the value of counter EMF across the armature of the motor. In this starter the shunt field rheostat is shorted during acceleration by the normally closed contact 3*A* across the field rheostat so that the motor cannot be started with a weakened field by the rheostat setting. The sequence of operation for this starter is as follows. The main switch is closed. The start button is pressed down energizing the coil of contactor *M*. All *M* contacts close. *M* in the armature circuit closes with R_1, R_2, and R_3 in the circuit. M_a (auxiliary) contact shorts the start button so that the operator need not hold down the start button that is normally open. When the counter EMF reaches about 50 percent of full value, relay 1*AR* will operate. Contact 1*AR* energizes contactor 1*A* which closes contact 1*A*, shunting out the first step of resistance R_1. When the counter EMF across the armature reaches about 75 percent of full value, relay 2*AR* operates closing contact 2*AR* which energizes contactor 2*A*. Contact 2*A* closes cutting out the second step of starting resistance R_2. When the counter EMF reaches about 90 percent of full value, relay 3*AR* operates with contact 3*AR* closing and energizing contactor 3*A*. Contact 3*A* shunts out the last step of starting resistance R_3 and the motor is now running with all armature starting resistance shunted. Also normally closed contact 3*A* across the shunt field rheostat opens thereby cutting in the rheostat to its preset value for the proper speed. For smaller motors up to 3 to 5 hp the relays may be replaced by contactors whose operation depends on the value of counter EMF across the armature.

The counter-EMF type of starter will accelerate the motor in relation to the motor load. It will also offer some overload protection by the possibility that the 3*AR* relay will open and thus cut in the R_3 step of resistance in the

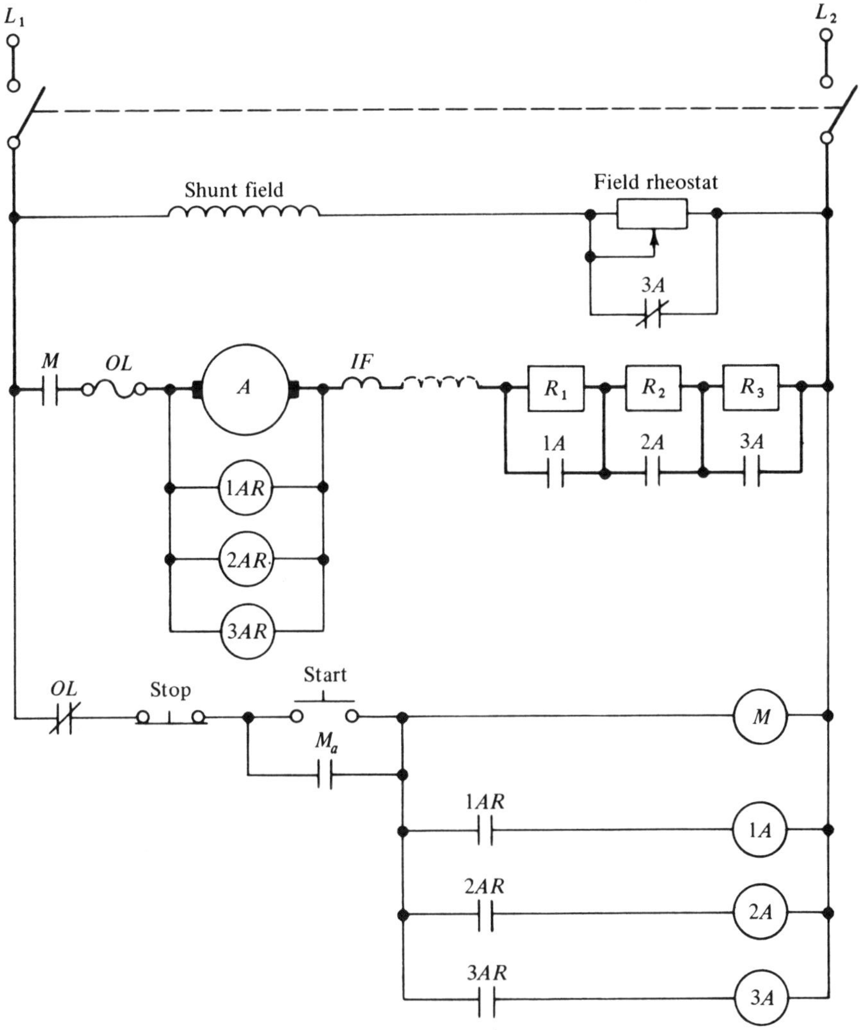

FIGURE 3-17. Schematic diagram of counter EMF push button starter.

armature. One disadvantage of this type starter is the number of relays and contactors that are continuously energized while the motor is running. Another disadvantage is that if the armature resistance is used for speed control, as the counter EMF decreases, more resistance of the starter would be added in the armature circuit.

Many types of *definite-time* starters have been developed, all of which have contactors closing at definite, progressively longer time intervals. Some of the types are mechanical timer, dashpot relay, inductive time delay, and neo timer, which utilizes neon tube. Figure 3-18 shows a *definite-time* starter

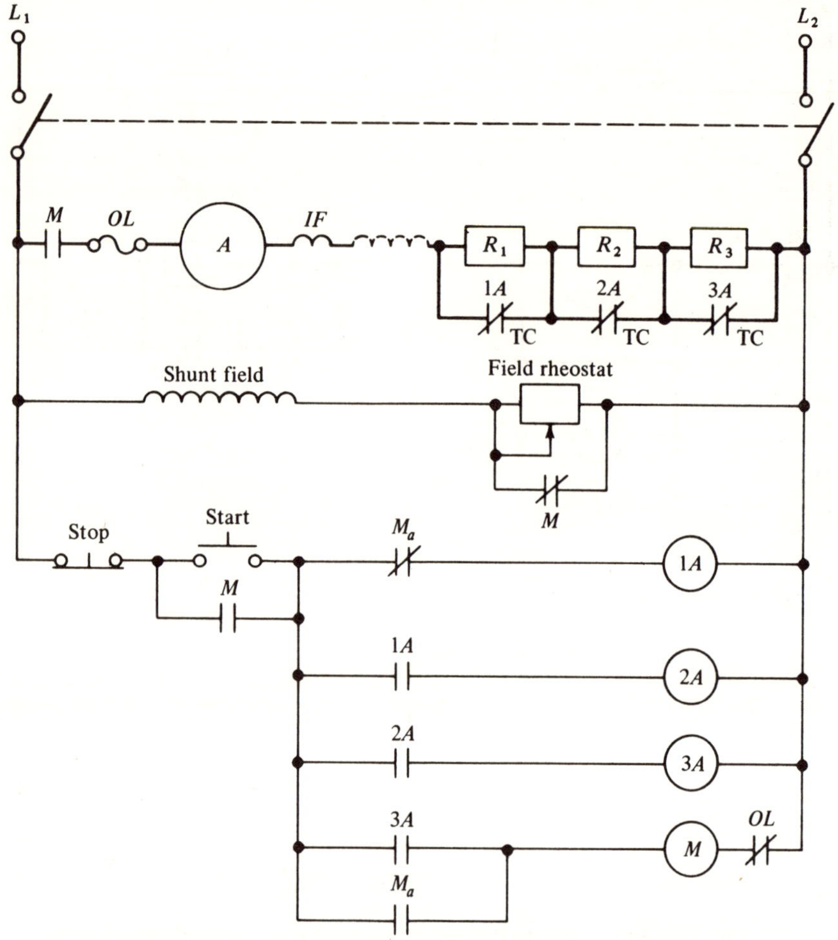

FIGURE 3-18. Schematic diagram of time-delay contactors push button starter.

that uses definite-time delay contactors without relays. The contactors are made with copper sleeves in which currents are induced in accordance with Lenz's law when the coil is de-energized, delaying the operating of normally closed contacts. The time is adjusted by the number of sleeves inserted. The circuit operates as follows: The start button is depressed energizing coil 1*A* through normally closed auxiliary contact M_a, and contactors 2*A*, 3*A*, and *M* in succession. These contactors have negligible delay in the energizing action, and the start button must be momentarily depressed long enough for contactors 1*A*, 2*A*, 3*A*, and *M* to pick up. The normally open contact M_a will keep the *M* coil energized after the start button is released and the main contact *M* will close the armature circuit with all resistance R_1, R_2, and R_3 in the circuit. The normally closed contact M_a opens when *M* is energized and coil 1*A* is de-energized. The closing of normally closed contact 1*A* is delayed by current flow in the copper sleeves in accordance with Lenz's law. Contactors 2*A* and 3*A* are de-energized successively with definite time delay cutting out R_2 and R_3. The normally open contact *M* across the field rheostat opens cutting in the field rheostat to its preset value.

In this type of starter when the motor is running, only contactor *M* is energized. The contacts 1*A*, 2*A*, and 3*A* across the armature resistors R_1, R_2, and R_3 are open before they carry current, nor do they interrupt current. Therefore, the contact gaps are short and not much power is required to operate them. The overload relay in this starter as well as in all automatic starters is a relay whose normally closed contact in the *M* coil circuit simply opens on a sustained overload.

Automatic starters eliminate any human error in the starting of motors. The stations can be placed conveniently for the machine operator. Another advantage of automatic starters is in the control of motors for reversing, braking, jogging; such controls can easily be added to the basic starter. The subject of control of motors is discussed in the next section. The subject of starters and controllers covered here is a small part of the subject and the references should be consulted for a more complete study.

Example 3-9. A 240-V, 20-hp shunt motor has a rated armature current of 70 A. The starter has three steps of resistance added in the armature circuit as shown in Figures 3-17 and 3-18. The armature circuit resistance including brushes is 0.25 ohm. The maximum in-rush currents are to be limited to 175 percent of rated value and successive resistances are shunted when the current drops to rated value. Calculate the values of R_1, R_2, and R_3.

From Equation (3-5), solving for *R*(added) yields

$$R(\text{added}) = \frac{V_T - E_g}{I_a} - R_a. \tag{3-5a}$$

Solution:

Step 1: $R_1 + R_2 + R_3$ in armature circuit:

$$R_1 + R_2 + R_3 = \frac{240 - 0}{70 \times 1.75} - 0.25 = 1.96 - 0.25 = 1.71 \text{ ohms.}$$

Step 2: E_g when R_1 is shunted:

$$E_g = V_T - I_a(R_a + R_1 + R_2 + R_3) = 240 - 70(0.25 + 1.71)$$
$$= 240 - 137.5 = 102.5 \text{ V}.$$

Step 3: $R_2 + R_3$ in armature circuit after R_1 is shunted:

$$R_2 + R_3 = \frac{240 - 102.5}{70 \times 1.75} - 0.25 = 1.12 - 0.25 = 0.87 \text{ ohm}.$$

Step 4: E_g when R_1 and R_2 are shunted:

$$E_g = 240 - 70(0.25 + 0.87) = 240 - 78.5 = 161.5 \text{ V}.$$

Step 5: R_3 in armature circuit after R_1 and R_2 are shunted:

$$R_3 = \frac{240 - 161.5}{70 \times 1.75} - 0.25 = 0.64 - 0.25 = 0.39 \text{ ohm}.$$

Step 6: value of R_2:

$$R_2 = (R_2 + R_3) - R_3 = 0.87 - 0.39 = 0.49 \text{ ohm}.$$

Step 7: R_1 value:

$$R_1 = (R_1 + R_2 + R_3) - (R_2 + R_3) = 1.71 - 0.87 = 0.84 \text{ ohm}.$$

The values of the counter EMF, the armature voltages, and the percentages of rated voltage across armature for the steps above are as follows:

Just before R_1 is shunted, $E_g = 102.5$ V, $V_a = 120$ V or 50 percent.
Just before R_2 is shunted, $E_g = 161.5$ V, $V_a = 161.5$ V or 74.5 percent.
Just before R_3 is shunted, $E_g = 195.2$ V, $V_a = 212.7$ V or 88.5 percent.

3-8 CONTROL OF DIRECT-CURRENT MOTORS[1],[6],[7]

In the preceding section only starting of motors was considered. Starting is one operation of controlling motors. Other control operations are reversing direction of rotation, jogging, dynamic braking, plugging, and speed control. Most of these control functions are best performed in connection with automatic starters except speed control which is usually a manual operation.

Figure 3-19 is a manual controller for shunt field control with a four-point starter. This is often called a compound starter and, in fact, the first operation is the starting operation with full-shunt field. After the motor has been accelerated to normal speed, the speed may be increased by cutting-in resistance in the shunt field with the built-in field rheostat. The controller has two arms—one the armature resistance arm, similar to the arm of the four-point starter, and the other the field resistance arm. The field resistance arm

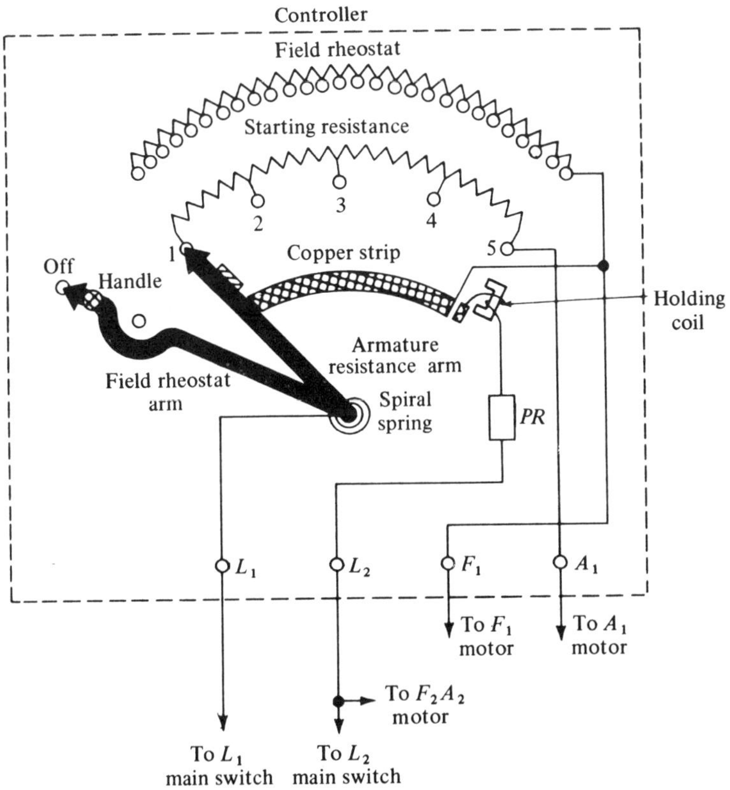

FIGURE 3-19. Manual controller for shunt field control with four-point starter.

pushes the armature resistance arm through the five positions after which it is held by the holding coil just as is the four-point starter. After the armature resistance arm is held by the holding coil, the field rheostat arm may be moved counterclockwise to cut in resistance in the field circuit. To stop the motor the main switch is opened and the spiral spring returns both arms to the off position. As with the four-point starter, the size of the motor is limited. It should also be observed that the brush on the copper strip shunts out the field rheostat during the acceleration period.

The *drum controller* is a manually operated controller that utilizes the same armature resistors for the starting operation and for speed control. The contacts are fingers that slide on contact strips which are fastened to a drum. Arc chutes and barriers are placed between each sliding contact. The arc chutes minimize the pitting of the contacts. Positions are definitely located by

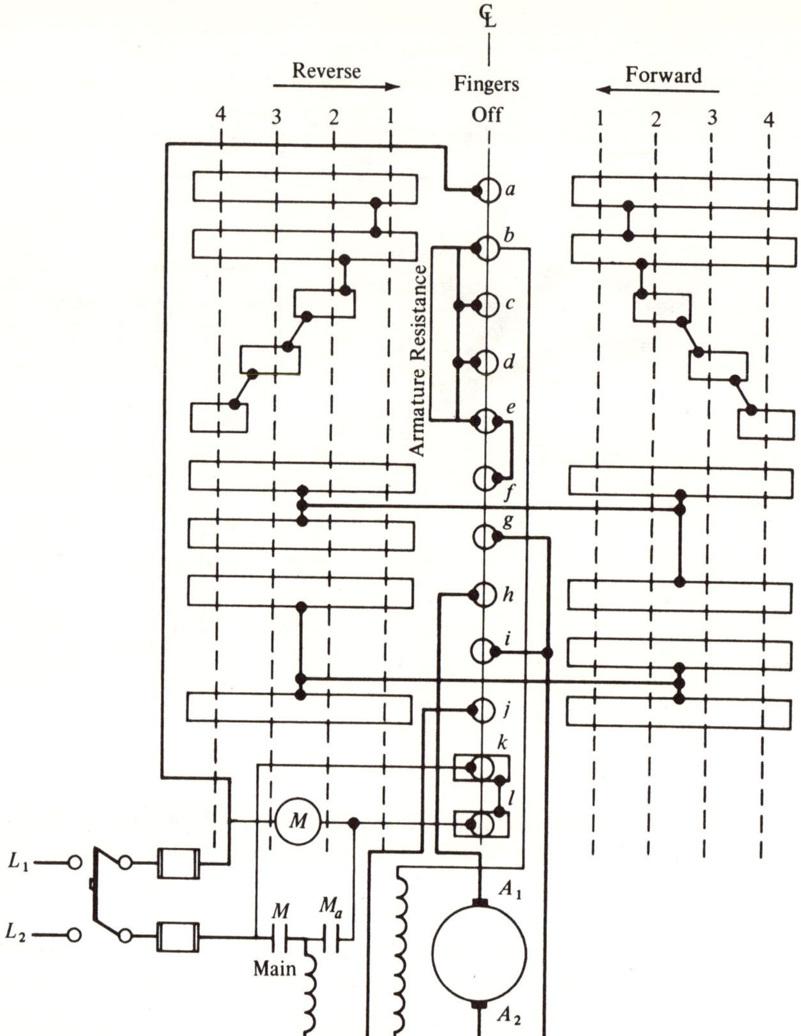

FIGURE 3-20. Manual reversing drum controller for shunt, compound, or series motor.

a roller that rides on a round notched wheel. The spring pressure on the roller positions the drum on which the contacts are placed. Figure 3-20 depicts a manual reversing drum controller for a compound motor. For a series motor the shunt field would be omitted, and for a shunt motor the series field would be omitted. The figure shows the drum laid out flat with the finger contacts located on the center line off position. This controller will reverse direction of rotation and slow down the speed by the addition of resistance in the armature circuit. The positions are shown by dotted lines with the position 1 being the slowest speed and position 4 being the maximum speed. Forward direction for rotation is indicated by motion of the drum in a clockwise direction. Moving the drum in a counterclockwise direction reverses the armature and the motor then rotates in a reverse direction. This controller provides *undervoltage protection.* At reduced line voltage the M contactor will open and disconnect line L_2 to the motor. Before the motor can be started again, the controller must first be moved to the off position for contactor M to pick up. On this position finger contacts k and l are placed over two short strips on the drum. On this position coil M is energized which connects the motor to line L_2 through contact M. M_a shunts out contacts k and l thereby keeping the circuit in readiness to move to position 1 forward or reverse. When position 1 is across the contactors, the circuit is as follows: L_1 to finger a to finger b through the drum segments which are connected together. From finger b through the full armature resistance to finger e, to finger f which is over the sixth contact segment. This segment is connected to the segment that is over finger h. Finger contact h is connected to A_1 (motor armature). A_2 is connected to fingers g and i. On forward 1, the i segment is on finger i. This segment is connected to the j segment and also to finger j which is connected through the series field of the motor to L_2. The shunt field is connected to the j finger lead and to finger b. This completes the motor connection for a short shunt compound motor with all armature resistance added in the armature circuit. By moving the controller arm to position 2 a step of resistance is cut out. On position 3 another step of resistance is cut out and, finally, on position 4 all the resistance is cut out and the motor operates at normal speed. A *field rheostat* can be and usually is placed in series with the shunt field to increase the motor speed.

The drum controller as described above is limited by size and by functions such as quick stopping, quick reversing, quick starting, and jogging. The drum controller is manually operated and the functions are controlled by the operator. Automatic controllers with push buttons or with master switches are used for larger motors and also perform the operational functions automatically to eliminate manual errors. Automatic push button controllers for quick reversing, quick stopping, jogging, and shunt field speed control are used with shunt and compound motors. For hoists, cranes, or mill drives, series motors are almost always used and for such operations the master-switch method is used.

The method used for reversing the direction of rotation of a motor can be accomplished in two ways: (a) Reversing the direction of the flux in the field or (b) reversing the direction of current flow through the armature. The latter is the method usually used since it is difficult to reverse the field(s) because of the high inductance of shunt fields. The schematic diagram of

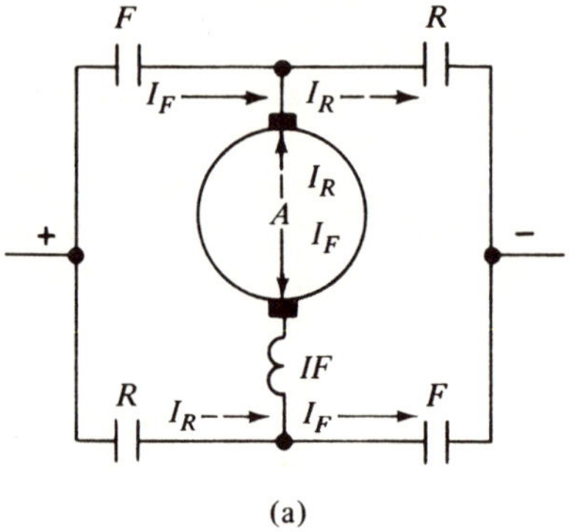

(a)

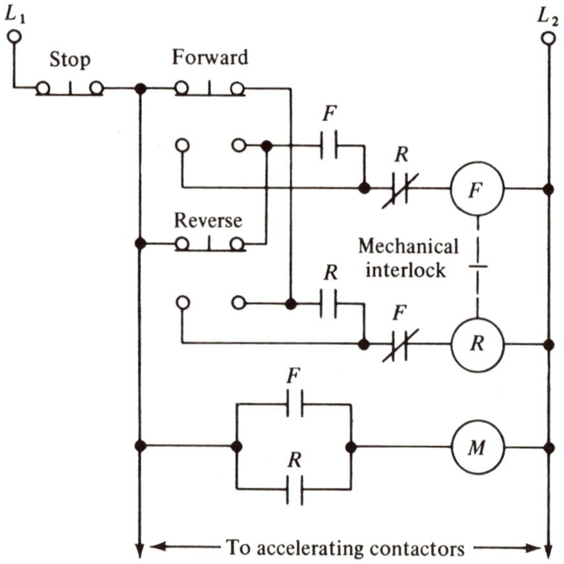

(b)

FIGURE 3-21. **(a) Schematic diagram showing how current flow in armature of motor is reversed. (b) Schematic diagram of control circuit for armature reversing of (a).**

Figure 3-21(a) shows the forward and reverse contacts for reversing the current through the armature. Figure 3-21(b) shows the control circuit having both electrical interlock and mechanical interlock. Electrical interlock is accomplished by the normally closed contacts *R* and *F* in the control circuit so that both *F* and *R* contactor coils cannot be energized at the same time. Mechanical interlock is accomplished on the *F* and *R* contacts in Figure 3-21(a) so that only one pair can be closed at any time. A complete schematic diagram for automatic reversal of a compound motor is shown in Figure 3-22. The timing of the closing of contactors 1*A*, 2*A*, and 3*A* are timed by dashpot relays *TR*1, *TR*2, and *TR*3. For smaller motors, time-closing auxiliary contacts can be used. In Figure 3-22 if the motor is running in the forward direction and the reverse button is depressed, all *M*, 1*A*, 2*A*, 3*A*, *TR*1, *TR*2, and *TR*3 contacts open. *R* contactor coil is energized and armature current will be reversed with R_1, R_2, and R_3 in the armature circuit. Actually the motor will come to rest quickly and proceed to accelerate in the reverse direction in regular fashion.

In the reversing controller that we have been discussing—reversing of the armature while the armature is turning in its respective direction—the armature comes very quickly to a stop. If instead of the armature accelerating normally in the opposite direction, it remains at rest, the process is called *plugging*. To do this a resistance must be placed in the armature circuit. In Figure 3-23(a) the plugging resistance with a normally closed *P* contact is shown. The plugging contactor is operated by a plugging relay. One type is attached to the shaft of the motor and when the motor is turning, the plugging relay is energized, but the coil of the plugging contactor is not energized since a normally closed *M* contact is in series with this circuit. Whenever the stop button is operated or if used with a reversing controller, whenever the *M* contactor is de-energized, the normally closed plugging contact *P* will open and insert the plugging resistance. In the automatic controller of Figure 3-22 without the added plugging resistance, the current flow would be almost double the starting current because the voltage appearing across the armature is practically doubled at the beginning of the plugging process. For this reason plugging is usually connected with any automatic reversing circuit.

Dynamic braking is a second method to obtain quick stopping. In dynamic braking a resistance is placed across the armature and the inertia of the motor rotation results in generator action with the energy being dissipated in the dynamic braking resistance. Figure 3-23(b) illustrates the schematic diagram of this method. As shown in the figure, part of the accelerating resistance is part of the load resistance.

Regenerative braking is another method for quick stopping of a motor.

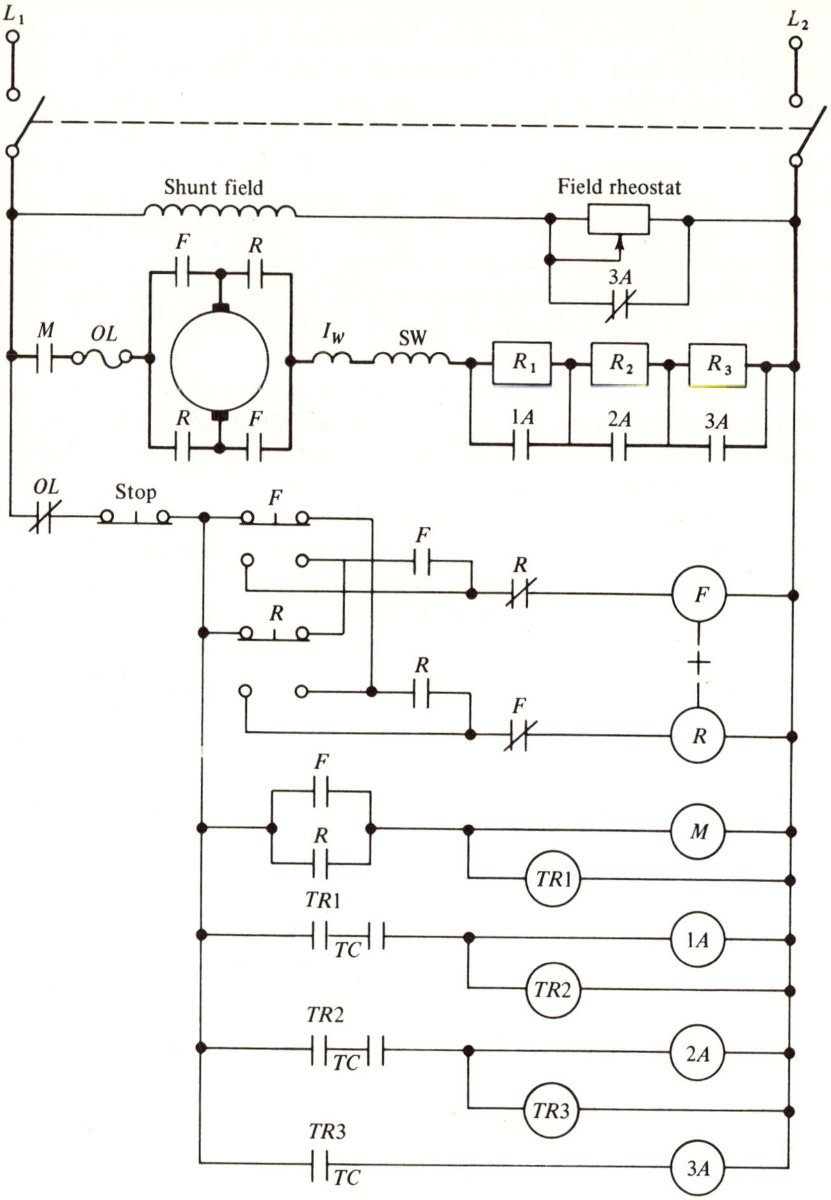

FIGURE 3-22. Schematic diagram for automatic reversing controller for compound motor using time delay relays.

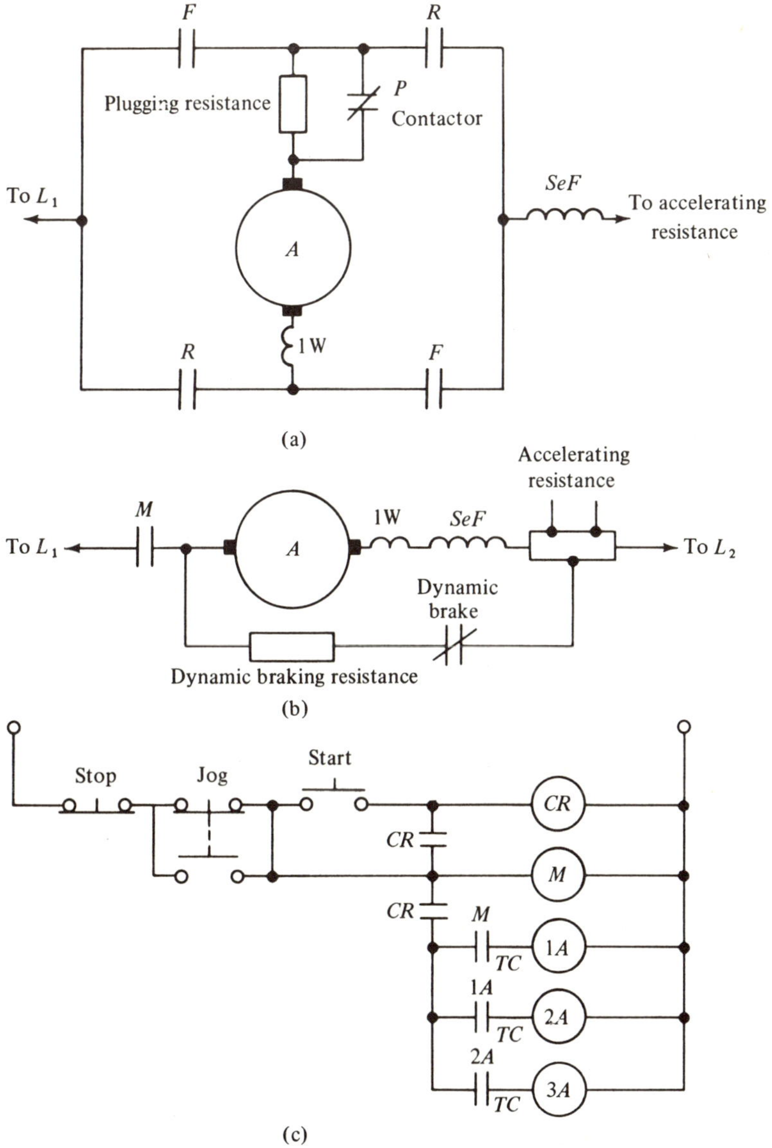

FIGURE 3-23. (a) Basic plugging resistance braking schematic diagram. (b) Basic dynamic braking schematic diagram. (c) Jogging diagram for jogging in one direction.

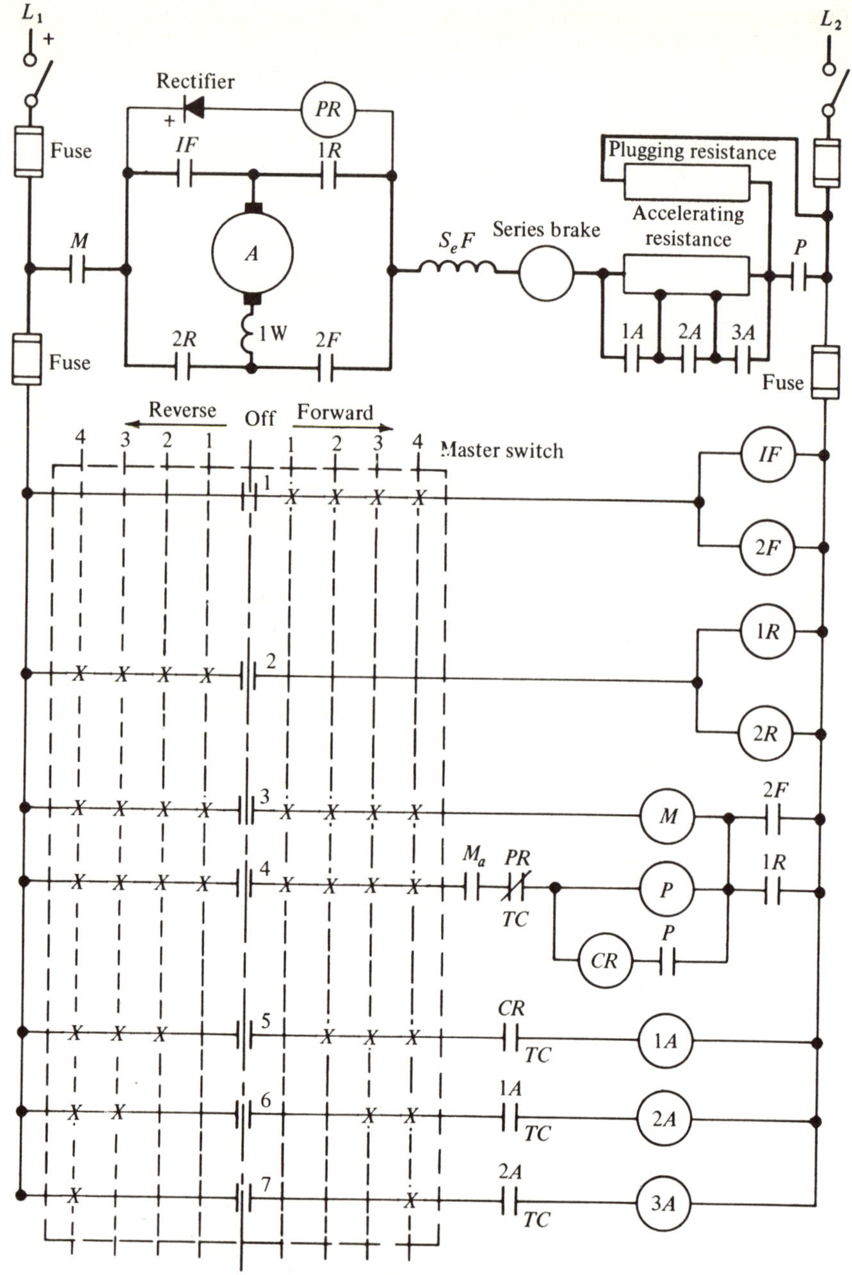

FIGURE 3-24. **Schematic diagram for reversing controller for a series motor. (Note: 1*F* and 1*R*, 2*F* and 2*R*, 2*F* and 1*R* contactors are interlocked mechanically.)**

This method is somewhat similar to dynamic braking but instead of the inertia generator action dissipating the energy in a resistance, the energy is fed into the line during the regenerative braking period.

Another control item for motors is *jogging*. Jogging is the moving of the motor by small amounts; it is needed in many tool and textile operations where alignment is essential. It is accomplished by a jogging button that starts the motor rotation with all the accelerating resistors in the armature circuit. Figure 3-23(c) shows how the jogging button is connected. As can be seen, if the jogging button is tapped quickly, the motor will rotate a very small amount.

Master switches may be of the *drum* type or the *cam* type. The drum type is similar to the drum controller with the contact fingers and segments placed on a drum. In cam switches the individual switches operate by cams that are moved by a lever. In both cases the master switches operate contactors that function directly in the drum controller. The operation of a motor by master switches is manual and is used on equipment such as hoists, cranes, freight elevators, or locomotives. Figure 3-24 shows the schematic diagram for a reversing controller for a series motor. A reverse plugging resistor is added to the accelerating resistors. The contactors and the *CR* relay are time-delay closing types and may be either dashpot time delay or variable magnetic-circuit time delay. The plugging relay *PR* is in series with a diode that passes current only in one direction (left to right). The *x*'s indicate the contacts that are closed. For instance, if the lever is moved to the forward 4 position, master switch contacts 1, 3, 4, 5, 6, and 7 are closed. Contact 1 energizes contactors 1*F* and 2*F*. Contact 3 energizes contactor *M*, which starts the motor with all resistance in the armature circuit including the plugging resistance. Contact 4 energizes *P* contactor through M_a contact and normally closed contact of relay *PR*. The plugging resistor is immediately cut out. Relay *CR* is energized by *P* contact. After a time delay, *CR* contact closes and energizes 1*A* contactor through master contact 5. Master contacts 6 and 7 are also closed and contactors 2*A* and 3*A* are energized in sequence and cut out the accelerating resistance. If the master is moved to off, all contacts open and the motor is stopped by the series solenoid brake. If, however, the master were moved rapidly to any reverse position, contactors 1*R* and 2*R* would be energized instead of 1*F* and 2*F* and the direction of current flow through the armature would be reversed. The voltage across the plugging relay is now equal to the impressed voltage plus the counter EMF of the motor armature. This voltage value activates the plugging relay *PR* and opens the normally closed *PR* contact. Contactor *P* is deactivated adding the plugging resistance to the accelerating resistance to bring the motor to a rapid stop from the forward direction. When the motor is at standstill, the voltage across the plugging relay *PR* is equal to the line voltage, *PR* is deactivated, and the motor accelerates in the reverse direction in a normal manner.

The following example illustrates the calculation of the value of *plugging resistance.*

Example 3-10. In Example 3-9 find the value of plugging resistance if the inrush current is to be held to 175 percent of the full-load value.

Solution:

The total voltage at instant plugging takes place as follows:

$$V_P = V + E_g = V + (V - I_{FL}R_a)$$
$$= 240 + (240 - 70(0.25)) = 462.5 \text{ V},$$

$$R_P = \frac{V_P}{1.75 I_{FL}} - R_a - R_{\text{acc}}$$

$$= \frac{462.5}{1.75 \times 70} - 0.25 - 1.71$$

$$= 3.78 - 0.25 - 1.71 = 1.82 \text{ ohms}.$$

The following example illustrates the calculation for a *dynamic braking resistance.*

Example 3-11. For the motor of Examples 3-9 and 3-10 what value of dynamic braking resistance is needed (peak current 225 percent rated)?

Solution:

$$E_g = V - I_{FL}R_a$$
$$= 240 - 70(0.25) = 240 - 17.5 = 222.5 \text{ V},$$

$$R_{DB} = \frac{E_g}{I_{\text{peak}}} = \frac{222.5}{70(2.25)} = 1.42 \text{ ohms}.$$

Dynamic braking is usually used to stop a motor quickly that operates in one direction only. In Example 3-9, R_2 and R_3 could be used, so that $1.42 - 0.84$ would yield a value of $0.58\ \Omega$.

3-9 SUMMARY

Operating characteristics of shunt and compound generators is expressed by percent voltage regulation. The formula is given in Equation (3-1),

$$\text{percent voltage regulation} = \frac{V_{NL} - V_{FL}}{V_{FL}} \times 100.$$

The terminal voltage of a shunt generator drops off from no load to full load with a voltage regulation of 8 to 15 percent. The undercompound generator has a lower voltage regulation than the shunt generator. The flat compound generator has a zero voltage regulation, while the overcompound generator has a negative percentage voltage regulation. The series generator has a low no-load voltage that is equal to the residual voltage and increases to a maximum value, then drops off. The reasons for drop in voltage as load is increased are: (a) Armature resistance drop (I_aR_a), (b) drop due to armature reaction, and (c) drop due to decreased excitation.

The magnetization curve, or the saturation curve, is the no-load voltage plotted against the field current with the generator operated at rated constant speed. From this curve the building-up process can be explained. For build up, the following factors are necessary:

(1) There must be residual magnetism (residual voltage);

(2) the resistance of the shunt field circuit must be less than the critical value—that is, the slope of the resistance line must be less than the slope of the straight-line portion of the magnetization curve;

(3) the armature and field circuits must be connected in proper relation to each other;

(4) the speed of the prime mover must be high enough (the name-plate speed should be used).

The operating characteristics of shunt and cumulative compound motors are shown in Figure 2-5. The speed of a shunt or compound generator was given by Equation (3-2),

$$S = \frac{E_g}{K\phi}.$$

The speed characteristic is expressed by percent speed regulation, and the formula was given in Equation (3-3)

$$\text{percent speed regulator} = \frac{S_{NL} - S_{FL}}{S_{FL}} \times 100.$$

A shunt motor is considered a constant-speed motor with a speed regulation not over 10 percent. Cumulative compound motors have a higher speed regulation and are classified as variable-speed motors. Both types have a definite no-load and full-load speed value. The torque characteristic for a shunt motor is a straight line, while the characteristic of a compound motor is concave.

Speed control of shunt and cumulative compound motors may be accomplished by adding resistance in the shunt field circuit or adding resistance in the motor armature circuit. Shunt field control yields speeds above normal rated values. Armature circuit added resistance decreases the speed below normal with a reduction of efficiency. The curves in Figure 3-9 show these relationships. For field control the speed regulation is good, but for armature control the percentage speed control increases.

The speed-torque curves of Figure 3-11 show that for low values of torque the speed is relatively high and for the series motor the speed is extremely high. The warning, therefore, is that series motors must not be operated at light load. On the other hand series motors have high starting torque and for this reason are used for hoists and lifting applications in which the load is solidly connected to the load.

The Ward Leonard system of speed control varies the armature voltage of the controlled motor. This is done with a motor-generator set with an exciter. This system offers smooth and stepless speed adjustments. The cost is comparatively high because three extra machines are needed. If a dc power supply is available, the exciter for the generator field is not needed. The overall efficiency with the controlled motor is low and around 60 to 70 percent. Silicon-controlled rectifiers and other electronic voltage supplies, and magnetic amplifiers with silicon diodes are replacing the Ward Leonard systems in present installations.

The basic principle of starting motors is to add resistance in the armature circuit at start which is cut out in steps as the speed of the motor accelerates. This is done manually with three- or four-point starters, drum controllers, or four-point compound starters. The last two are controllers. In the drum controller the same resistance is used for starting and speed control. Manual starter circuits are shown in Figures 3-13 to 3-16. Figures 3-17, 3-18, and 3-22 are automatic starter circuits with the last also performing a reversing of rotation function. Figures 3-19, 3-20, and 3-24 are manually operated circuits that perform control operations as well as the starting operation.

Dynamic braking utilizes the kinetic energy of the motor to dissipate the energy in a resistance as heat to stop a motor quickly. Figure 3-23(b) illustrates this principle schematically. Regenerative braking utilizes the kinetic energy by putting the energy back to the line. Plugging is another quick stopping method whereby the motor armature is reversed until it comes to a stop. A resistance, commonly called a *plugging resistor*, must be inserted in the armature circuit during the braking period. This is accomplished with plugging relays. Jogging is an operation by which the motor may be rotated by very small amounts. A scheme for this is shown in the circuit diagram of Figure 3-23(c).

Starting and control of dc motors is a large field for study. The information in this chapter is intended to give some basic principles. For further study the references for this chapter should be consulted.

Problems

1. A 25-kW, 250-V shunt generator has a no-load voltage of 262 V. (a) Calculate voltage regulation for rated load. (b) If the terminal voltage is 254 V, calculate the output assuming the load characteristic is a straight line.
2. A 50-kW, 250-V overcompound generator has a regulation of −7.5 percent. (a) Calculate the no-load voltage. (b) If it is operating at 240 V, calculate the load assuming a straight-line load characteristic.
3. A 10-kW, 125-V generator has a regulation of 6 percent. It is operating at rated speed at 120 V. Assuming straight-line load characteristic, calculate the load it is supplying.
4. Two generators, one 10 kW, 125 V with 6 percent voltage regulation, and the second 15 kW, 125 V with 10 percent voltage regulation, are operating in parallel at 128 V. (a) Calculate the load of the 10-kW generator. (b) Calculate the load of the 15-kW generator.
5. A 240-V shunt motor has an armature resistance of 0.18 ohm and a shunt field resistance of 120 ohms. At full load when the current is 89 A, the speed is 1150 rpm. At no load the current input is 7.0 A. Assume a 3-V brush drop and that the field flux remains unchanged. (a) Calculate the no-load speed. (b) Calculate the percent speed regulation.
6. A 5-hp, 120-V long-shunt compound motor has a series field resistance of 0.08 ohm, armature resistance of 0.22 ohm, and a shunt field resistance of 120 ohms. Rated speed is 1750 rpm and the input current is 40 A. At no load the input current is 4 A and the field flux is reduced 90 percent. (a) Calculate the no-load speed. (b) Calculate the speed regulation (assume a 2-V brush drop).
7. A 10-hp, 240-V, 1150-rpm motor has an armature resistance of 0.20 ohm,

an interpole winding resistance of 0.06 ohm, a series field resistance of 0.08 ohm, and a shunt field resistance of 160 ohms. The speed at no load is 1380 rpm when the motor draws 5 A. (a) Calculate rated speed when the motor current is 38 A and the field flux increases 15 percent (assume a 4-V brush drop). (b) Calculate the percent speed regulation.

8. A 15-hp, 240-V, 55-A, 850-rpm long-shunt compound motor has a field resistance of 120 ohms. The armature circuit resistance including series field and interpole winding is 0.18 V. The no-load speed is 1060 rpm at a motor current of 5 A. What percentage of the no-load flux is the full-load flux (use a 4-V brush drop)?
9. The 25-hp shunt motor of Problem 5 is operated with a 1.25-ohms resistance (R_{as}) added in the armature circuit. (a) Calculate the speed at which it will operate. (b) Calculate the power loss in horsepower in the added resistance. (c) Calculate the resistance that must be added to operate at 800 rpm.
10. In Problem 9, if 4.0 ohms is shunted across the armature and the current through the armature is 40 A, (a) calculate the speed. (b) If the torque requirement is increased so that the current through the armature is 60 A, calculate the speed.
11. The motor of Problem 6 is operated with 1.4 ohms added in the armature circuit with rated armature current. Assuming no change of field flux, calculate (a) the operating speed of the motor. (b) Calculate the hp power loss in added resistance. (c) Calculate the value of added resistance R_{as} if the speed is to be 1400 rpm.
12. In Problem 11, 3.3 ohms is shunted across the armature with 20 A through the armature. The flux drops 11 percent from rated load flux to no load. Assume a 2-V brush drop throughout. Calculate the speed at which motor operates.
13. The shunt motor of Problems 5 and 9 is to operate at 85 percent of rated speed and 65 percent of rated full-load torque. Calculate the value of armature resistance to be added to the armature circuit. Assume same brush drop of 3 V throughout and no change in flux at various loads.
14. A 25-hp, 240-V, 1150-rpm series motor has an armature resistance of 0.12 ohm, a series field resistance of 0.055 ohm, and a rated current of 89 A. Use a 3-V brush drop throughout and assume that the flux varies directly as the field current (straight-line portion of magnetization curve). (a) Calculate the speed at a current of 50 A. (b) Calculate the resistance that must be added to reduce the full-load speed to 750 rpm (with same I_a or torque).
15. The series motor of Problem 14 is operated at 135 A. Because of satura-

tion, the flux only increases 10 percent above value at rated current. (a) Calculate the speed at the new conditions. (b) If the developed torque is 118 lb-ft at rated load and current, what is the developed torque at the new conditions? (c) Calculate the value of power developed.

16. In Problem 14 calculate the stray power loss at rated load if the torque developed is 118 lb-ft.
17. In the series motor of Problem 14, the current is reduced 25 percent from rated value while the flux is reduced 15 percent with a diverter across the series field. (a) Calculate the value of the diverter resistance. (b) Calculate the new speed. (c) Calculate the value of developed torque.
18. A 600-V series railway motor has an armature resistance of 0.12 ohm, a series field resistance of 0.08 ohm, and a commutating field winding resistance of 0.05 ohm. The rated horsepower is 50, rated current is 68.0 A, and rated speed is 600 rpm (assume a brush drop of 8 V). (a) Calculate the developed torque and the speed at 35 A. (b) Calculate the developed torque and the speed at 80 A. Assume a straight-line field saturation and neglect any armature reaction.
19. In Problem 18, (a) calculate the value of a series field diverter to shunt 10 percent of the series field current. (b) Calculate the developed torque at 35 A with the diverter of (a). (c) Calculate the speed at 35 A with diverter of (a).
20. Calculate the stray power loss at rated load for Problem 18.
21. A 5-hp, 240-V shunt motor has an armature resistance of 0.28 ohm. The shunt field resistance is 160 ohms. The rated motor current is 20 A. (a) Calculate the value of the starting resistance for a four-point manual starter if the armature current is limited to 150 percent of rated value (assume a 4-V brush drop). (b) Calculate the resistance between the first and second studs if the arm is moved to the second stud when the armature current has reached rated value.
22. The motor of Problem 7, which is a 10-hp, 240-V, 1150-rpm compound motor with the following values: Armature resistance is 0.20 ohm; series field resistance is 0.08 ohm; interpole winding resistance is 0.06 ohm; shunt field resistance is 160 ohms; rated motor current is 38 A (assume 4-V brush drop). The maximum armature current allowed is 175 percent rated value. Calculate the values for three resistances: R_1, R_2, and R_3 as shown in Figures 3-17 and 3-18.
23. For the motor of Problem 22, and using the counter EMF schematic diagram of Figure 3-17, calculate the voltage values and the percentages at which relays 1*AR*, 2*AR*, and 3*AR* should pick up.
24. For the motor of Problem 22, calculate the value of a plugging resistor

with an allowable armature current of 175 percent of rated value. The accelerating resistors are to be used.

25. For Problem 22, if the motor is stopped by dynamic braking, calculate the value of a dynamic braking resistor. Calculate (a) if the resistor is placed across armature and (b) if R_1 and R_2 are to be used.

References

[1] Kosow, I. L., *Electrical Machinery and Control.* Englewood Cliffs, N.J.: Prentice-Hall, 1964.

[2] Siskind, C. S., *Electrical Machines—Direct and Alternating Current*, 2nd ed. New York: McGraw-Hill, 1959.

[3] Carr, C. C., *Electrical Machinery.* New York: Wiley, 1958.

[4] Dawes, C. L., *Electrical Engineering*, Vol I, *Direct Currents*, 4th ed. New York: McGraw-Hill, 1952.

[5] Ku, Y. H., *Electric Energy Conversion.* New York: Ronald Press, 1959.

[6] Siskind, C. S., *Electrical Control Systems in Industry.* New York: McGraw-Hill, 1963.

[7] *Electric Motor Control, Electrical Series.* Albany, N.Y.: Delmar Publishers.

Alternating-Current Synchronous Machines

4

An ac synchronous machine is a rotating machine whose average speed is proportional to the frequency of the generated voltage or of the applied voltage. The two basic parts of such machines are the *armature* whose excitation is always alternating current and the *field* whose excitation is usually direct current. The two basic parts are coupled by flux ϕ and torque T is produced or applied by the shaft. If torque is applied to the shaft and the dc field is excited, the machine is an ac generator or, as it is commonly called, an *alternator*. Other names sometimes used for the alternator are *synchronous generator* and *synchronous alternator*. If torque is produced at the shaft when the dc field and the ac armature are excited, the machine is called a *synchronous motor*. An alternator and a synchronous motor are identical machines and the term *synchronous dynamo* may refer to either. When the *synchronous dynamo* is driven by a prime mover, such as a steam turbine, the frequency of the generated voltage is determined by the speed it rotates and the machine is an *alternator* or *synchronous generator*. When a voltage source is applied to the dynamo, the speed at which it rotates is determined by the frequency of the source and the machine is a *synchronous motor*.

The basic difference between the dc dynamo and the ac synchronous machine is that in the ac synchronous machine the armature is the stator and the field rotates, while in the dc dynamo the field is stationary and the armature rotates. This change is necessary for a number of reasons, for example, increased capacity, higher operating voltage, and constructional advantages.

This chapter will discuss the basic construction, frequency, and speed relationships, the dc field, the armature, the voltage regulation, synchronous reactance, efficiency, operation, and control of ac synchronous machines.

4-1 ELEMENTARY SINGLE-PHASE GENERATOR[1]–[3]

Section 1-2 (Generator Action) and Figure 1-6(a) show how a single turn rotating at a constant speed in a uniform field generates an alternating voltage. Figures 1-6(b) and 1-6(c) show how a sinusoidal voltage is generated. It would not be practical to have a two-pole generator with one coil in one set of slots. Figure 4-1(a) shows a section of a four-pole, single-phase alternator with a rotating armature. Figure 4-1(b) shows a section of a four-pole single-phase alternator with a rotating field. The rotating-field structure is a salient pole rotor. High-speed alternators usually have circular field (non-salient pole) rotors. Figures 1-11 and 4-5 show this type in which the field is induced by radial coils on the periphery of the rotor. The directions of generated voltage are indicated on the figures. In Figure 4-1(b), by the use of Fleming's right-hand rule, the direction of rotation of the conductors is opposite to the direction in which the field rotates. Because the conductors are stationary, the relative direction of their rotation is opposite to that of the field rotation.

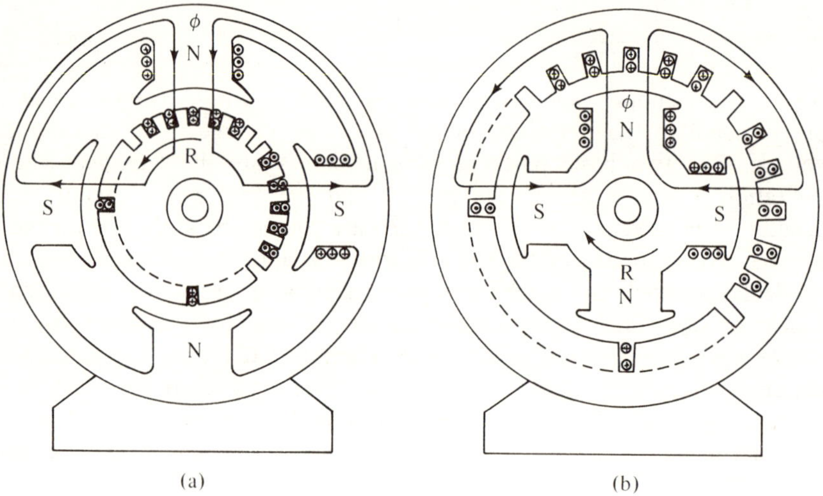

FIGURE 4-1. (a) Cross section of four-pole, single-phase alternator with stationary salient pole field. (b) Cross section of four-pole, single-phase alternator with rotating salient pole field.

The formula for the average generated voltage for a single coil was given in Equation (1-1) as

$$E_{\text{ave}} = \frac{\phi}{t}\ \text{V}.$$

This is the average generated voltage as depicted in Figure 1-7(c) where ϕ, the flux, is in webers. Since 10^8 lines (maxwells) equal one weber, the formula becomes

$$E_{\text{ave}} = \frac{\phi}{t} \times 10^{-8}\ \text{V}.$$

Because the ac waveshape of alternator voltages are sinusoidal by design of the flux distribution in the air gap and by the addition of distributed coils in the armature slots for circular armature rotors, the time for one-quarter cycle, the time for the flux to change from zero to maximum, becomes $1/4f$. Substituting this value for time t, the average generated voltage E_{ave} becomes

$$E_{\text{ave}} = 4f\phi \times 10^{-8}\ \text{V}. \tag{4-1}$$

In this equation f is the frequency in hertz and ϕ is the flux in lines. The *form factor*, which is the ratio of the effective value to the average value of a sine wave, is 1.11 and the effective value of sine-wave voltage E is $1.11 \times E_{\text{ave}}$; for a generator of N coils in series, the effective generated voltage E of an alternator becomes

$$E = 4.44fN\phi \times 10^{-8}\ \text{V}. \tag{4-2}$$

The value of frequency f is dependent on the speed at which it rotates. Speed is measured in revolutions per minute (rpm) and frequency in hertz, the number of cycles per second. A two-pole alternator making 60 rps or 3600 rpm generates a 60-Hz frequency. By formula

$$f = \frac{(\text{rpm})P}{120}\ \text{Hz}. \tag{4-3}$$

Alternators have *distributed windings.* The coils are distributed in the slots around the periphery of the armature. The individual coil voltages are out of phase with each other by the number of electrical degrees between centers of the slots. Therefore the coil voltages must be added as phasors. Figure 4-2 shows a belt or group of five coils in series in slots that are 18 electrical deg apart. For a four-pole winding the slots are nine space degrees apart. In Figure 4-2, E_T is found by graphical solution to be 0.905 times the coil voltage times the number of coils. This value of 0.905 is called the *distribution factor* k_d. The analytical solution for the distribution factor is

$$k_d = \frac{\sin(sd^\circ/2)}{s \sin(d^\circ/2)} \tag{4-4}$$

where s is the number of slots per pole (in a polyphase alternator this is the number of slots per phase per pole); and d° is the number of electrical degrees between adjacent slots (center to center).

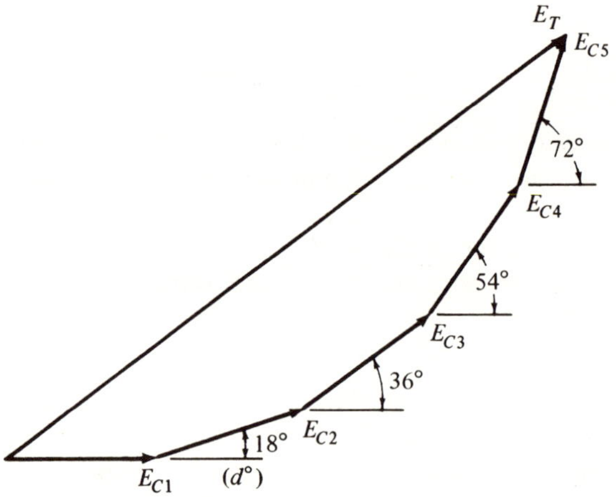

FIGURE 4-2. **Phasor diagram for coil EMFs and total EMF generated in armature windings of Figure 4-1(a) and (b).**

Example 4-1. A single-phase alternator has 4 poles, 40 slots, and a distributed winding using 5 slots per pole. Calculate the distribution factor.

Solution:

$$d^\circ = \frac{2 \times 360^\circ}{40} = 18^\circ,$$

$$s = 5 \qquad \text{(from problem)},$$

$$k_d = \frac{\sin(sd^\circ/2)}{s \sin(d^\circ/2)} = \frac{\sin(5 \times 18^\circ/2)}{5 \times \sin 18^\circ/2} = \frac{\sin 45^\circ}{5 \sin 9^\circ},$$

$$k_d = \frac{0.707}{5(0.1565)} = 0.905.$$

In alternators the coil span often is less than 180 deg. This is done to save on the amount of copper in a winding, for the ends of the coil have no function other than to connect the coil sides. By shortening the coil ends, copper can be saved without too much sacrifice of generated voltage. In Figure 4-3 are shown the phasor voltages of a coil having a span of 160 deg. The voltages are added graphically for equal coil-side voltages E_{cs1} and E_{cs2} which are equal. The sum is numerically equal to $0.985 \times E_{cs1}$; or for the

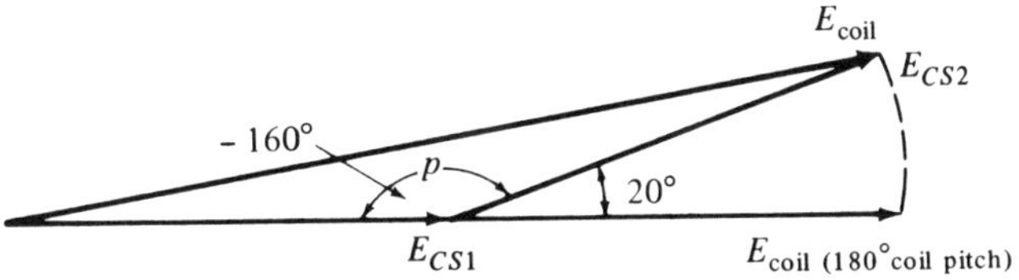

FIGURE 4-3. Phasor diagram showing coil EMF for coil with 180° span and 160° span.

coil, E_{coil} is equal to 0.985 times the coil voltage if the coil were full pitch of 180 deg. This value of 0.985 is called the *pitch factor* k_p and from the figure is

$$k_p = \sin\frac{p^\circ}{2}, \tag{4-5}$$

where p° is the coil pitch in electrical degrees.

Example 4-2. Calculate the pitch factors for the following: (a) 72 slots, 8 poles, coil span from slot 1 to 8; (b) 36 slots, 6 poles, coil span from slot 1 to 6; (c) 144 slots, 12 poles, coil span from 1 to 10.

Solution:

$$\text{(a) } p^\circ = \frac{7}{9} \times 180^\circ = 140^\circ \quad \text{and} \quad k_p = \sin\frac{140^\circ}{2} = 0.940;$$

$$\text{(b) } p^\circ = \frac{5}{6} \times 180^\circ = 150^\circ \quad \text{and} \quad k_p = \sin\frac{150^\circ}{2} = 0.965;$$

$$\text{(c) } p^\circ = \frac{9}{12} \times 180^\circ = 135^\circ \quad \text{and} \quad k_p = \sin\frac{135^\circ}{2} = 0.924.$$

Equation (4-2) for the generated voltage of an alternator with the pitch factor k_p and the distribution factor k_d becomes

$$E = 4.44 f N \phi k_d k_p \times 10^{-8}\ \text{V}. \tag{4-6}$$

The terms are as previously defined.

Example 4-3. A single-phase alternator has 24 slots; 4 poles; speed, 1800 rpm; 6 turns per coil; ϕ per pole is 1.20×10^6 lines; coil span is from slot

1 to 7, and so on; 5 slots are used per pole; and the coils are whole coiled. Calculate the generated voltage.

Solution:

$$k_p = \sin\frac{p^\circ}{2}; \qquad p^\circ = \frac{6}{6}\times 180^\circ = 180^\circ; \qquad k_p = \sin 90^\circ = 1.$$

$$k_d = \frac{\sin(sd^\circ/2)}{s\sin(d^\circ/2)}; \qquad d^\circ = \frac{180^\circ}{6} = 30^\circ$$

$$= \frac{\sin(5\times 30^\circ/2)}{5\sin 15^\circ} = \frac{\sin 75^\circ}{5(0.2585)} = \frac{0.966}{1.293} = 0.748.$$

$$E = 4.44 f N \phi k_p k_d \times 10^{-8}; \quad f = (1800)\frac{4}{120} = 60 \text{ Hz}$$

$$= 4.44(60)(20\times 6)(1.2\times 10^6)(1)(0.748)\times 10^{-8} = 276 \text{ V}.$$

Single-phase alternators are not widely used; they are used for emergency systems, portable systems, and the generators for automobiles. It is obvious that the coils around the periphery of the armature are not fully effective as seen in the last example. Polyphase alternators can make use of the full circumference of the armature and for the same size are considerably higher in rating. The balance of this chapter is concerned with polyphase machines.

4-2 POLYPHASE ALTERNATORS[3],[4]

The polyphase alternator or ac generator may have a two-phase, a three-phase, or a six-phase armature winding. The field winding is the same as for the single-phase alternator just described. The two-phase armature winding has two belts of coils displaced by 90 deg. The three-phase armature winding narrows the belt width to 60 deg. Degrees as referred to here are in electrical degrees and shall be such unless otherwise stated hereafter. Similarly, a six-phase alternator armature has a 30-deg belt width. Since practically all commercial power is generated as three phase, this type will be considered here.

It was mentioned before that in alternator construction the armature winding is placed in slots around the periphery of the stator and the dc field is on the rotor. This change is made for a number of reasons some of which are as follows:

(1) Higher operational voltages are possible. A stationary armature can be better insulated because size and weight are not as limited as with a rotating field. Insulation between collector rings and brushes is

not needed except for the dc field which operates at low voltages of $62\frac{1}{2}$ to 500 V. Voltage ratings range from 125/216Y to 14,400 V, three phase.

(2) Construction is more easily accomplished for the complex armature connections for three-phase windings on a stationary structure than on a rotating armature.

(3) The armature winding can be more securely braced on a stationary structure than on a rotating structure. It is seen in Figures 4-1(a) and 4-1(b) that the teeth have more strength because as they are made deeper, the distance between the teeth at the bottom of the slot becomes wider.

(4) The air-gap flux by the field MMF is increased due to reduced reluctance. Increased slot depth actually increases the iron cross section between the slots. This reduces leakage flux and reduces armature reaction of the armature flux.

FIGURE 4-4. **Cross section of a four-pole, 36-slot, 3-phase alternator with salient pole field.**

(5) Ventilation of the armature can be improved by providing cooling slots for forced air or inert gas through it and size is not as critical to the stator as it would be to the rotor. Section 1-5 on armature conductors in the first chapter showed an armature winding with the three phases in Figure 1-17 and the complete full pitch winding in Figure 1-18 for a 4-pole armature with 36 slots. In the first figure each phase was shown to consist of four groups or belts of coils, each group containing three coils. Figure 1-18 showed the complete winding, which is a simplex lap winding. Here the various connections are made more easily on a stational part than on a revolving part. Phase-A terminals are A_s and A_e,

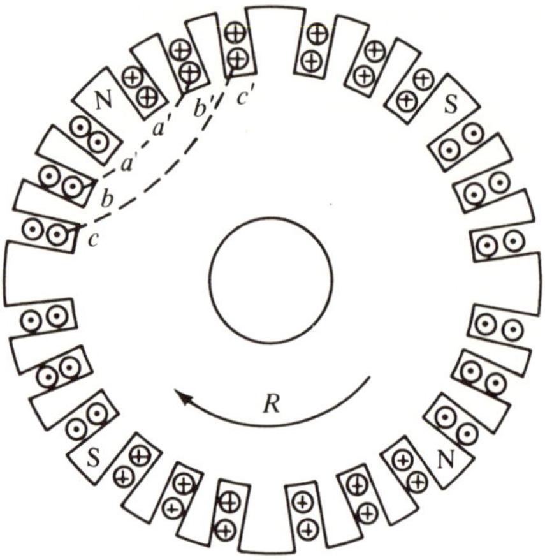

(a)

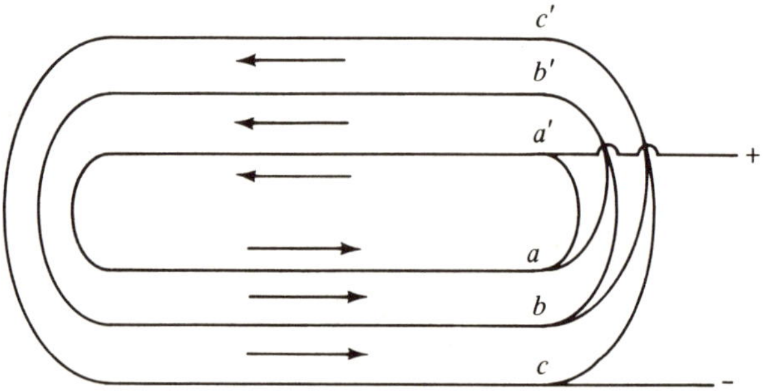

(b)

FIGURE 4-5. (a) Cross section of a cylindrical four-pole rotating field for the alternator of Figure 4-4. (b) Coils for one pole of a cylindrical field rotor.

phase-B terminals are B_s and B_e, while phase-C terminals are C_s and C_e. Figure 4-4 showed a cross section of the stator and salient-pole type of rotor field for the 4-pole, 36-slot armature winding of Figure 1-18. The directions of the generated EMF were shown. The slip rings were not shown but were needed to conduct dc current to the field windings.

Figure 4-5(a) shows a section of a *cylindrical rotor* for the same alternator. Each pole is induced by three radial coils *a-a′*, *b-b′*, and *c-c′* as shown in the sketch of Figure 4-5(b) for one induced north pole. The wire for these coils is usually bare rectangular wire. Three coils are shown here but more are usually used, especially for a two-pole (3600 rpm) rotor. The spacing of the coils is so designed that the MMF has a sinusoidal shape in the air gap.

The instantaneous values of generated voltage generated in the alternator of Figure 4-4 and the winding of Figure 1-18 are shown in Figure 4-6.

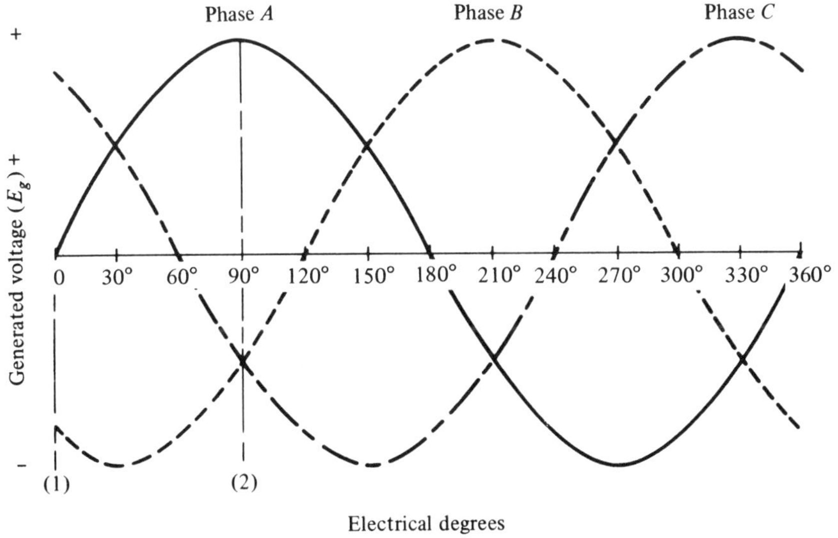

FIGURE 4-6. Generated voltage waves (sinusoidal) for generator of Figure 4-4.

Zero degrees is shown as position (1) on Figure 4-4. The instantaneous value generated voltage of phase A is zero; phase B is (−)0.866 of the maximum, or equal to $E_{\max} \sin(\theta - 120 \text{ deg})$; and phase C is 0.866 of the maximum or equal to $E_{\max} \sin(\theta + 120 \text{ deg})$. When the field turns through 45-space deg, the phase voltages advance 90 electrical deg as shown by position (2). The instantaneous EMF of phase A is equal to $E_{\max} \sin 90$ deg or $E_{\max}$. For phase B, e is equal to E_{max} sin (90 deg − 120 deg) or (−)0.5E_{max}. For phase C, e is equal to E_{max} sin (90 deg + 120 deg) or (−)0.5E_{max}. The instantaneous algebraic sum of the three generated voltages at any point is equal to zero.

The following example illustrates the use of Equations (4-2) through (4-6) as applied to three-phase alternators.

Example 4-4. The three-phase alternator of Figure 4-4 has 36 slots. Each coil has 8 turns. The speed is 1800 rpm and the flux per pole is 1.18×10^6 lines. As mentioned this winding has full-pitch coils. Calculate the generated voltage per phase.

Solution:

$$k_p = \frac{\sin p^\circ}{2} = \frac{\sin 180^\circ}{2} = 1;$$

$$k_d = \frac{\sin (sd^\circ)/2}{s \sin d^\circ/2}, \qquad s = \frac{36}{3 \times 4} = 3 \text{ and } d^\circ = \frac{720}{36} = 20^\circ;$$

$$k_d = \frac{\sin (3 \times 20^\circ)/2}{3 \sin 20^\circ/2} = \frac{\sin 30^\circ}{3 \sin 10^\circ} = 0.960;$$

$$f = \frac{(P)(\text{rpm})}{120} = \frac{4 \times 1800}{120} = 60 \text{ Hz};$$

$$\frac{E}{\text{phase}} = 4.44 f N \phi \times 10^{-8} \text{ V}, \qquad N = \frac{36 \times 8}{3} = 96 \frac{\text{turns}}{\text{phase}}$$

$$\frac{E}{\text{phase}} = 4.44(60)(96)(1.18 \times 10^6)(10^{-8}) = 302 \text{ V}.$$

4-3 THREE-PHASE ALTERNATOR CONNECTIONS[3],[4]

Three-phase alternators or synchronous generators are usually connected in *wye* (star) as shown in Figure 4-7(a). One reason for the predominance of the wye connection is the saving of insulation. For instance, a 14,400-V (line-to-line) alternator has a phase voltage of 8300 V. Another reason is the advantage of the neutral that often is brought out and used as a ground. In the figure A', B', and C' are connected together to form the neutral. Figure 4-8(a) shows the phasor diagram for the wye connection. In this connection the phase voltages are 60 deg apart so that the line voltage is numerically equal to 1.7312($\sqrt{3}$) times the phase voltage. The phasor diagram is for the balanced load condition that the utility companies strive to attain. Also, the line current is the same as the phase current. The angle between

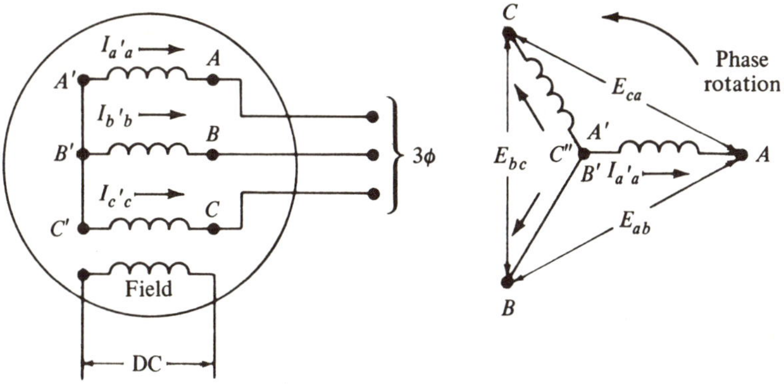

(a)

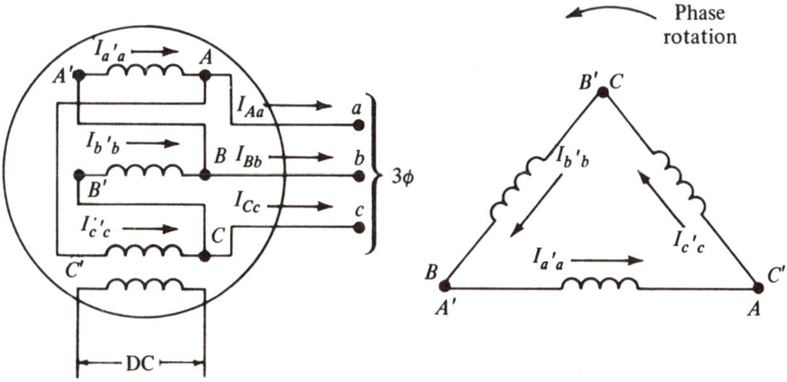

(b)

FIGURE 4-7. **(a) Schematic connection diagram for three-phase *wye* or star connected alternator. (b) Schematic connection diagram for three-phase *delta* connected alternator.**

the phase current and phase voltage is indicated by β. This angle is larger than the load power-factor angle θ as shown in Figure 4-14(c). In the phasor diagram the following are the generated voltage expressions:

$$\overline{E}_{ca} = \overline{E}_{cc'} + \overline{E}_{a'a},$$
$$\overline{E}_{ab} = \overline{E}_{aa'} + \overline{E}_{b'b},$$
$$\overline{E}_{bc} = \overline{E}_{bb'} + \overline{E}_{c'c}.$$

Figure 4-7(b) depicts the delta connection for the same machine. Figure

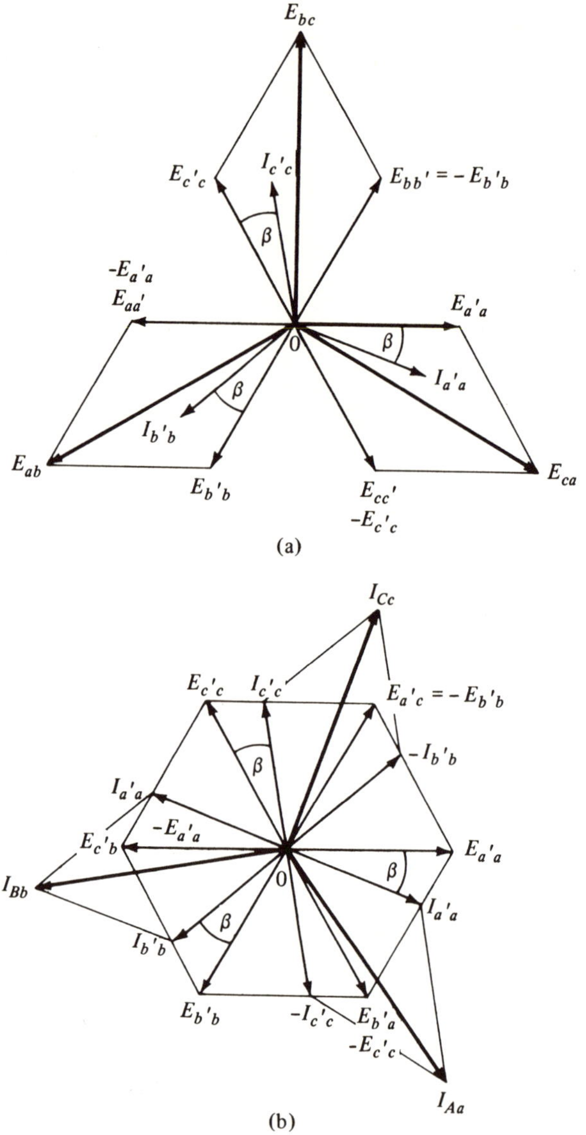

FIGURE 4-8. **(a) Phasor diagram for *wye* connected alternator of Figure 4-7(a). (b) Phasor diagram for *delta* connected alternator of Figure 4-8(b).**

4-8(b) shows the phasor diagram and the line voltage is equal to the phase voltage. The following show the generated voltage relationships:

$$-\overline{E}_{a'a} = \overline{E}_{c'b} = \overline{E}_{c'c} + \overline{E}_{b'b},$$
$$-\overline{E}_{b'b} = \overline{E}_{a'c} = \overline{E}_{a'a} + \overline{E}_{c'c},$$
$$-\overline{E}_{c'c} = \overline{E}_{b'a} = \overline{E}_{b'b} + \overline{E}_{a'a}.$$

Here the line current is numerically equal to 1.7312($\sqrt{3}$) times the phase current. The following are the expressions for the line currents:

$$\overline{I}_{Aa} = \overline{I}_{a'a} + \overline{I}_{cc'} = \overline{I}_{a'a} - \overline{I}_{c'c},$$
$$\overline{I}_{Bb} = \overline{I}_{b'b} - \overline{I}_{a'a},$$
$$\overline{I}_{Cc} = \overline{I}_{c'c} - \overline{I}_{b'b}.$$

In both the wye connection and the delta connection—since phase voltages, phase currents, and phase angles are the same for each phase—the power generated is given by the formula,

$$P_g = 3(E_p I_p \cos \beta) = 3 E_l I_l \cos \beta. \quad \textbf{(4-7)}$$

Example 4-5. If the alternator in Example 4-4 is connected in wye as in Figure 4-7(a) and the phase current is 64 A, (a) calculate the value of line generated voltage; (b) calculate the value of line current; (c) calculate the value of generated power for a balanced load with an angle θ whose cosine is 0.80 lagging.

Solution:

(a) $E_l = 1.7312 E_p = 1.7312 \times 302 = 523$ V;

(b) $I_l = I_p = 64$ A (given);

(c) $P_g = 3 E_p I_p \cos \theta = \dfrac{3(302)(64)(0.80)}{1000} = 46.4$ kW.

Alternate:

$$P_g = 3 E_l I_l \cos \theta = 1.7312 \times 523 \times 64 \times \frac{0.80}{1000}$$
$$= 46.4 \text{ kW}.$$

Example 4-6. If the alternator of Example 4-5 is connected in delta, (a) calculate the value of line-generated voltage; (b) calculate the value of line current; (c) calculate the value of generated power.

Solution:

(a) $E_l = E_p = 302$ V;

(b) $I_l = 3 I_p = 1.7312 \times 64 = 111$ A;

(c) $P_g = 3 E_l I_l = 1.7312 \times 302 \times 111 \times \dfrac{0.80}{1000}$

$= 46.4$ kW.

4-4 FREQUENCY AND SPEED[3]

An alternator is a rotating machine whose frequency of the generated voltage is proportional to the speed. A single coil rotating in a field produced by a north pole and a south pole such as shown in Figures 1-6(a) and (b) generates one cycle in one revolution. Therefore, if the coil is turning at one revolution per second (rps), the generated voltage would go through one cycle each second or the frequency is one hertz (one cycle per second). If the speed of the coil is 25 rps, the frequency is 25 Hz. *Rotational speed* is given in revolutions per minute (rpm) so that rpm/60 is equal to the speed in rps. The frequency of the generated voltage of a two-pole alternator is equal to the rpm/60. In Figures 4-1(a), 4-1(b), and 4-4 which show four-pole alternators, in one revolution of the rotor field two cycles of generated EMF result. If the speed of this rotor is 1500 rpm or 25 rps, the frequency of the generated voltage is twice the frequency of the two-pole alternator or 50 Hz. From this it is obvious that frequency (f) is proportional to the speed (rpm/60) and pairs of poles ($P/2$) or, as given in Equation (4-3),

$$f = \frac{\text{rpm}}{60} \times \frac{P}{2} = \frac{(\text{rpm})(P)}{120}. \tag{4-8}$$

At the usual commercial frequency of 60 Hz, a two-pole alternator revolves at exactly 3600 rpm and the four-pole alternator turns at 1800 rpm. The following examples illustrate the pole, speed, and frequency relationship of alternators or synchronous motors.

Example 4-7. A 16-pole alternator is to be driven by a hydroturbine at 450 rpm. Calculate the frequency of the alternator generated voltage.

Solution:

$$f = \frac{(\text{rpm})(P)}{120} = \frac{(450)(16)}{120} = 60 \text{ Hz}.$$

Example 4-8. A 25-Hz alternator is turning at a speed of 214 rpm. How many poles does the rotor have?

Solution:

$$P = \frac{120(f)}{\text{rpm}} = \frac{120(25)}{214} = 14 \text{ poles}.$$

Example 4-9. A 50-Hz, 18-pole alternator must be driven at what constant speed?

Solution:

$$\text{rpm} = \frac{120(f)}{P} = \frac{120(50)}{18} = 333\tfrac{1}{3}\ \text{rpm}.$$

4-5 ARMATURE MAGNETOMOTIVE FORCE[1],[5],[6]

In Section 2-5 the effect of armature magnetomotive force (MMF) in a dc dynamo was seen to distort the air-gap flux and because of magnetic circuit saturation, the magnitude of the net flux in the air gap was decreased. Brush position controls the magnitude and direction for a given value of armature current. In an alternator that has no brushes, the effect of armature MMF depends on the power factor of the load or the relationship between the generated voltage and the armature current. This is the angle β on the phasor diagrams of Figure 4-8(a) and (b). Figure 4-9(a) shows the three-phase currents of the alternator of Figure 4-4 with the currents of the three phases in phase with the generated voltages. Figure 4-9(b) shows both the armature flux ϕ_a and the resulting MMF curve when the rotor is in position (1). Half of the winding is shown; for the entire winding, two MMF sine waves are present.

Coil sides in slots a and b, belt B and C, yield an MMF rectangle represented by base $a'b' \times \sin 60 \text{ deg} \times I_{\max}$. Coil sides in slots c and d yield the MMF rectangle $c'd' \times \sin 60 \text{ deg} \times I_{\max}$. The coil sides in slots e and f yield the MMF rectangle $e'f' \times \sin 60 \text{ deg} \times I_{\max}$. Adding the three rectangles yields the stepped curve. Because of fringing of the flux in the armature slots, the corners are rounded off and the sine wave as shown results. Salient poles are shown in the figure. Actually, turboalternators are high speed and use cylindrical rotors so that the MMF around the periphery of the rotor as shown in Figure 4-5(a) is uniform. In either the case of a salient pole or a cylindrical rotor field, the armature MMF curve is a sine wave.

In Figure 4-9(c) the rotor field is advanced 30 deg as shown by position (2) on Figure 4-9(a). Adding the MMF rectangles as in Figure 4-9(b) results in an identical armature MMF curve but advanced 30 deg from that of Figure 4-9(b). Examination shows that the relationship of the armature MMF is advancing in space and in step with the revolving rotor field. A further examination of the figures shows that the armature flux is in quadrature (90 deg) with the main flux. This is the case for the phase current in phase with the generated voltage.

For a zero power-factor leading when the phase armature current leads the phase generated voltage by 90 deg, the armature MMF adds directly to the field MMF. For a zero lagging power factor when the phase armature current lags the phase generated voltage by 90 deg, the armature MMF subtracts directly from the field MMF. This relationship is shown in Figure

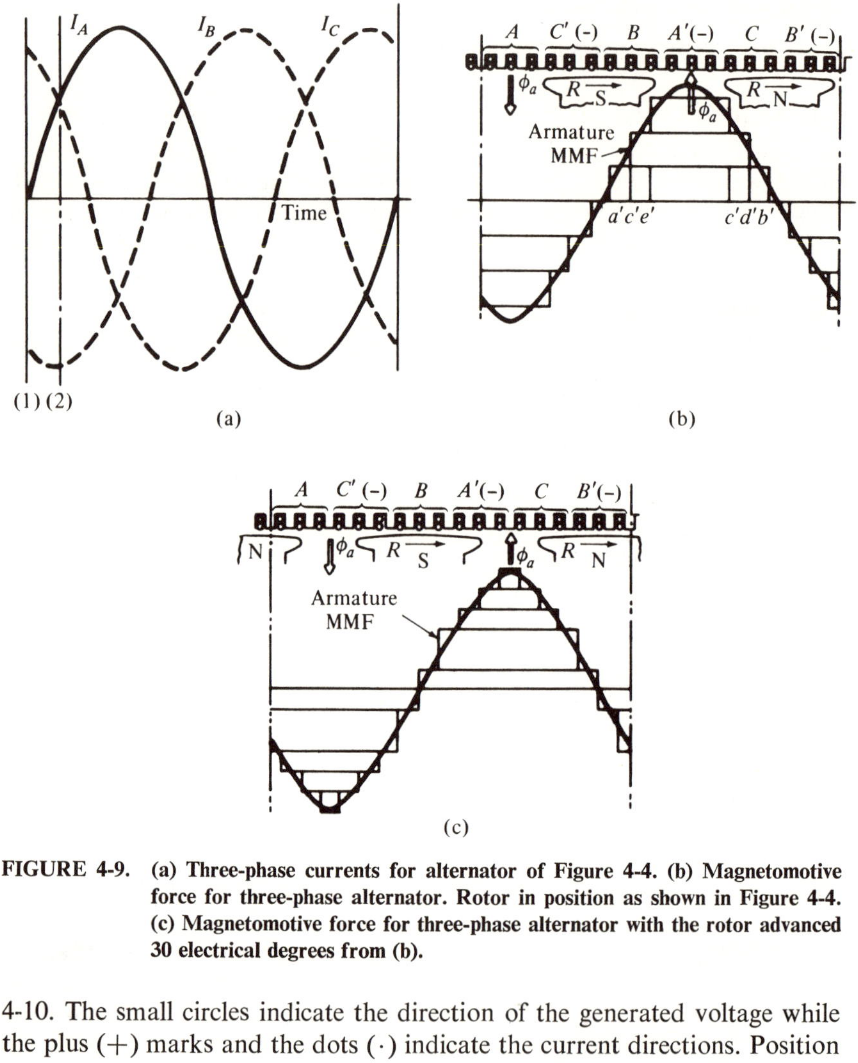

FIGURE 4-9. (a) Three-phase currents for alternator of Figure 4-4. (b) Magnetomotive force for three-phase alternator. Rotor in position as shown in Figure 4-4. (c) Magnetomotive force for three-phase alternator with the rotor advanced 30 electrical degrees from (b).

4-10. The small circles indicate the direction of the generated voltage while the plus (+) marks and the dots (·) indicate the current directions. Position (1) indicates the current in phase with the generated voltage. The rotor field is at position (2) as indicated in Figure 4-4. The current directions are indicated by (2) and the right-hand rule for direction of flux shows that the flux by the armature MMF adds directly to the dc rotor flux. Position (3) shows the case of a 90-deg lagging current and the flux subtracts directly from the main flux.

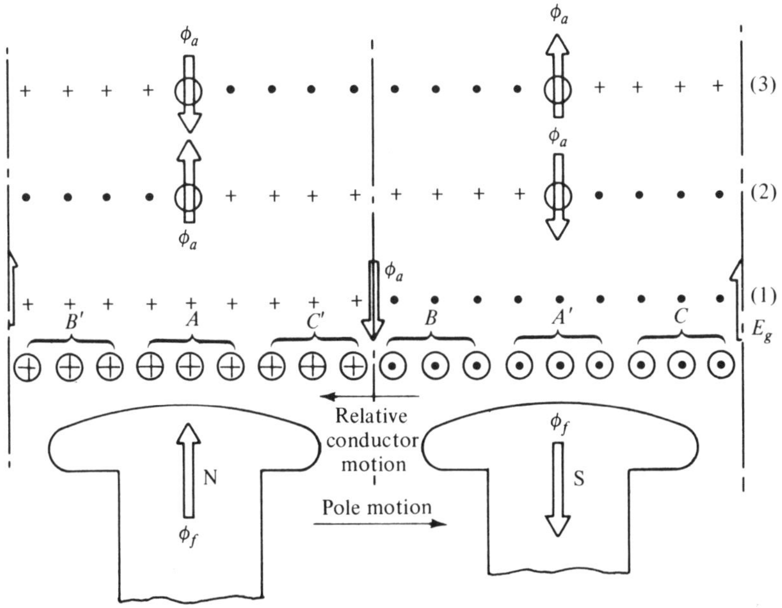

FIGURE 4-10. **Armature flux for unity power factor, zero power factor lead, and zero power factor lag of three-phase alternator. Note: (1) unity power factor; (2) zero power factor leading; (3) zero power factor lagging.**

The phasor diagrams for the three relationships, on a per phase basis, unity power factor, 90-deg leading power factor, and 90-deg lagging power factor are shown in Figures 4-11(a), (b), and (c). These figures show how the voltage of armature reaction, the reaction of the armature MMF with the dc field MMF, reacts with the generated voltage. For unity power-load factor, it reacts approximately 90 deg with the generated voltage. For 90-deg lagging power factor it subtracts from the generated voltage; for the 90-deg leading power factor it adds to the generated voltage. In these diagrams no armature resistance or reactance is considered.

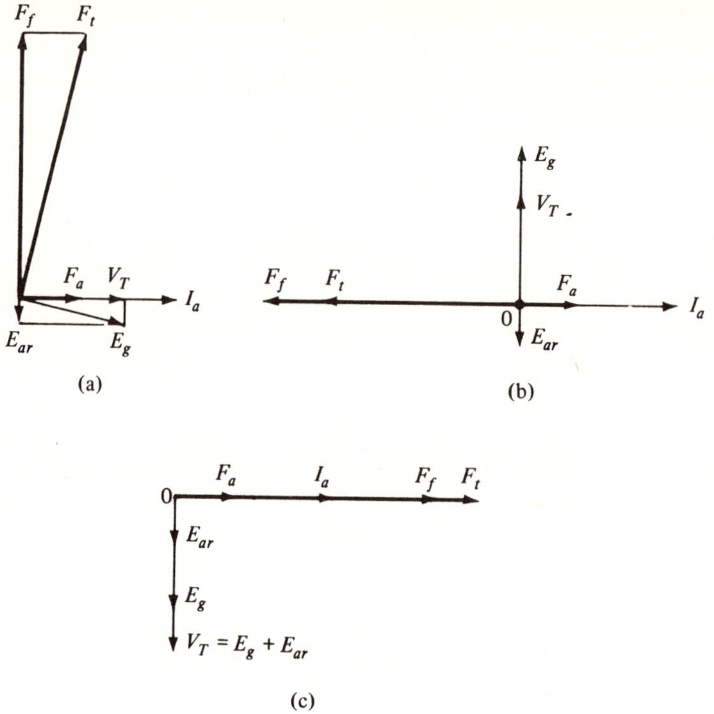

FIGURE 4-11. **Phasor diagrams showing magnetomotive force, voltages, and currents on a per phase basis for (a) unity power factor, (b) 90° lagging power factor, and (c) 90° leading power factor.**

4-6 THE ROTATING MAGNETIC FIELD[1],[3],[4]

It was shown in the preceding section that when a synchronous machine is operating as an ac generator, the armature conductors produce MMF that revolves with a fixed relationship to the rotor dc field. The corollary of the alternator armature MMF is when the same winding is energized by a three-phase ac source. Actually when a polyphase machine winding is connected to a polyphase source, a revolving magnetic field is induced. A graphical analysis as shown by Figures 4-12(c) through 4-12(h) shows the field at various times. Figure 4-12(c) shows a section of a two-pole machine with currents indicated for position (1) on Figure 4-12(b) or for zero time. The wave diagrams show that the current in phase A is zero, or $I_{max} \sin \omega t$. The current in phase B is $(-)0.866 \times I_{max}$ or equal to $I_{max} \sin(\omega t - 120°)$. In phase C the current is $0.866 \times I_{max}$ or equal to $I_{max} \sin(\omega t + 120°)$. Assuming that there is no saturation of the iron, the flux ϕ is proportional to the current and ϕ_A is zero. ϕ_B is $0.866 \times \phi_{max}$ as shown and ϕ_C is $0.866 \times \phi_{max}$ as shown. Adding the phasors, ϕ_B and ϕ_C, ϕ_R is $1.5 \times \phi_{max}$; ϕ_B and ϕ_C are 60 deg apart and the resultant flux ϕ_R is directly downward. The induced polarities are indicated on the sketch.

The next figure, Figure 4-12(d), is shown as position (2) on Figure 4-12(b). This is an advance of 30 deg from position (1) and for a 60-Hz supply represents a time advance of 1/720 (0.00139) sec. I_C is $0.5I_{max}$ or ϕ_C is $0.5\phi_{max}$, as shown, perpendicular to coils CC'. Similarly, I_A is $0.5I_{max}$ and ϕ_A is $0.5\phi_{max}$ and perpendicular to coils AA'. I_B is at maximum negative value and ϕ_B is equal to ϕ_{max} as shown perpendicular to coils BB'. Adding the phasors, ϕ_A, ϕ_B, and ϕ_C, ϕ_R equal in magnitude of $1.5\phi_{max}$ is the phasor sum and advanced 30 deg from position (1). Because the winding is two pole, space and electrical degrees are the same. For a four-pole winding, the resultant flux ϕ_R would move only 15 space deg.

For position (3) on Figure 4-12(b) the total time advance is 1/360 sec or 0.00278 sec for a 60-Hz supply. ϕ_C is zero and the total of ϕ_A and ϕ_B yields ϕ_R, also equal to $1.5\phi_{max}$ and advanced 60 deg from position (1). Figure 4-12(e) shows the phasor diagram. Figure 4-12(f) shows the resultant flux ϕ_R displaced 90 deg. This is position (4) on Figure 4-12(b), which is 90 deg or 1/240 (0.00417) sec from position (1). Figure 4-12(g) shows the resultant flux ϕ_R for position (5), which is 270 deg or 3/240 (0.0125) sec later than position (1). The flux is also advanced 270 deg. The last illustration is shown in Figure 4-12(h) for a time of 7/480 (0.0146) sec for the 60-Hz supply, or 315 deg from position (1). The resultant flux ϕ_R is still $1.5\phi_{max}$ and displaced 315 deg from position (1).

This study shows the resultant flux as always equal to 1.5 times the maximum flux ϕ_{max} per phase. This flux rotates at a synchronous speed, as given in Equation (4-3a), of

$$\text{rpm} = \frac{120f}{P}. \qquad \textbf{(4-8a)}$$

Stator windings of synchronous motors, synchronous capacitors, and polyphase induction motors are wound similar to the alternator stator winding and in each a revolving flux is produced in the stationary winding. The value of 1.5 times ϕ_{max} per phase for the resultant flux is somewhat less because of saturation of the iron and the varying reluctance around the periphery of salient poles or slot spacing of circular rotors.

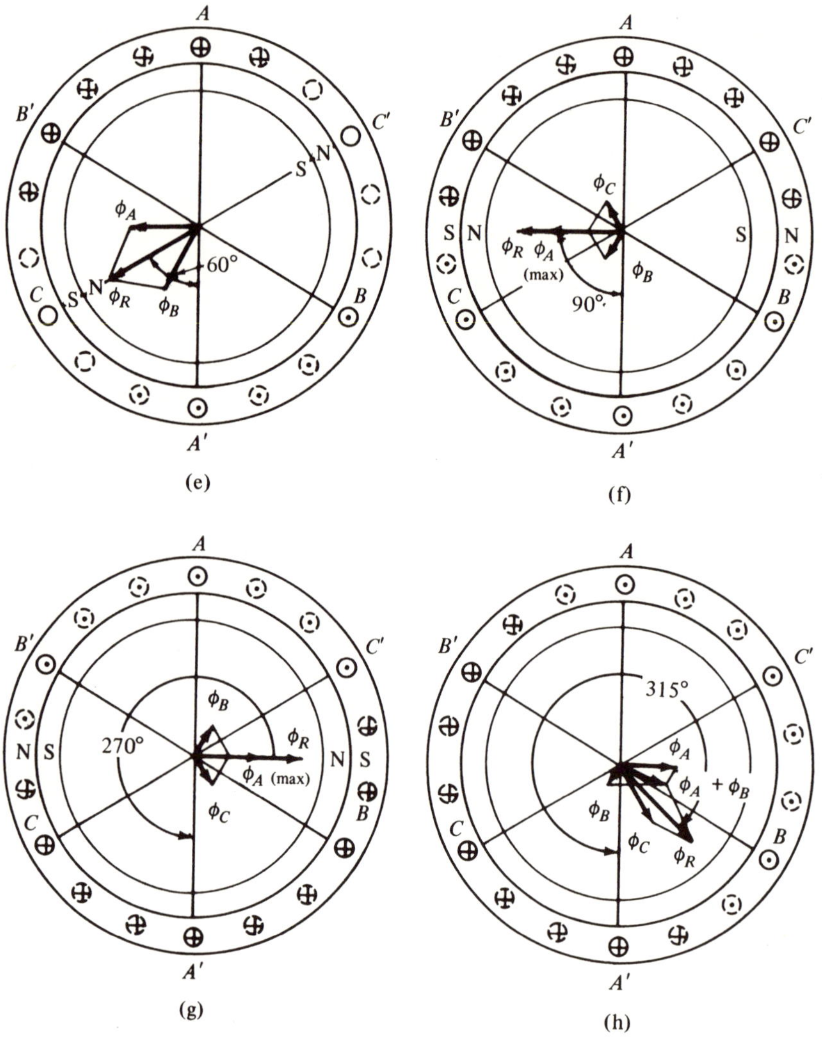

FIGURE 4-12. (a) Schematic diagram of a three-phase *wye* connected two-pole, single-layer, 18-slot armature winding. (b) Wave diagrams (sine) of three-phase currents in armature of (a). (c) Magnetic flux for armature at position (1) of (b). (d) Magnetic flux for armature at position (2) in (b). (e) Magnetic flux for armature at position (3) in (b). (f) Magnetic flux for armature at position (4) in (b). (g) Magnetic flux for armature at position (5) in (b). (h) Magnetic flux for armature at position (6) in (b).

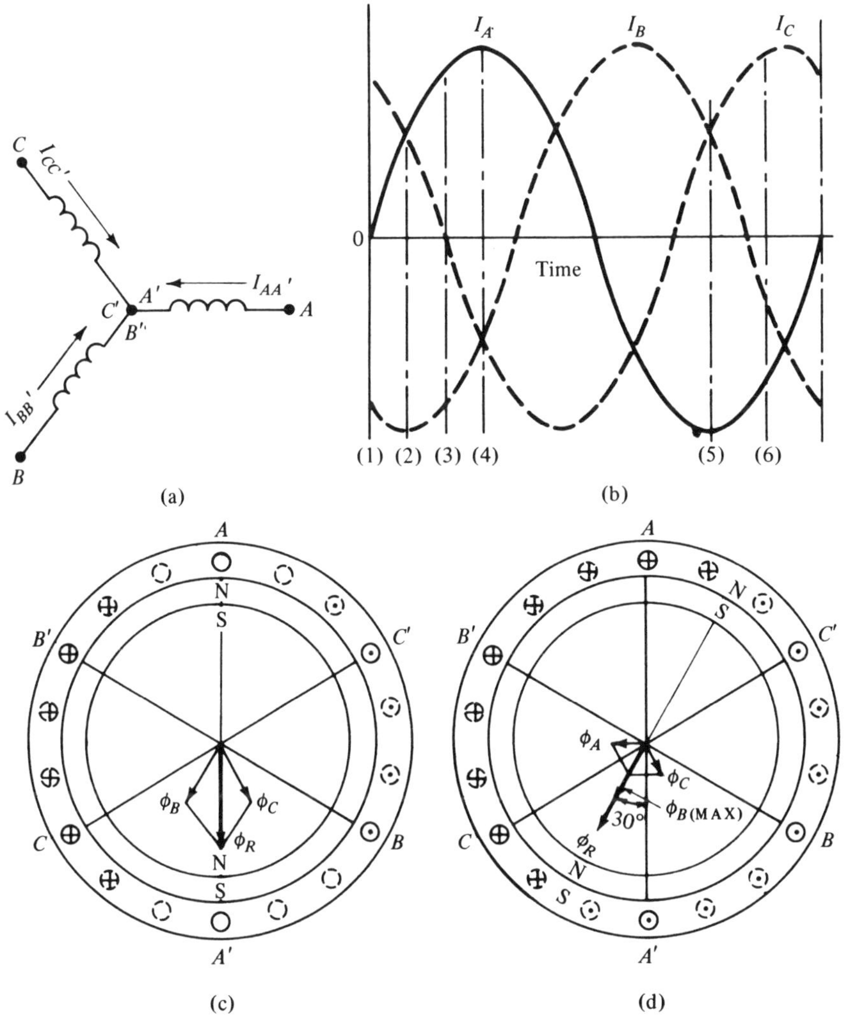

FIGURE 4-12. (*continued*).

4-7 SYNCHRONOUS REACTANCE[1],[2],[5],[7],[8]

The alternator is similar to the separately excited shunt generator since the alternator is separately excited by dc current. In the separately excited dc generator the voltage at the terminals under load is reduced due to armature resistance drop I_aR_a and armature reaction. In the alternator there are three factors causing a drop (or increase) in terminal voltage under load. These factors are (a) armature resistance drop, I_aR_a; (b) leakage reactance drop, I_aX_a; and (c) armature reaction drop. Phasor relationships are unlike the dc counterpart and calculations must be made by the use of phasor algebra. Power factor is determined by the load and the load voltage may possibly be larger than the generated voltage. Because alternators are, for the most part, three phase, calculations are made on a phase basis and the wye connection is used unless otherwise specified.

Considering armature resistance and leakage reactance as the factors causing a drop in voltage at the terminals under load, the equivalent diagram is shown in Figure 4-13(a). The phasor diagrams for unity power-factor

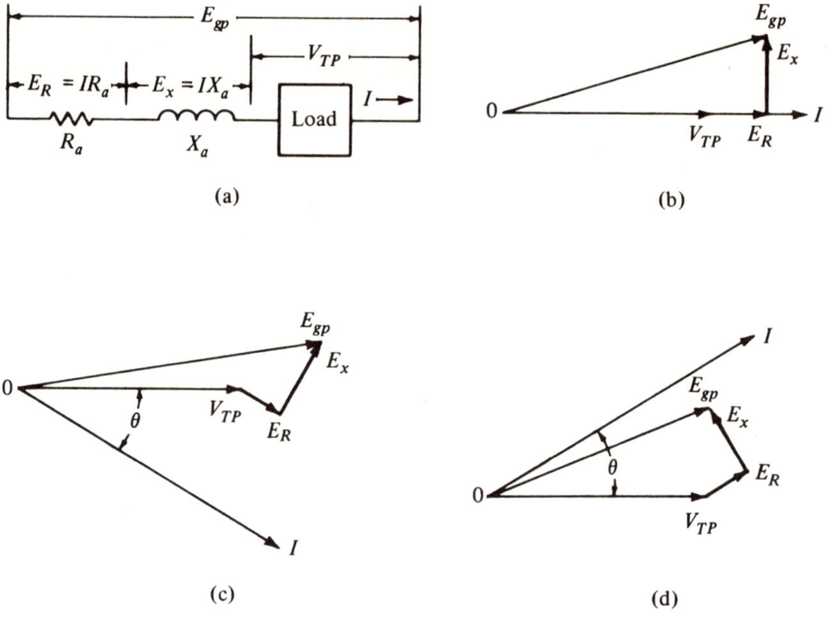

FIGURE 4-13. (a) Equivalent circuit of one phase of alternator without armature reaction. (b) Phasor diagram for (a) with unity power factor load. (c) Phasor diagram for (a) with lagging power factor phase angle. (d) Phasor diagram for (a) with leading phase angle (leading power factor).

load, lagging power-factor load, and a leading power-factor load are shown in Figures 4-13(b), 4-13(c), and 4-13(d), respectively. The circuit shows a resistance R_a that represents the effective resistance per phase. The dc (ohmic) resistance can readily be measured between terminals and the ohmic resistance per phase is one half this resistance. To obtain the effective ac resistance the ohmic value is multiplied by a factor of 1.25 to 1.75. The value of leakage reactance can readily be obtained with an ammeter, voltmeter, and wattmeter. The following example will illustrate the calculations.

Example 4-10. A 2400-V, three-phase, 60-Hz, 250-kVA alternator was tested and the average resistance (dc) across each pair of terminals was found to be 1.12 ohms. The leakage reactance per pair of terminals was found to be 4.90 ohms. Calculate (a) the armature resistance drop at full load; (b) the armature leakage reactance drop at full load; (c) the generated voltage per phase for full load, unity power factor; (d) the generated voltage for full load, 80 percent power factor lagging; (e) the generated voltage for full load, 80 percent leading power factor.

Solution:

$$I_a \text{ (rated)} = \frac{\text{kVA} \times 1000}{\sqrt{3}V_L} = \frac{250 \times 1000}{1.7312(2400)} = 60.1 \text{ A};$$

$$R_a/\text{phase} = 1.12 \times 1.5/2 = 0.84 \text{ ohm}.$$

(a) $E_R = I_aR_a = 60.1 \times 0.84 = 50.5$ V;

(b) $E_X = I_aX_a = 60.1 \times (4.90/2) = 147.2$ V;

(c) See phasor diagram, Figure 4-13(b),

$$E_{gp} = \sqrt{(V_{TP} + I_aR_a)^2 + (I_aX_a)^2} \qquad V_{TP} = \frac{2400}{\sqrt{3}} = 1386 \text{ V},$$

$$E_{gp} = \sqrt{(1386 + 50.5)^2 + (147.2)^2}$$
$$= \sqrt{(1436.5)^2 + (147.2)^2} = 1438 \text{ V}.$$

(d) See phasor diagram, Figure 4-13(c),

$$E_{gp} = \sqrt{(V_{TP}\cos\theta + I_aR_a)^2 + (V_{TP}\sin\theta + I_aX_a)^2}$$
$$= \sqrt{(1386 \times 0.8 + 50.5)^2 + (1386 \times 0.6 + 147.2)^2}$$
$$= \sqrt{(1008 + 50.5)^2 + (830 + 147.2)^2} = 1441 \text{ V}.$$

(e) See phasor diagram, Figure 4-13(d),

$$E_{gp} = \sqrt{(V_{TP}\cos\theta + I_aR_a)^2 + (V_{TP}\sin\theta - I_aX_a)^2}$$
$$= \sqrt{(1386 \times 0.80 + 50.5)^2 + (1386 \times 0.60 - 147.2)^2}$$
$$= \sqrt{(1058.5)^2 + (682.8)^2} = 1260 \text{ V}.$$

Referring to Figures 4-11(a), (b), and (c), it shows that the armature reaction voltage (E_{ar}) lags the armature current by 90 deg for each load power factor. In Figure 4-14(a) is the equivalent phase diagram showing this armature reaction voltage as caused by another reactance (X_{ar}). In each of

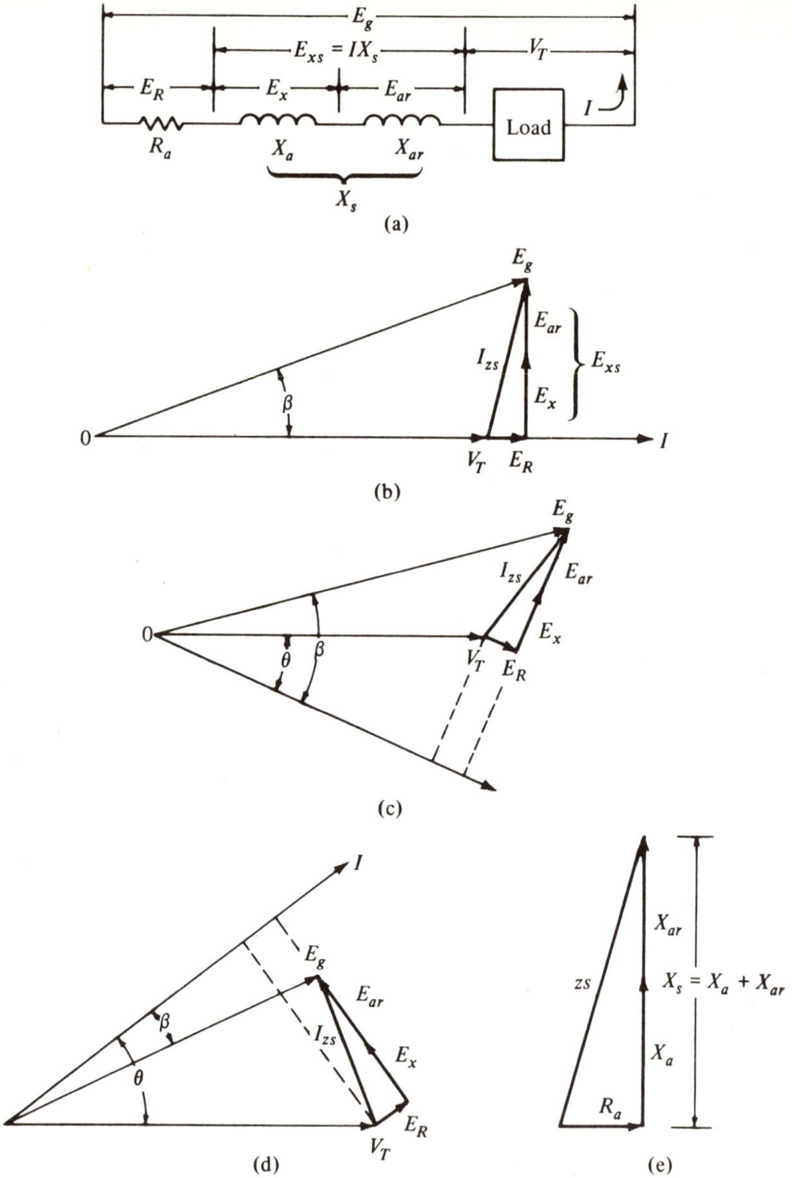

FIGURE 4-14. **(a) Equivalent circuit diagram with armature reaction. (b) Phasor diagram for (a) at unity power factor (PF) load. (c) Phasor diagram for (a) with lagging PF load. (d) Phasor diagram for (a) with leading PF load. (e) Synchronous impedance triangle for one phase.**

the phasor diagrams for unity, lagging and leading power factors the armature reaction voltages are in phase with the leakage reactance drops. In Figure 4-14(b) these two voltages drops are added because they are in phase with each other and it is shown as the synchronous reactance voltage E_{XS} drop. In the impedance triangle of Figure 4-14(e) the armature reactance X_a and the fictitious armature reaction reactance X_{ar} are added together and their sum is called *synchronous reactance* X_S. The *synchronous impedance* Z_S is the hypotenuse of the triangle and is almost numerically equal to the synchronous reactance X_S since the armature resistance R_a is small compared to the synchronous reactance X_S. The following example shows how this synchronous reactance X_S is used.

Example 4-11. In the previous Example 4-10 the synchronous reactance was calculated to be 8.5 Ω. Calculate (a) the generated voltage per phase for unity power-factor load; (b) the generated voltage per phase for 0.80 power-factor load lagging; (c) the generated voltage per phase for 0.80 power-factor load leading.

Solution:

From Example 4-10, $I_a = 60.1$ A, $V_{TP} = 1386$ V, and $R_{a(\mathrm{ac})} = 0.84$ ohm/phase.

(a) See phasor diagram, Figure 4-14(b),

$$E_{gp} = \sqrt{(V_{TP} + I_aR_a)^2 + (I_aX_S)^2}$$
$$= \sqrt{(1386 + 50.5)^2 + (60.1 \times 8.5)^2}$$
$$= \sqrt{(1436.5)^2 + (511)^2} = 1524 \text{ V}.$$

(b) See phasor diagram, Figure 4-14(c),

$$E_{gp} = \sqrt{(V_{TP}\cos\theta + I_aR_a)^2 + (V_{TP}\sin\theta + I_aX_S)^2}$$
$$= \sqrt{(1386 \times 0.8 + 50.5)^2 + (1386 \times 0.60 + 511)^2}$$
$$= \sqrt{(1058)^2 + (1341)^2} = 1707 \text{ V}.$$

(c) See phasor diagram, Figure 4-14(d),

$$E_{gp} = \sqrt{(V_{TP}\cos\theta + I_aR_a)^2 + (V_{TP}\sin\theta - I_aX_S)^2}$$
$$= \sqrt{(1386 \times 0.80 + 50.5)^2 + (1386 \times 0.60 - 511)^2}$$
$$= \sqrt{(1058)^2 + (830 - 511)^2} = 1104 \text{ V}.$$

The formulas as given for (a), (b), and (c) are taken from the phasor diagrams and hence are not listed as given formulas. However, the (a) formula always holds for unity power-factor load, while the (b) formula is for lagging power factor, and the (c) formula is for leading power factor. For the leading power-factor solution the generated voltage is less than the terminal voltage.

To find the synchronous reactance, three tests are necessary. The first test is a *resistance test.* The dc resistance is measured between each pair of terminals. Assuming the wye connection, the dc resistance is half of this resistance. A factor of 1.25 to 1.75 is applied for the effective ac resistance.

The second test is an *open-circuit* test to obtain a saturation curve of

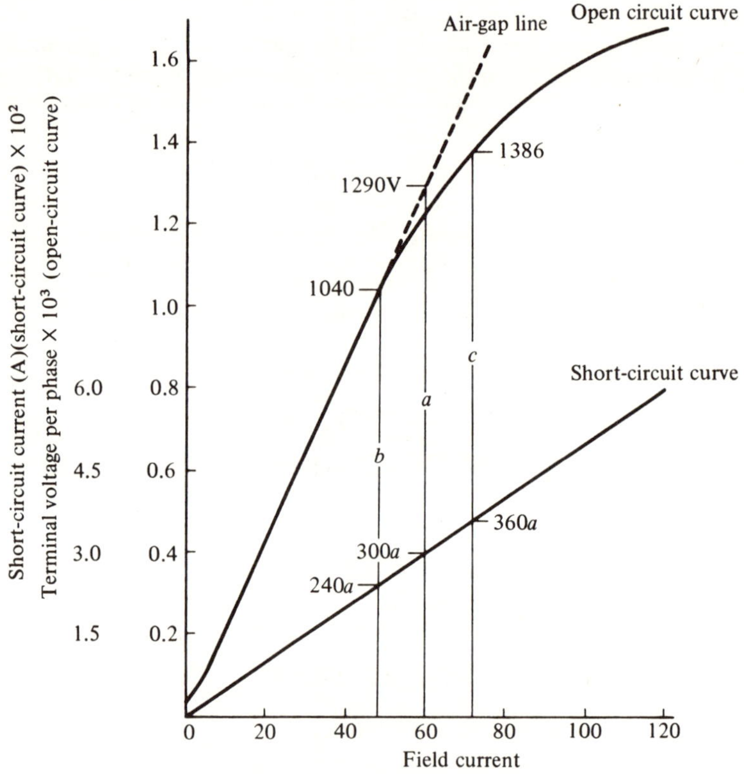

FIGURE 4-15. Open-circuit (saturation) curve and short-circuit current–field current curve for 1000 kVA, 3 phase, 2400 volt, 60 Hz alternator.

open-circuit voltage–field (dc) current. In this test the terminal voltage at no load is recorded for increasing values of dc field current.

The third test is a *short-circuit* test. Ammeters are inserted in the lines and the ends are tied together. In other words the ammeters are connected in wye. The field current is varied. This test must be performed with great care because low values of field current are used. A rheostat in the field circuit of high resistance is used to keep the current values at not more than twice the rated current value. This curve of short-circuit current–field current is a straight line.

Figure 4-15 shows a typical open-circuit curve and a short-circuit current–field current curve for a 1000-kVA, 2400-V, three-phase alternator. The following example will show the method of calculating the value of synchronous impedance for the alternator.

Example 4-12. The average ohmic resistance between terminals of an alternator was measured and found to be 0.38 ohm. The open-circuit test and the short-circuit test for the 2400-V three-phase alternator rated at 1000 kVA for a 0.80 power factor and 60 Hz are shown on Figure 4-15. (a) Calculate the synchronous reactance using the air-gap method. (b) Calculate the synchronous reactance using rated-current value. (c) Calculate the synchronous reactance using rated-voltage value.

Solution:

(a) The air-gap line is an extension of the straight-line portion of the open-circuit curve. Using any value of field current, divide the value of terminal voltage by the value of short-circuit current. In this example a value of 60 A was used for field current. For 60 A a terminal voltage per phase is found to be 1290 V and the short-circuit current is 300 A.

$$Z_S = \frac{V(\text{air-gap line})}{I_{SC}} = \frac{1290}{300} = 4.30 \text{ ohms},$$

$$R_{a(\text{ac})} = \frac{0.38}{2} \times 1.5 = 0.285 \text{ ohm},$$

$$X_S = \sqrt{Z_S{}^2 - R_a{}^2} = \sqrt{(4.30)^2 - (0.285)^2} = 4.29 \text{ ohms}.$$

(Notice that Z_S could have been used for the value of X_S since R_a is less than one percent of Z_S.)

(b) Rated current of the alternator:

$$I_L = \frac{\text{kVA(rated)} \times 1000}{\sqrt{3}\, E_L} = \frac{1000 \times 1000}{1.732 \times 2400} = 240 \text{ A},$$

$$I_{SC} = I_L = 240 \text{ A},$$
$$I_F = 48 \text{ A},$$
$$V(\text{at } 48 \text{ A}) = 1040 \text{ V (from curve } b),$$
$$Z_S = \frac{V(\text{from curve})}{I_{\text{rated}}} = \frac{1040}{240} = 4.33 \text{ ohms},$$
$$X_S = \sqrt{(4.33)^2 - (0.285)^2} = 4.32 \text{ ohms}.$$

(c) Rated voltage:

$$\text{rated voltage/phase} = \frac{2400}{\sqrt{3}} = 1385 \text{ V},$$
$$I_f \text{ for voltage/phase of } 1385 = 68 \text{ A},$$
$$I_{SC}(\text{from curve } c) \text{ at } 68 \text{ A} = 360 \text{ A},$$
$$Z_S = \frac{1385}{360} = 3.85 \text{ ohms},$$
$$X_S = (3.85)^2 - (0.285)^2 = 3.84 \text{ ohms}.$$

Methods for (a) and (b) yield just about the same values for synchronous reactance. The last method (c) yields a lower value because the voltage value is taken beyond the straight-line portion of the saturation curve. It is lower because at rated voltage there is some saturation in the iron. Methods of (b) and (c) do not require the plotting of an open-circuit–field-current curve. The methods of (a) and (b) yield very nearly equal results. In any event these methods yield a value of generated voltage that is higher than is actually the case, so that the value of voltage regulation that is discussed in the next section will yield pessimistic values.

These methods of calculating synchronous impedance are based on assumptions that the value of armature reaction does not depend on saturation; that the reluctance is constant for all power factors; that the net magnetic flux is constant for all loads for the same values of field excitation; and that the flux due to armature reaction is proportional to the armature current. For the cylindrical rotor the error of the assumptions is not as great as for the salient pole rotor. More exact methods for calculating the synchronous reactance may be found in the references since these methods are beyond the scope of this discussion. The Potier triangle method, the slip method, the ANSI (formerly ASA) method are some of the methods to yield more precise values of generated voltage. An actual load test or tests will yield exact results, but because of power limitations for large generators they are not always practical.

4-8 ALTERNATOR VOLTAGE REGULATION AND CONTROL[3],[4],[6]

For dc generators, voltage regulation is defined by the expression as given in Equation (3-1):

$$\text{percent voltage regulation} = \frac{V_{NL} - V_{FL}}{V_{FL}} \times 100,$$

where V_{NL} is the no-load voltage and V_{FL} is the full-load or rated-load voltage. The speed was held constant.

For alternators the speed is always held constant to maintain a fixed value of frequency. However, in alternators the power factor of the load has a great influence on value of regulation. For an alternator, voltage regulation is the voltage change in percent from rated load to no load at a specified power factor. Equation (3-1) is used. The no-load voltage is the generated voltage as calculated using armature resistance and synchronous reactance. Phase voltages are used rather than line-to-line voltages, although it is immaterial because the $\sqrt{3}$ would divide out in the expression of Equation (3-1). The expression in terms of the generated voltage per phase and the terminal voltage per phase for percent voltage regulation becomes

$$\text{percent voltage regulation} = \frac{E_{gp} - V_{TP}}{V_{TP}} \times 100, \tag{4-9}$$

where E_{gp} is the generated voltage per phase and V_{TP} is the terminal voltage per phase.

Example 4-13. Calculate the voltage regulation for the 250-kVA, 3-phase, 60-Hz, 2400-V alternator of Examples 4-10 and 4-11 at (a) unity power factor; (b) 0.80 lagging power factor; (c) 0.80 leading power factor.

Solution:

(a) $$\text{percent } VR = \frac{E_{gp} - V_{TP}}{V_{TP}} \times 100$$

$$= \frac{1524^* - 2400/\sqrt{3}}{2400\sqrt{3}} \times 100$$

$$= 9.95\% \text{ (unity power factor);}$$

(b) $$\text{percent } VR = \frac{1707^* - 1385}{1385} \times 100$$

$$= 23.1\% \text{ (0.80 power-factor lag);}$$

---

* Values of E_{gp} from Example 4-11.

$$\text{(c) percent } VR = \frac{1104^* - 1385}{1385} \times 100$$

$$= -20.4\% \text{ (0.80 power-factor lead).}$$

In this example there is a variation of 32.1 to −20.4 percent in voltage regulation or a total variation of 43.5 percent from 0.80 power-factor lagging to 0.80 power-factor leading. Figure 4-16 shows typical terminal-voltage–armature-current curves for an alternator.

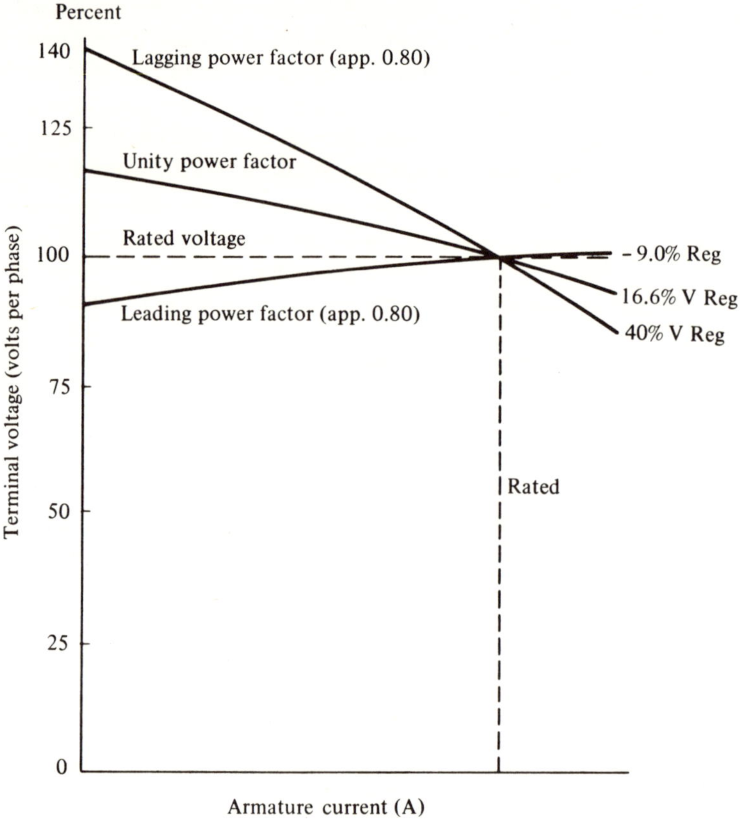

FIGURE 4-16. Typical terminal-voltage–armature-current curves for an alternator at leading, unity, and lagging power factors.

The wide variation of terminal voltages due to the variation of the power factors of ac loads makes it especially necessary to employ *automatic regu-*

lators to maintain uniform terminal voltages. Such regulators are feedback devices. Usually a transformer across the alternator output will sense a change in voltage. This change of voltage is transformed to a signal to add resistance in the field circuit if the voltage is high or to drop resistance in the field circuit. The field circuit of the exciter is changed since a small change of resistance in the exciter field is amplified to the field of the alternator itself. If the alternator does not have its own exciter, the regulator must control the alternator field and must handle larger values of power. Such rheostats must be motor driven if the alternator is of any appreciable size.

4-9 EFFICIENCY AND LOSSES[2]–[4],[8]

The efficiencies of synchronous machines are calculated for rated output conditions by determination of the losses and by the same formula (2-12c) that is given in Chapter 2, as follows:

$$\text{percent efficiency} = \left(1 - \frac{P_l}{P_o + P_l}\right) \times 100,$$

where P_l is the sum of the power losses and P_o is the rated power output. The losses are as follows:

(a) stray power or rotational losses group:
- (1) friction and windage losses
- (2) brush friction loss
- (3) core loss
- (4) ventilation and cooling loss

(b) electrical losses group:
- (5) armature loss ($I_a^2R_a$)
- (6) field loss ($I_f^2R_f$)
- (7) brush contact loss
- (8) rheostat loss

(c) (9) exciter losses

(d) (10) stray load loss

These losses are determined as noted in the following discussion.

(1) There are friction and windage losses.

(2) Brush friction loss is combined and determined by driving the machine with a calibrated motor or a dynamometer at rated speed.

Synchronous machines operate at constant average speed so that there is no variation as in dc machines. If a cooling gas is used, this loss is deducted.

(3) The core loss, made up of hysteresis and eddy current losses in the iron, is found by driving the machine with a calibrated motor or dynamometer at rated speed and with normal excitation. The loss from (1) and (2) is deducted to yield the core loss. The voltage at the terminals must correspond to the calculated internal voltage. This internal voltage is calculated by correcting the rated terminal voltage for the resistance drop only.

(4) Ventilation and cooling loss is the power required to circulate the cooling gas if a coolant medium is used.

(5) The armature loss $I_a{}^2R_a$ for a three-phase machine is three times the $I_a{}^2R_a$ corrected to 75°C for a 40°C temperature rise machine. This may also be found by driving the short-circuited alternator with a calibrated motor with rated armature current. The friction losses of (1) and (2) are deducted.

(6) The field loss is the dc field current squared times the ohmic resistance of the field corrected to 75°C. This may also be calculated by the product of the field current and the field voltage at the heated or normal operating temperature.

(7) Brush contact loss is an electrical loss of I^2R, where R is the resistance of the brushes, brush contact, and collector rings. This loss is low and usually is omitted without any appreciable change of the efficiency value.

(8) The rheostat loss is usually omitted and charged to plant operation. If added, it is I^2R, where I is the field current and the R is the hot value of rheostat resistance.

(9) Exciter losses are the losses in the exciter to supply the dc field excitation. These losses are usually omitted and are charged to plant operation.

(10) The stray load loss is made up of the various losses due to eddy current in copper and additional core loss due to the distribution of the magnetic flux; and for dc machines it is taken as one percent of the output for machines larger than 150 kVA.

Efficiency of a synchronous machine depends on the power output and therefore in specifying the efficiency at rated kVA; the power factor at which it operates must also be specified. The following example shows that at unity power factor, the efficiency is higher than at some given power factor.

Example 4-14. A 1000-kVA, 2400-V, 3-phase, 1200-rpm, 60-Hz alternator was tested as follows:

> Resistance between terminals corrected to 75°C, 0.30 ohm. Calibrated motor input to alternator without field and driven at 1200 rpm, 26.2 kW. Calibrated motor input to alternator with field excitation for open-circuit voltage of 1430 V/phase or 2480 V (line) was 50.4 kW.

The dc field operates at 125 V and field current is 64 A. (Use an ac resistance factor of 1.25.) Calculate the efficiency at unity power-factor load and for a 0.80 power-factor load (lag).

Solution:

Losses:

friction, windage, and brush friction	26.2 kW
core loss (50.4 to 26.2)	24.2 kW

armature electrical loss:

$$I_a = \frac{1000 \times 1000}{\sqrt{3} \times 2400} = 240 \text{ A}$$

$$R_a/\text{phase} = \frac{0.30}{2} \times 1.25 = 0.1875 \text{ ohm}$$

$$I_a^2 R_a = 3(240)^2 \frac{0.1875}{1000} = \qquad 32.4 \text{ kW}$$

field loss (125 × 64)/1000	8.0 kW
stray load loss, 1 percent of 1000	10.0 kW
total losses	100.8 kW

$$\text{efficiency (unity power factor)} = \left(1 - \frac{P_l}{P_o + P_l}\right) \times 100$$

$$= \left(1 - \frac{100.8}{1000 + 100.8}\right) \times 100$$

$$= 90.85\%$$

$$\text{Efficiency (0.80 power factor)} = \left(1 - \frac{100.8}{1000 \times 0.8 + 100.8}\right) \times 100$$

$$= 88.8\%$$

4-10 PARALLEL OPERATION OF ALTERNATORS[3],[4]

The operation of alternators in parallel is obviously necessary for continuity of service, efficiency of operation, and for scheduled maintenance. Alternators are built in large sizes for commercial use with 1000-MW units not uncommon. Stations within a system are tied together and systems are tied together to form grids covering large areas. Basically, many alternators are operating in parallel commercially, and load is continually being shifted. Alternators are put on the line and taken off the line continuously. This section is concerned with the basic process of paralleling two alternators and the shifting of load between them.

The requirements for alternators to operate in parallel may be stated as follows:

(1) The voltages must have the same effective values.

(2) The waveforms must be the same, hence the sine wave is the standard waveform.

(3) The frequencies must be the same.

(4) The phase sequence of each must be the same.

(5) Each alternator must be in phase opposition with the bus or the first alternator. This is similar to dc operation where generators and batteries are connected (+) to (+).

(6) The overall speed load characteristic of alternator and prime mover should have drooping characteristics.

The first five requirements are necessary for the first step of paralleling before the breaker of the incoming alternator may be closed. Figure 4-17 shows the schematic diagram of two alternators connected to a load bus by triple-pole single-throw switches. The switch of alternator I is closed and alternator II is to be paralleled or *synchronized* as the process is commonly called. This operation uses the dark-lamp method in which lamps are connected across the switch terminals, a set across each phase. Two lamps in series are used for the double voltage by phase opposition that at instances will be across the lamps. For voltages higher than lamp voltages, step-down transformers need to be used, with the high side across the terminals and the low side connected to the lamps. A double-pole double-throw switch is used with a voltmeter to check on the first requirement. As for the lamps, for high voltages a potential transformer has to be used with the voltmeter. The synchronizing procedure is given below.

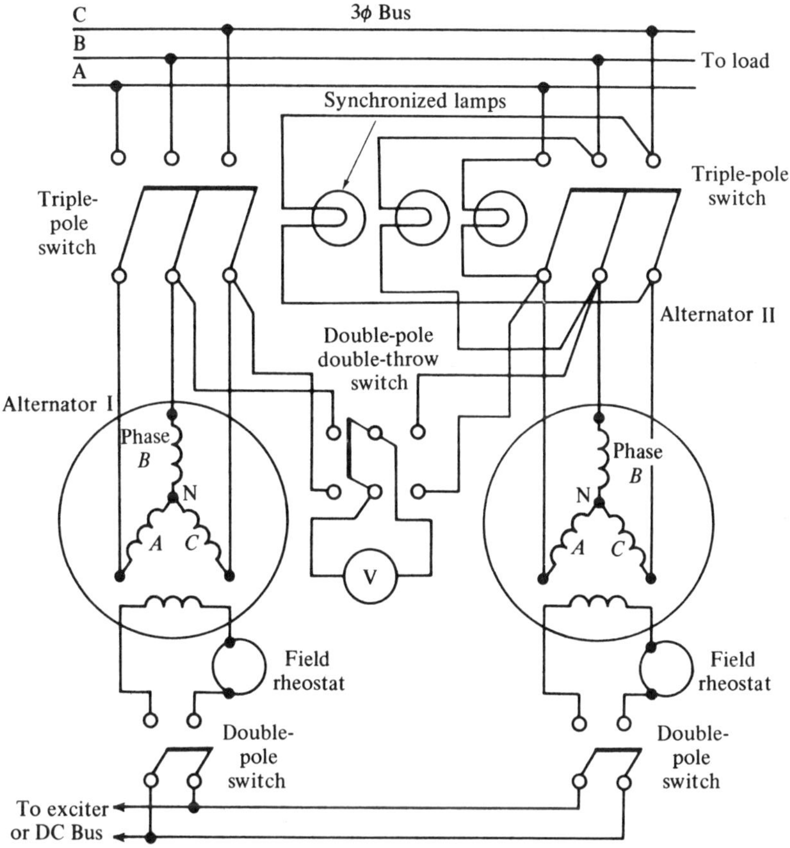

FIGURE 4-17. Connection diagram for paralleling two alternators with the dark lamp method.

Alternator II is brought up to speed and its field rheostat is adjusted so that the voltage of alternator II is equal to the voltage of alternator I. The three sets of synchronizing lamps should flicker in unison. If they do not, the phase sequence of both are not the same and any two leads of alternator II will have to be interchanged. The lamps now will flicker in unison, indicating that both have the same phase sequence but that the frequency of alternator II is not the same as I. The speed of the prime mover of alternator II must be changed until the lamps do not flicker. At this point the field rheostat may have to be adjusted so that the terminal voltages are equal. If the lamps stay bright, the speed must be changed until the lamps are dark.

At this point the voltages are equal, both alternators are in phase opposition, both have the same phase sequence, and both have the same frequencies. The switch of alternator II may now be closed, putting it in parallel with alternator I. Alternator I is now paralleled but not delivering any load. Figure 4-18(a) shows the phasor diagram of the alternator terminal voltages.

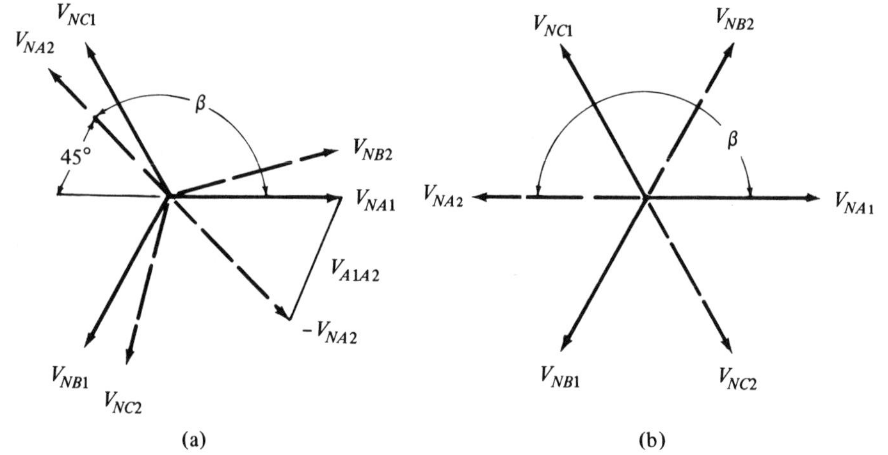

FIGURE 4-18. **(a) Phasor diagram for terminal voltages of two alternators out of phase opposition by 135°. (b) Phasor diagram for terminal voltages of two alternators in phase opposition.**

The solid lines are for alternator I and the broken lines for alternator II. The diagram shows that alternator II is out of phase with I by the angle β or approximately 135 deg. The voltage across the *A*-phase lamps is the phasor difference as shown by V_{A1A2}. A voltage of the same magnitude is across the *B* and the *C* leg of the switch. In Figure 4-18(b) the phase opposition is 180 deg as shown by angle β and the voltages across each leg are zero. It is at this instant that the switch is closed as stated before.

It is somewhat difficult to detect the zero lamp voltage due to the nature of incandescent lamps. If the lamps are connected as shown in Figure 4-19(a), the lamps will all be equally but partially bright, which may be as difficult to perceive as the middle of the dark period. Another scheme for connecting the lamps is shown by Figure 4-19(b). This is the rotating-lamp method. By examination of Figure 4-19(b) and the phasor diagrams, the proper moment to close the switch is when the middle lamp is dark and the two other sets are of equal brilliance. This is most difficult to perceive for the *B*-phase

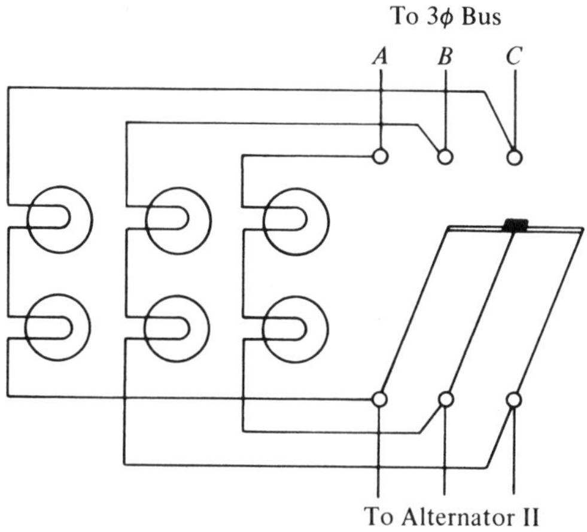

(a)

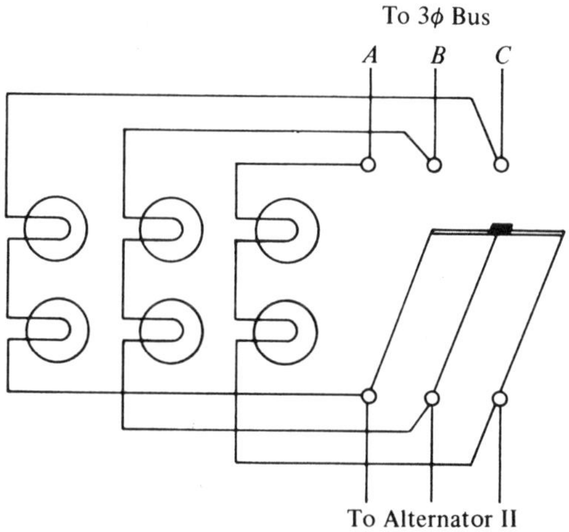

(b)

FIGURE 4-19. **(a) Synchronizing lamp connections for all-bright lamp method. (b) Synchronizing lamp connections for rotating lamp method.**

lamps must be dark. If mistakenly the *A*-phase or the *C*-phase lamps are taken as the dark lamps, this would be the worst time to close the switch.

The synchroscope is an instrument that best tells the instant to close the switch of the incoming alternator. It is a single-phase instrument and matches the frequencies on one phase. Figure 4-20 shows the face of the instrument.

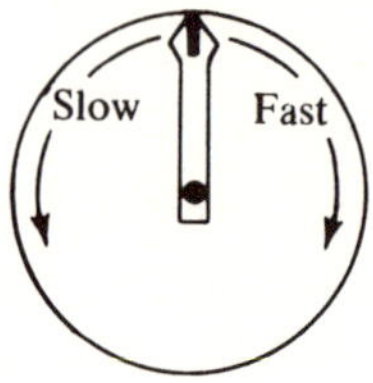

FIGURE 4-20. Sketch of face of a synchroscope.

When the pointer is still, the frequencies are the same but when the pointer is stationary in the upright position, the phase opposition is 180 deg and this is the correct instant to close the switch. For a new connection, lamps are needed to indicate the correct phase sequence. A combination of the dark lamps and the synchroscope is usually used for the synchronizing procedure.

After synchronizing is completed, shifting of load is accomplished by manipulation of the governor setting of the turbine drive or speed adjustment of the prime mover. The governor setting of alternator II is advanced to increase the speed, while the setting of alternator I is lowered. This will cause alternator II to take load while I gives up some of its load. This is an exacting procedure for an isolated situation of two alternators, for the frequency and the voltage of the combination must be kept constant. Field rheostat adjustment of synchronized alternators does not change the voltage to any great extent and load shifting is accomplished by speed adjustment or governor setting of the prime mover.

The last requirement of drooping overall speed–load characteristics is necessary for stable load output of an alternator. Figure 4-21 shows the speed load characteristics of two alternators with their prime movers. Each has a drooping characteristic. It is obvious that if two alternators with the characteristic of *A*1 are in parallel, each will carry the same load. The characteristic of *A*1 by increasing the governor setting will raise the characteristic as shown by the broken line, *A*1*A*. This effect is similar to adjusting the shunt field rheostat in a dc generator to raise its terminal voltage–load characteristic. Unless the governor setting of *A*2 is lowered, both the voltage and the frequency will increase. The horizontal dash line, indicated by operating speed, will be raised. The alternators need not have the same number of poles. One could be a four-pole machine whose synchronous speed is 1800 rpm and the other might be an eight-pole machine whose synchronous speed is 900 rpm.

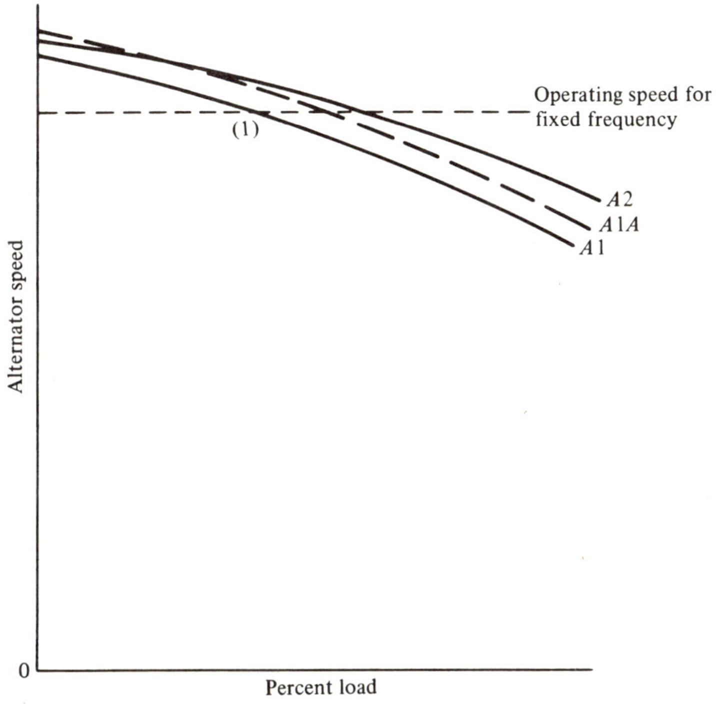

FIGURE 4-21. Alternator speed–percent load characteristics of combination of alternator with its drive.

Another facet of parallel operation of alternators is the setting of dc field current by their respective rheostats. When in the synchronizing procedure, the alternators are not in phase opposition; a synchronizing current flows between them through the synchronizing lamps. When paralleled alternators are delivering load, the field rheostats should be set so that each alternator operates as nearly as possible at the same power factor as that of the load. The settings should be such that each alternator current is a minimum; then the synchronizing current is at a minimum value. To overexcite an alternator causes its current to lag the generated voltage. As was noted in the section on armature reaction, for zero lagging power factor, the armature MMF demagnetizes the main flux. So the current will lag the generated voltage to reduce the net flux to keep it constant for constant terminal voltage. Under-excitation causes the current to lead the generated voltage. This synchronizing current flow may cause excessive heating in an alternator, especially if it is operating at rated power output. In the next section on synchronous motors *V* curves are discussed and shown in Figure 4-26. These apply equally to alternators, except that excitation above normal for an alternator causes it to have lagging power factor. Below normal, the power factor is leading. *Normal excitation* is that value at which the operation is at *unity power factor.*

4-11 SYNCHRONOUS MOTORS, GENERAL THEORY

Synchronous motors operate at an average constant speed, the synchronous speed, at all times. The speed in rpm, solving Equation (4-3) for the speed, yields the formula given in Equation (4-8a),

$$\text{rpm} = \frac{120f}{P}.$$

The terms are the same as previously defined. The synchronous speed, therefore, depends on the frequency of the source and the number of poles in the motor. Speed regulation is zero.

As stated before the excitation for a synchronous motor is from two sources. One source is an ac supply to the armature. The ac supply is also the power supply to the motor. The other source for excitation is the dc to the field. In special cases of small synchronous motors the dc excitation may be permanent or induced magnets. The dc excitation can be varied. If the excitation is of such value that the motor operates at unity power factor, this is normal excitation. If the field current is increased above normal, the motor will operate at a leading power factor. If the excitation is below normal, the operation is at a lagging power factor.

The properties of constant speed and power-factor adjustment are advantages for the synchronous motor over other types. These motors are used for such applications as drives for blowers, compressors, centrifugal pumps, mills, and printing presses, where speed must be constant. Also, they are used for direct drives at lower speed application for power-factor improvement. They are less costly than other types in the 50 to 500 hp sizes in low-range speed below 500 rpm. They also operate, in most cases, at higher efficiencies than other types of motors.

The synchronous motor itself does not produce any starting torque. A squirrel cage is built into the pole faces to start it as an induction motor and when the speed is near the synchronous speed, the field is energized and the rotor will fall into its rated speed and operate as a synchronous motor. The squirrel cage is also known as the *amortisseur winding*, which serves to prevent hunting when the load is suddenly changed. This winding, therefore, serves to start the motor and to keep the speed from oscillating when load is changed in its operation.

The fact that no starting torque is developed by this motor is explained by Figures 4-22(a), 4-22(b), and 4-22(c). In Figure 4-22(a) with the north

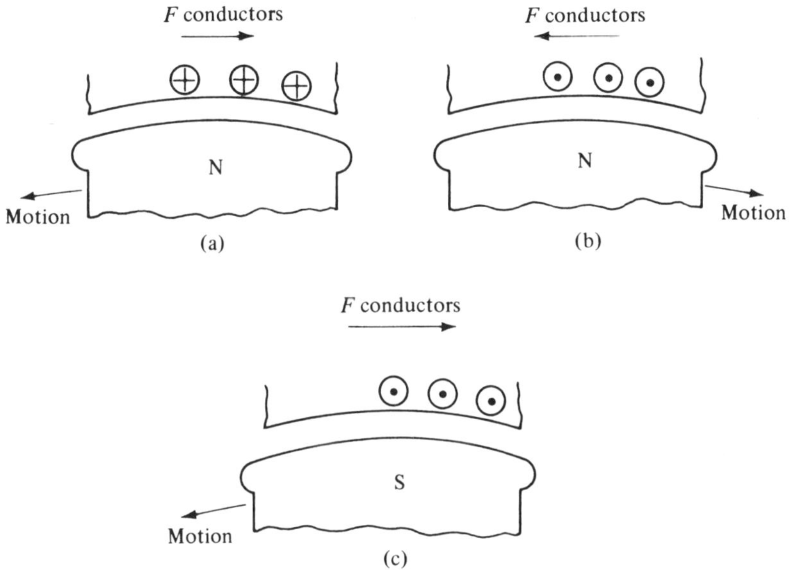

FIGURE 4-22. (a) Motion of a rotor pole at rest when current is applied to armature (current away from observer). (b) Motion of same rotor pole at rest when current is reversed (2nd half of cycle). (c) Motion of same pole at rest when current is reversed as in (b), and pole has advanced 180° from (a).

pole stationary and current as shown in the conductors, a force is exerted on them as shown. This direction can be determined by the left-hand motor rule. The conductors cannot move because they are in slots in the stator, so a force will cause the rotor pole to move in the opposite direction as shown. In Figure 4-22(b), which is 180 deg (1/120 sec, for a 60-Hz supply) later, the current in the conductors is reversed and the motion is opposite to that in the (a) figure. In another half cycle the force and motion is again as in (a) and the rotor will only try to oscillate, but because of inertia and counter torque, it will stand still. If in some way the poles were made to rotate so that they would rotate through 180 electrical deg, the forces and motion would be in the same direction. Figure 4-22(c) shows that the motion is in the same direction as in (a). Therefore, if the rotor is started, it will continue to rotate at synchronous speed to a point where the developed torque can no longer overcome the counter torque. At this point the rotor will stop rotating without any drop in the average constant speed prior to stalling. The rotor rotates at average synchronous speed up to the stalling point and then stops abruptly.

The synchronous motor action is actually one of magnetic action. In Section 4-6 it is shown how a three-phase supply to a properly wound armature stator produces a rotating magnetic field. It is known that opposite magnetic poles are attracted to each other by a force. Therefore, the south poles of the rotor are attracted to the induced north poles, and the north

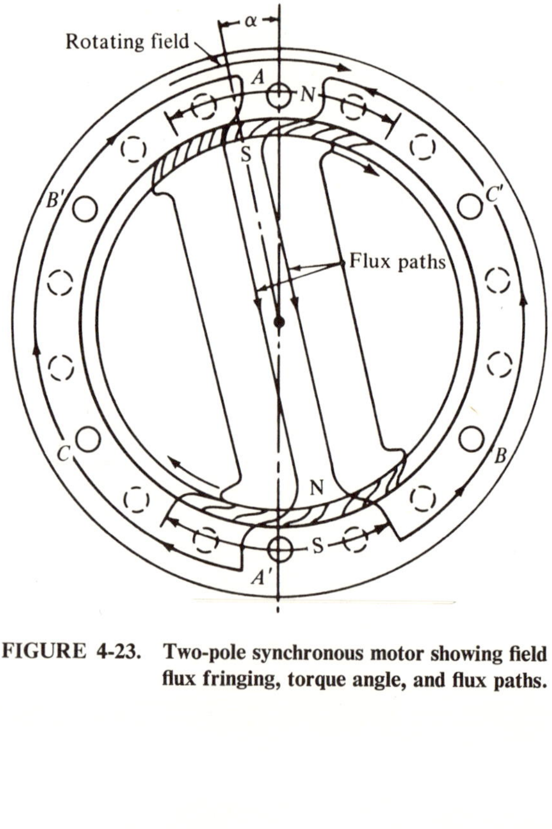

FIGURE 4-23. Two-pole synchronous motor showing field flux fringing, torque angle, and flux paths.

poles are attracted by the induced south poles. If the rotor poles are made to rotate at near synchronous speed, the rotor poles will lock in step with the rotating field of the stator. The action is shown in Figure 4-23. The *rotating field* of the stator rotates at synchronous speed and the two-pole rotor will follow at the same speed. The poles do not align on the center lines due to the counter torque of the load and the mechanical losses of the motor itself. The rotor follows the stator magnetic field and lags by an angle α, the *torque angle*. The greater the load torque, the larger will be the torque angle. When the counter torque gets too great, the force of attraction between the poles no longer can hold the rotor in step with the stator field. The torque angle is always in electrical degrees. In Figure 4-23, which is two pole, the torque angle is the same as the angle between pole centers.

In dc motors as load increases, the armature current must increase; for torque is proportional to the magnetic flux and the armature current ($T = K\phi I_a$). For a shunt motor the armature current I_a was given by the formula in Equation (3-5a),

$$I_a = \frac{V - E_g}{R_a}.$$

For a synchronous motor, the formula is quite similar,

$$\bar{I}_a = \frac{\bar{V} - \bar{E}_g}{\bar{Z}_a}, \qquad \textbf{(4-10)}$$

where

V is the terminal voltage (a phasor quantity) per phase,
E_g is the generated voltage (a phasor quantity) per phase,
Z_a is the impedance of the armature per phase (also Z_S).

To increase I_a in the shunt motor the value of E_g decreases by a decrease in speed. In the synchronous motor the speed does not change and therefore to increase $(\bar{V} - \bar{E}_g)$ is a different problem. The armature resistance is low compared to the synchronous reactance, so for practical purposes the synchronous impedance is equal to synchronous reactance. This means that the armature current lags the net voltage $(\bar{V} - \bar{E}_g)$ by almost 90 deg; 90 deg is used in the phasor diagrams of Figures 4-24(a), (b), and (c). The first diagram (a) shows the condition for light load, with $\bar{E}_g$ approximately equal to the impressed voltage $\bar{V}$. The phasor voltage of the difference $(\bar{V} - \bar{E}_g)$ is small as is the torque angle α. In the (b) figure for medium load the voltage difference $(\bar{V} - \bar{E}_g)$ must increase so that the armature current $\bar{I}_a$ may increase in accordance with Equation (4-8). The value of generated voltage remains approximately the same since excitation of the dc field remains constant, but the position of the generated voltage phasor changes increasing the

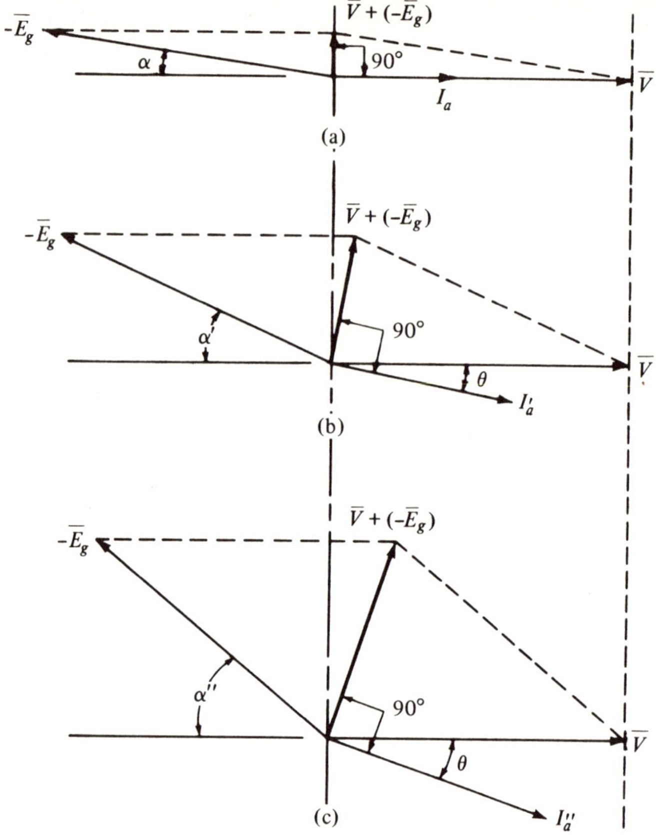

FIGURE 4-24. (a) Phasor diagram showing light load with normal excitation for a synchronous motor. (b) Phasor diagram as in (a) for a medium load. (c) Phasor diagram as in (a) for a heavy load.

torque angle α'. Hence the magnitude of $(\bar{V} - \bar{E}_g)$ increases and tips to the right causing the armature current I_a' to increase in magnitude and lag the terminal voltage by the angle θ. Because the rotor slips back as the load is increased, the new torque angle (α') increases causing both $\bar{V} + (-\bar{E}_g)$ and I_a to become larger. Figure 4-24(c), for a heavy load, shows further increases in the voltage difference $(\bar{V} - \bar{E}_g)$, the armature current I_a'', the phase angle θ, and the torque angle α''.

The characteristic that overexcitation causes the synchronous motor to operate at a leading power factor, and underexcitation results in lagging power-factor operation may be shown by phasor diagrams with the same assumptions that were used in Figures 4-24(a), (b), and (c). Figures 4-25(a),

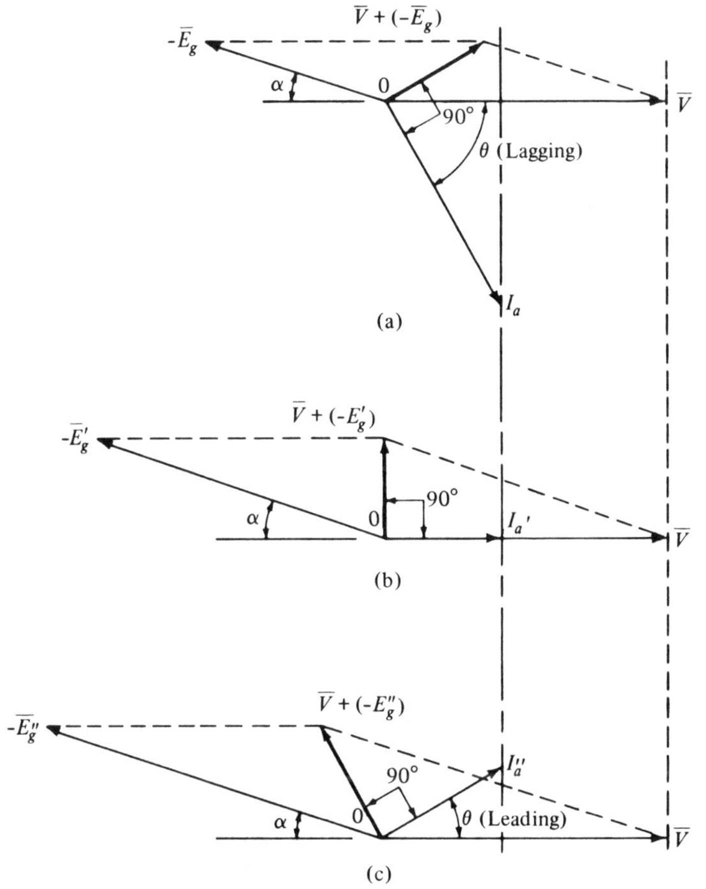

FIGURE 4-25. (a) Phasor diagram of synchronous motor showing lagging power factor at ordinary load and under excited. (b) Phasor diagram of the same motor at same load with normal excitation and unity power factor. (c) Phasor diagram of same motor at same load with overexcitation and leading power factor.

(b), and (c) illustrate how the phase angle θ for a constant load changes with a change in excitation. These are based on the assumption that the losses are constant, therefore the power input is constant and this power is given by the expression, $P = VI(\cos\theta)$. The torque angle α for a constant load does not change. In Figure 4-25(a) the value of the generated voltage $\overline{E}_g$ is less than for unity power factor because the field excitation is reduced, the difference $(\bar{V} - \bar{E}_g)$ tips to the right, and the armature current lagging the voltage difference by approximately 90 deg also lags the terminal voltage by the angle θ. In Figure 4-25(b) the value of generated voltage is approximately equal to the terminal voltage and the difference of the two voltages $[\bar{V} + (-E_g)]$ leads the terminal voltage and the armature current by 90 deg. The armature current and the terminal voltage are in phase for unity power factor. The magnitude of the armature current is equal to $I_a \cos\theta$ of that shown in Figure 4-25(a). In Figure 4-25(c) the value of generated voltage is greater than that given in (a) or (b) since the value of excitation is increased above normal. The voltage difference $(\bar{V} - \bar{E}_g'')$ increases from that given in Figure 4-25(b) and tips to the left. The armature current lags the voltage difference by 90 deg and leads the terminal voltage by angle θ. These phasor diagrams are all on a phase basis. For a three-phase synchronous motor, the power input is given by the expression, $P = 3VI\cos\theta$.

The operation of synchronous motors is facilitated by the use of V curves. These are curves of armature current or simply load current versus field current for various values of constant load. Figure 4-26 shows the V curves for a 500-hp, 2300-V, three-phase synchronous motor. The curves are shown for no load, half load, and full load. A more complete set might also include one-quarter, three-quarter, and one and one-quarter load values. They have the shape of the letter V and hence are called V *curves*. The bottom of the curve represents the unity power-factor values. The same figure also shows power factor plotted for the three-load values plotted against power factor. It should be noticed that the input rating, the line current, and the power factor are also given. At full-load operation this motor may operate at 0.80 power factor. From the V curve the excitation for this power-factor leading is 67.5 A (dc). The power input is 400 kW and it would draw $(500 \times \sin\theta)$ 300 leading reactive kVA. Thus operated at full load with an excitation of 67.5 A the motor would neutralize 300 lagging reactive kVA. At the same load increasing the excitation current to 75 A changes the power factor to 0.715 lead and a current input of 140 A according to the V curve. At full load and this power factor, the motor would draw 558 kVA and 393 reactive kVA leading. The motor would be operating at overload and overheating would occur. Neglecting losses, this synchronous motor at no load could operate at 500 reactive kVA. *Synchronous motors* designed to operate at no load at zero leading power factor are called *synchronous condensers*. In the design the shaft can be quite small and bearings would also be much lighter. Such a machine floats on the line and the value of reactive kilovolt amperes depends on the value of field excitation. In all respects it behaves as a capacitor and could be called a *synchronous capacitor*. The following examples illustrate

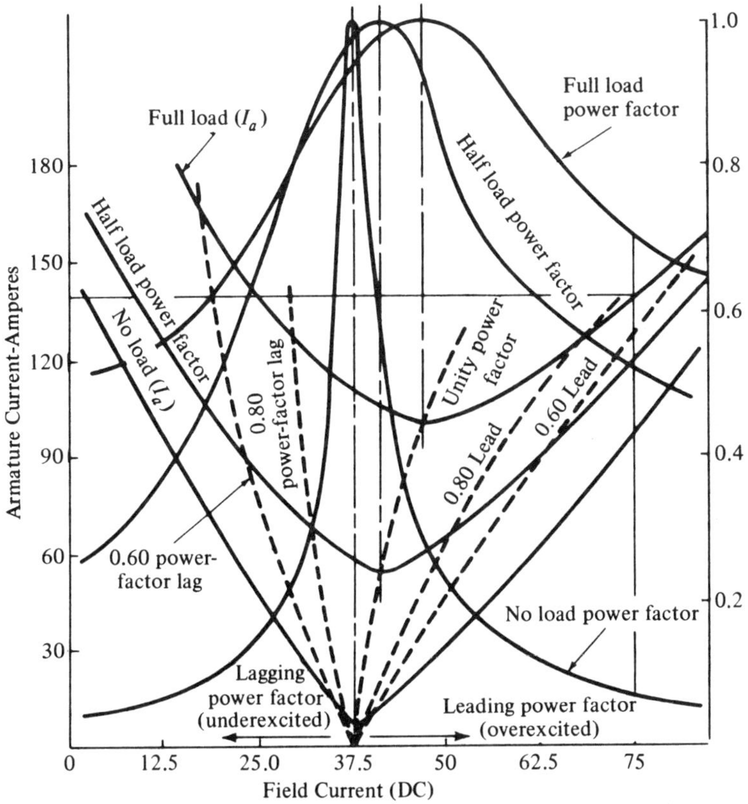

FIGURE 4-26. V **curves** ($I_a - I_f$) **for a typical synchronous motor.**

how power-factor improvement is accomplished with synchronous motors.

Example 4-15. The V curves are for a 500-hp, 2300-V, 3-phase, 1200-rpm, 125-A synchronous motor whose input is rated at 500 kVA at a power factor of 0.80. A plant has a 1000-kVA load at a power factor of 0.70 lagging. If the motor is added and operated at full load with a leading power factor of 0.80, (a) calculate the value of dc field excitation at which the motor must operate. (b) Calculate the overall kVA with the motor installed and the new plant power factor.

Solution:

(a) From the full-load power-factor curve, $I_f = 67.5$ A,
from the V curve at 125 A, $I_f = 67.5$ A.

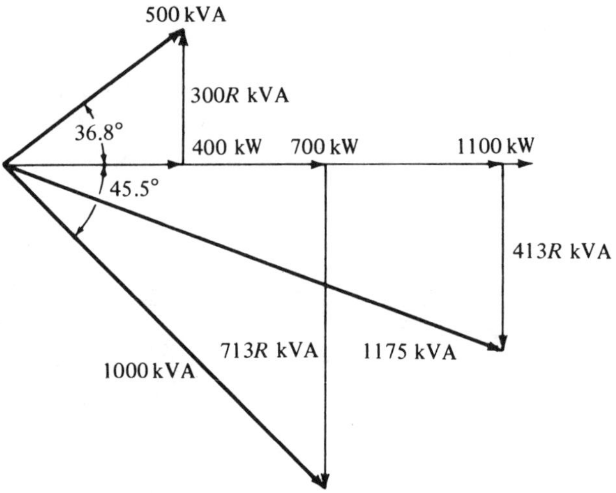

FIGURE 4-27. Phasor diagram for Example 15.

(b) See the phasor diagram of Figure 4-27:

original load, 1000 kVA at 0.70 lag power factor	$700 - j713$
motor load, 500 kVA at 0.80 lead power factor	$400 + j300$
total kVA	$1100 - j413$
	$1175\lfloor 20.55°$
power factor = cos (−20.55°)	0.935 lag
or 1100/1175	0.935 lag

Note that the original load is 700 kW with 713 lagging reactive kVA. The new total load is 1175 kVA, 1100 kW with 413 lagging reactive kVA.

Example 4-16. If the motor operates at no load as a synchronous capacitor is added together with a motor load of 400 kW at a lagging power factor of 0.80 lagging (input), calculate the total kVA and power factor.

Solution:

See the phasor diagram of Figure 4-28:

original load (same as Example 4-15)	$700 - j713$
synchronous capacitor	$j500$
new motor load, 400 kW at 0.70 lag power factor	$400 - j407$
total kVA	$1100 - j620$
	$1264\lfloor 29.4°$
power factor (cos −29.4 deg)	0.87 lag

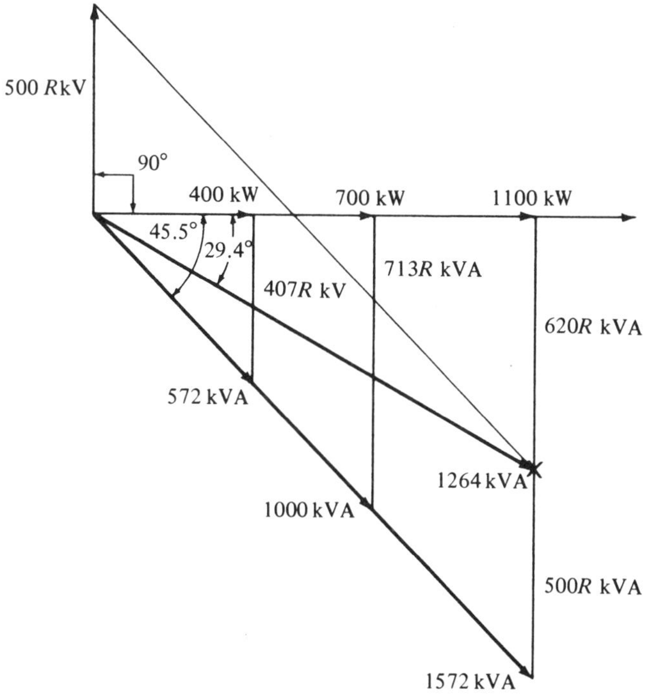

FIGURE 4-28. Phasor diagram for Example 16.

4-12 SUMMARY

The ac synchronous machine is a rotating machine whose average speed is proportional to the frequency of the generated voltage or of the applied voltage. The generator is known as an alternator, ac generator, synchronous generator, or synchronous alternator with the connotation of alternator the most common. The alternator generated voltage is given by the expression in Equation (4-2),

$$E = 4.44 f N \phi \times 10^{-8} \text{ V}.$$

This expression is for the basic single-phase alternator or for the generated voltage per phase if the coil side turns are all in one slot and the coil sides span 180 electrical deg. For an actual alternator that has the armature winding distributed around the periphery of the stator, the expression becomes that given in Equation (4-6),

$$E = 4.44 f N \phi k_d k_p \times 10^{-8} \text{ V}.$$

In this expression f is the frequency of the generated voltage which is given by the expression of Equation (4-3),

$$f = \frac{(\text{rpm})(P)}{120}\ \text{Hz}.$$

The distribution factor—a factor compensating for the fact that the winding is distributed, and for a polyphase winding, the coils are in groups or belts lying in adjacent slots—is given by the expression of Equation (4-4),

$$k_d = \frac{\sin (sd^\circ/2)}{s \sin (d^\circ/2)}.$$

The pitch factor, which compensates for the fact that the coil sides may be less than 180 deg between coil sides, is given by the expression in Eq. (4-5),

$$k_p = \sin \left(\frac{p^\circ}{2}\right).$$

In construction the synchronous motor and the alternator are identical. The usual synchronous machine has the armature on the stator and the field winding on the rotor. The field is excited by a dc supply, except in special cases where induced magnets or permanent magnets are used. The rotor for high-speed operation is of the cylindrical type and for medium and low-speed operation the salient pole is used. The change of the armature to the stator from the dc dynamo has many advantages, such as higher operational voltages, better facilities for the construction of armature windings on stators, more secure bracing of armature windings, increase of air-gap flux, and better ventilation facilities. In the rotor poles, amortisseur windings (squirrel cage, see Chapter 5) are placed in slots on the pole faces. In the alternator these windings serve to prevent hunting, which may be caused from minute variations in speed, for example, from reciprocating engines. In the synchronous motor these windings may be used for starting and to prevent speed oscillations (hunting) in operation from sudden variations in load.

Polyphase alternators are usually of the three-phase type. In this chapter the three-phase type was considered and, more specifically, the wye (Y) connection for alternators was considered. In calculations it is immaterial whether the connection is wye or delta, the wye-connection assumption yields the results desired a little more simply. Calculations are made on this basis in this chapter.

The power generated by an alternator or the counter power of a synchronous motor is given by the formula in Equation (4-7),

$$P_g = 3(E_p I_p \cos \beta) = \sqrt{3} E_l I_l \cos \beta.$$

In these formulas β is the phase angle between the generated voltage per phase E_{gp} and the current per phase. The difference between load phase angle θ and the generated voltage phase angle is shown in Figures 4-14(b), 4-14(c) and 4-14(d).

Armature MMF due to current in the armature windings is explained by Figures 4-9, 4-10, and 4-11. For a polyphase alternator it is sinusoidal in shape and for the stationary stator it moves in space in a fixed position with the main field of the rotating rotor which rotates at synchronous speed. The position of the armature MMF curve depends on the power factor of the load. For unity power factor it is practically in phase with the generated voltage, for zero lagging power, it lags the generated voltage by 90 deg and for zero leading power factor it leads the generated voltage by 90 deg. Therefore for unity power factor the armature MMF has a cross-magnetizing effect on the main field MMF, for a zero lagging power factor it has a demagnetizing effect, and for zero leading power factor it has a magnetizing effect.

In an alternator the rotating field of the rotor cuts the armature conductors, thereby inducing generated voltages in these conductors. The current flow in the armature conductors produces a rotating armature MMF and hence armature flux. If current is to flow from a polyphase source in the same windings, an armature MMF and hence magnetic flux that rotates around the stationary armature must be produced. The same Figures 4-9, 4-10, and 4-11 could be used to show this fact. The only difference in the figures is that if a three-phase supply is applied to the same winding, the current directions and, therefore, the MMF and flux directions would also be reversed. However, the series of illustrations of Figure 4-12 show that a truly rotating magnetic field is produced in the armature winding.

The terminal voltage of an alternator drops from the generated voltage value (no-load) to full-load value due to armature resistance drop, leakage reactance drop, and armature reaction. The resistance is determined by measurement between each pair of terminals. Since a wye connection is assumed, the resistance per phase is one half the average value multiplied by a factor of 1.25 to 1.75 for the effective armature resistance per phase. Since the drop due to leakage reactance and the drop due to armature reaction are in phase, the reactance of the two drops are combined and give the term, *synchronous reactance* (X_S). The synchronous reactance is calculated from two tests. An open-circuit test to yield a saturation curve and a short-circuit test are performed. The independent variable is field current (dc). The synchronous impedance is equal to the voltage per phase divided by the short-circuit current. The voltage may be taken at rated, or the short-circuit current may be taken at rated, or a value on the air-gap line divided by the short-circuit current for the same field current value. From the value of synchronous impedance the value of synchronous reactance is calculated as per phasor diagram of Figure 4-14(e). The calculations for unity, lagging, and leading power factors are made by use of phasor diagrams of Figures 4-14(b), (c), and (d). The equations are as follows:

$$\text{unity power factor } E_g = \sqrt{(V_T + I_aR_a)^2 + (IX_S)^2},$$
$$\text{lagging power factor } E_g = \sqrt{(V_T \cos\theta + I_aR_a)^2 + (V_T \sin\theta + I_aX_S)^2},$$
$$\text{leading power factor } E_g = \sqrt{(V_T \cos\theta + I_aR_a)^2 + (V_T \sin\theta - I_aX_S)^2}.$$

This method of calculating the synchronous reactance and the generated voltage yields a high value for the no-load voltage. This method is often called the pessimistic method. The Potier triangle and the ASA methods yield more realistic results.

Voltage regulation for an alternator may be calculated by the following formula (Equation (4-7)),

$$\text{percent voltage regulation} = \frac{E_{gp} - V_{TP}}{V_{TP}} \times 100.$$

Typical terminal-voltage–armature-current curves are shown in Figure 4-16. For leading power factor, regulation may be negative and quite high at lagging power factor. The power factor is determined by the load conditions.

Efficiencies of synchronous machine are quite high. Machines over 50 kVA may be as high as 92 to 94 percent. Efficiencies are calculated by determining the losses. The losses are the stray power losses consisting of friction and windage loss, brush friction loss, core loss, and ventilation and cooling loss. The electrical losses are made up of the armature loss $I_a^2R_a$, field loss $I_f^2R_f$, brush contact loss, and rheostat loss. The last two are usually not included. The exciter losses are usually charged to plant operation. The stray load loss taken as one percent of the output for machines larger than 150 kVA.

Alternators are operated in parallel, taken off the line and put on the line for continuity of service, economical operation, maintenance, and operation at optimum efficiencies. The basic diagram is shown by Figure 4-17. The dark-lamp method with a synchroscope is generally used. Before the alternator may be connected, the voltage must equal that of the other alternator or bus, the waveform must match, the frequency must be the same, the phase sequence must be the same, and the alternator must be in phase opposition. For operation and load stability, the speed-load characteristic of the alternator with its driver must have drooping characteristics. To shift load the speed setting of the prime mover must be altered.

Synchronous motors have no speed regulation because they operate at an average constant speed. The speed is given by the formula given in Equation (4-8a) as

$$\text{rpm} = \frac{120f}{P}.$$

The motor is doubly excited. Alternating current to the armature, which provides a rotating magnetic field and to the power source, and dc to the rotor field are the two sources of excitation. When a synchronous motor is overexcited, it will operate at a leading power factor; when underexcited, it will operate at a lagging power factor. Normal excitation is when it operates at unity power factor. Such motors operate at high efficiencies and may be used to operate a rotating load and improve power factor by overexcited operation. Some motors are designed to operate without any mechanical load to provide leading reactive kVA. Such machines are called *synchronous condensers*. Figure 4-26 shows V curves (armature current–field current) and power-factor–field-current curves for a 500-hp motor. The calculations for power-factor correction follows typical ac calculations with the use of phasor diagrams.

Problems

1. Make a table showing synchronous speeds for 60-Hz, 50-Hz, and 25-Hz frequencies for alternators from 2 poles to 30 poles for every pair of poles.
2. A single-phase alternator has 32 slots, 4 poles, a speed of 1800 rpm, 3 turns per coil, and a flux per pole of 1×10^6 lines; the coil span is 7 (1 to 8); all the slots are used. Calculate the generated voltage.
3. In Problem 2, the winding is altered so that 4 slots per pole are used and the turns per coil are doubled and a full-pitch winding is used. For the same generated voltage as in Problem 2, what must be the flux per pole?
4. In Problem 3, (a) what number of turns per coil are needed for the same generated voltage with approximately the same value of flux per pole as in Problem 2? (b) Calculate the actual new value of flux per pole.
5. A three-phase generator has 48 slots, 4 poles, and 8 turns per coil; the armature winding is whole coiled with a flux per pole of 5.1×10^6 lines; the coils have a slot pitch of 10 (1-11, 2-12, and so on). Calculate the generated voltage per phase.
6. A 240-rpm, 30-pole, 450-slot, three-phase alternator has a flux per pole of 2.1×10^6 lines; the coil span is 13. Calculate the generated voltage per phase.

7. If the alternator of Problem 6 turns at a speed of 200 rpm, (a) what is the frequency, and (b) what is the value of flux per pole for the same generated voltage per phase?
8. A 20-pole, 360-rpm, 6900-V, three-phase alternator requires a generated voltage of 4750 V/phase. If the alternator has 240 slots, a whole coil winding with 8 turns per coil, and a coil pitch of 10, calculate the value of flux per pole required.
9. In Problem 8, (a) calculate the line generated voltage; (b) if the phase current is 83.5 A and the angle β between E_g and I is 42 deg, calculate the power generated by the alternator.
10. A synchronous motor is coupled to two alternators and runs at a speed of 600 rpm. One alternator is to furnish a 400-Hz supply, the second to furnish a 50-Hz supply. Calculate the number of poles of each alternator.
11. A motor drives two alternators at 300 rpm. One alternator has 10 poles and the other has 16 poles. What is the frequency of each alternator voltage?
12. A three-phase, 500-kVA, 2400-V alternator has a dc resistance between terminals of 0.8 ohm and a synchronous impedance of 6.25 ohms. Calculate (a) the generated voltage at unity power factor; (b) the generated voltage at 0.80 lagging power factor; and (c) the generated voltage at 0.80 leading power factor.
13. In the alternator of Example 4-12, using the synchronous reactance of the (b) method, calculate (a) the generated voltage for unity power factor, (b) the generated voltage for 0.80 lagging power factor, and (c) the generated voltage for 0.80 leading power factor.
14. A three-phase, 750-kVA, 2400-V alternator is short circuited at rated speed with the field current adjusted for rated line current. The short circuit is removed and with the same field current, the open-circuit line voltage is 1425 V. The ohmic resistance between stator terminals is 0.6 ohm. Calculate the generated voltage for 0.80 lagging power factor and for 0.80 leading power factor.
15. The alternator of Example 4-14 supplies a unity power-factor load of 750 kW at rated voltage. With the same excitation the no-load voltage is found to be 2880 V. From this data calculate the value of synchronous reactance and the generated voltage at 0.80 lagging power factor and for 0.80 leading power factor.
16. In Example 4-12 and Problem 13, when the alternator is operated at rated load of 1000 kVA and 0.80 lagging power factor, the load is removed and with the same dc excitation the terminal voltage is found

to be 3500 V (line). Calculate (a) the value of synchronous reactance, (b) the generated line voltage at unity power factor, (c) the generated voltage at 0.80 leading power factor.

17. Calculate the voltage regulation for (a) the unity power factor, (b) the 0.80 lagging power factor, (c) the 0.80 leading power factor for Problem 13.
18. For Problem 16 calculate the voltage regulation for (a) the unity power factor, (b) the 0.80 lagging power factor, (c) the 0.80 leading power factor.
19. A 250-kVA, 2400-V, three-phase, 1200-rpm, 60-Hz alternator has a voltage regulation of 13.5 percent at unity power factor. The ac resistance per phase is 0.09 ohm at 75°C. Calculate the regulation for (a) 0.70 lagging power factor and (b) 0.70 leading power factor.
20. A 1000-kVA, 6900-V, three-phase, 750-rpm, 25-Hz alternator has a voltage regulation of −8.50 percent at a leading power factor of 0.85. The ac resistance per phase is 1.8 ohm corrected for 75°C. Calculate the regulation for (a) the unity power factor and (b) 0.75 lagging power factor.
21. A 750-kVA, 2400-V, three-phase, 1200-rpm alternator was tested as follows: The average resistance between stator terminals corrected to 75°C was 0.396 ohm; friction and windage losses with calibrated motor were 15.5 kW; it was run at normal excitation of 45 A at 125 V dc with calibrated motor input of 30.5 kW. Use a 1.25 ac resistance factor. Calculate the efficiency at unity power factor and at 0.80 lagging power factor.
22. A 50-kVA, 575-V, three-phase, 1200-rpm alternator is operating at rated output at a lagging power factor of 0.75. A running light test as a motor was made at rated voltage with an input of 2560 W at a current of 8 A to find the rotation losses. The effective (ac) resistance between stator terminals is 0.6 ohm. Calculate the efficiency.
23. A plant has an average load of 750 kW at 0.75 lagging power factor, at 550 V, three phase. (a) Calculate rating of synchronous condenser to raise the plant power factor to unity; (b) calculate the rating of synchronous condenser if the power factor is raised to only 0.90 lagging power factor.
24. In Problem 23, the plant increases its load to 1000 kW by adding several synchronous motors taking 250 kW at a power factor of 0.80 lead. (a) Calculate the total kVA and (b) the new overall plant power factor.
25. A 2300-kW load at 0.76 lagging power factor is to be increased by adding two 400-hp synchronous motors. Each motor has an efficiency of 91

percent at rated output with the input at 0.80 power factor. (a) What is the new kVA if the two motors are operated at 0.80 leading power factor and full load? (b) What is the new overall power factor?

References

[1] ROBERTSON, B. L. and L. J. BLACK, *Electric Circuits and Machines*, 2nd ed. Princeton, N.J.: Van Nostrand, 1957.

[2] ROSENBLATT, J. and M. H. FRIEDMAN, *Direct and Alternating Current Machinery*. New York: McGraw-Hill, 1963.

[3] SISKIND, C. S., *Electrical Machines—Direct and Alternating Current*, 2nd ed. New York: McGraw-Hill, 1959.

[4] KOSOW, I. L., *Electrical Machinery and Control*. Englewood Cliffs, N.J.: Prentice-Hall, 1964.

[5] CARR, C. C., *Electrical Machinery*. New York: Wiley, Inc., 1958.

[6] DAWES, C. L., *Electrical Engineering*, Vol. II, *Alternating Currents*, 4th ed. New York: McGraw-Hill, 1947.

[7] FITZGERALD, A. E. and C. KINGSLEY, JR., *Electric Machinery*, 2nd ed. New York: McGraw-Hill, 1961.

[8] *Test Procedures for Synchronous Machines*, IEEE No. 115. New York: Institute of Electrical and Electronics Engineers, March, 1965.

Alternating-Current Induction Motors

5

An ac induction motor is defined by NEMA and ANSI (ASA) standards as a motor whose primary winding on one member (usually the stator) is connected to the power source and whose polyphase secondary winding or squirrel-cage secondary winding on the other member (usually the rotor) carries induced current. Unlike the dc motor and the synchronous motor in which there are electrical connections to the rotor through either a commutator or slip rings and brushes, currents are induced in the rotor through magnetic flux in the rotor winding in the same way currents are induced in a transformer secondary winding. The secondary of the induction motor rotates, however, and an air gap between rotor and stator is part of the magnetic circuit. The frequency of the induced currents in the secondary winding is not the same as the primary current when the motor is delivering power to a mechanical load.

The *induction motor* is an *asynchronous* machine whose average speed when operating normally is not proportional to the frequency of the system to which it is connected. As a motor, its speed is less than the synchronous speed a synchronous motor would have if connected to the same source and if it is driven at a speed greater than the synchronous speed, the induction motor will operate as an induction generator. The induction generator has few applications for emergency power supplies or regenerative braking application. A type of high-frequency generator operates on this principle. Because of the limited application of the induction machine as a generator it will not be considered here.

This chapter is concerned chiefly with the *polyphase* induction motor and, to a lesser extent, to the *single-phase* induction motor. The induction motor is the most widely used of all electrical motors because of its reliability, simplicity, ruggedness, comparable lower cost, and the variety of operating characteristics it may have.

5-1 ROTATING FIELD OF A POLYPHASE INDUCTION MOTOR[1]–[3]

The stator of the polyphase induction motor is similar to that of the synchronous machine. In the smaller sizes the stator is a cylindrical slotted core that is placed in the frame after the winding is completed in the slots. Figure 5-1 is a sketch of the cross section of such a machine. The winding

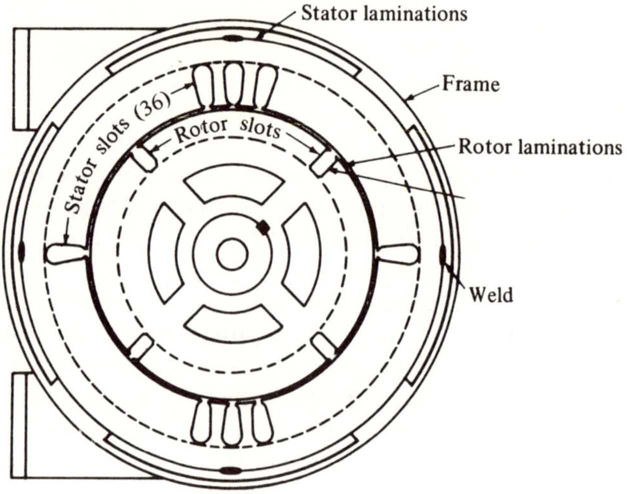

FIGURE 5-1. Cross section of magnetic circuits and structure of a polyphase induction motor.

connections are identical to that of the synchronous machine. The stator is made of laminated magnetic steel. The laminations are welded together. The slots usually are of the semiclosed type. The coils usually are not preformed but are made up of flexible insulated wire that is fed into the slots directly.

Figure 5-2(a) shows an enlargement of such stator slots. In one slot the cross-hatched area shows the irregular shape of the coils after they are in place. The partially closed slot is preferable for more uniform air-gap flux, which improves operation. In larger machines the stator core usually has the open slots so that preformed coils can be used. The slots are deeper than those of the synchronous machine to offset, to some degree, the fringing around the slots. Figure 5-2(b) shows a section of the open slot with the coil sides shown in the first slot.

In the preceding chapter in Section 4-6 it was shown that in a synchronous motor with a three-phase winding a revolving or rotating field is produced when the winding is properly connected. The electrical winding for a three-phase induction-motor stator is identical to that of a synchronous machine.

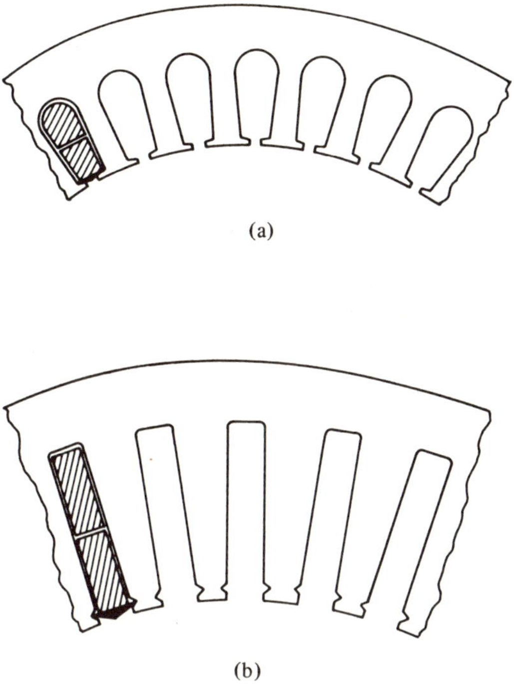

FIGURE 5-2. (a) Cross section of partially-closed stator slots. (b) Cross section of open and deep stator slots.

A revolving magnetic field by the stator is the first ingredient for the production of torque of a polyphase motor. In Figure 5-1, the stator has 36 slots. The winding as shown in Figure 1-18 (Chapter 1) would fit into the slots of this figure and a rotating field would be produced in this stator. To reverse the direction of rotation of the field, two of the terminals must be interchanged. The speed of the revolving field is given by the same expression used in Section 4-6—that is, Equation (4-8a),

$$\text{rpm} = \frac{120f}{P}.$$

5-2 ROTOR CONSTRUCTION OF AN INDUCTION MOTOR[1],[3];[4]

There are two types of rotors for induction motors. These are (a) *squirrel-cage rotor* and (b) the *wound rotor*. The type of induction motor is identified by the type of rotor it possesses. If a squirrel-cage rotor is used, the motor is known as a *squirrel-cage induction motor* (SCIM) and if a wound rotor is used, it is known as a *wound-rotor induction motor* (WRIM). The iron core is made of laminated steel punchings that are pressed directly onto a shaft in the smaller sized motor and on a spider structure for the larger sized motor.

The semiclosed slot in the rotor cross section of Figure 5-1 is the type slot that is used for the wound rotor. A polyphase winding that is practically always a three-phase wye-connected winding is placed in the slots. The three terminals are brought out to slip rings and brushes. A wye-connected rheostat is connected to the brushes so that the resistance of the winding can be increased to limit the starting current and at the same time increase the starting torque. Variation of the rotor resistance also varies the speed of

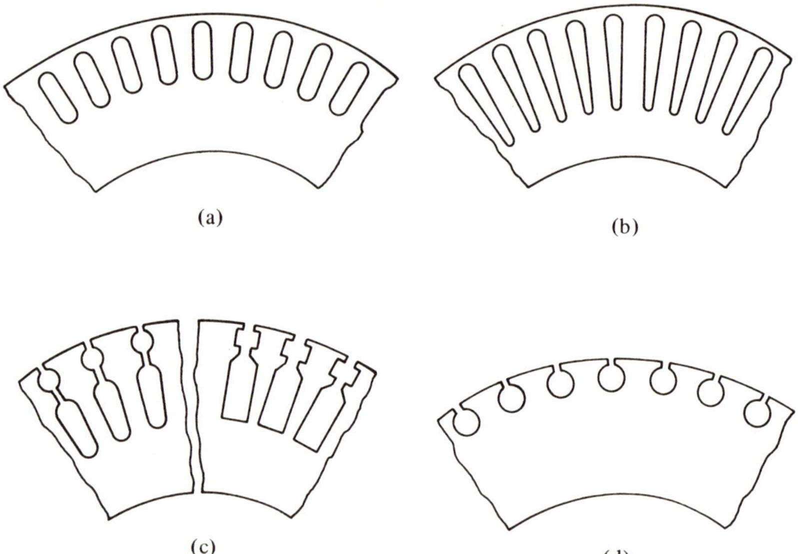

FIGURE 5-3. (a) Cross section of medium-depth rotor slots for a squirrel-cage rotor. (b) Cross section of deep rotor slots for squirrel-cage rotor. (c) Cross section of slots for double-cage squirrel-cage rotor. (d) Cross section of slots for high-resistance squirrel-cage rotor.

rotor. The wound rotor is discussed in more detail in Section 5-8. The winding must always be wound for the same number of poles as the stator winding.

Various shapes squirrel-cage rotor slots and bars are shown in Figures 5-3(a), 5-3(b), 5-3(c), and 5-3(d). In each type bare solid conductors fill out the slots. The ends of the rotor conductors are connected to end rings so that electric circuits are formed. Figure 5-4 shows a drawing of a squirrel-cage

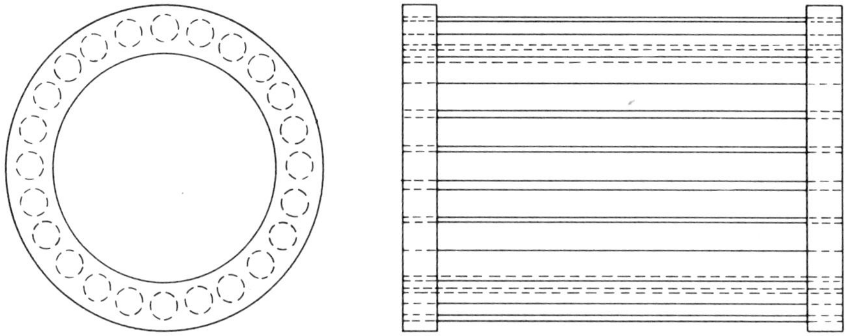

FIGURE 5-4. Squirrel-cage winding with round rotor bars.

winding with bars for the slots as shown in Figure 5-3(d). The squirrel-cage winding for smaller machines usually is molded by placing the laminated core in a mold and pouring the molten metal (usually aluminum) to form the bars and end rings in one operation. Fins are formed on the end rings for forcing cooling air through the motor. In the larger-sized motors copper bars are placed in the slots and the ends are brazed onto the end rings. The variety of shapes of the slots and bars of the squirrel-cage rotor offers a variety of motor characteristics.

The principle of operation may be demonstrated by use of a squirrel-cage rotor that fulfills the second requirement of a secondary winding. In Figure 5-5 the stator is shown as a rotating magnetic field as if the iron structure rotates at synchronous speed on the same shaft as the rotor. As the field rotates and the rotor is at standstill, voltages are induced in the rotor conductors. The right-hand generator rule may be used to determine the direction of the induced EMF in the bars. If the power factor is unity for the rotor winding, currents will flow in the same directions in the rotor conductors as shown for the induced EMF. By application of the left-hand motor rule, tangential forces are exerted on the rotor conductors in the same direction as the rotating magnetic field, and if the torque produced by the sum of these tangential forces is sufficient to overcome the counter torque of the motor and load, rotation of the rotor is in the same direction as the magnetic field.

In actual operation the rotor rotates, but never at the synchronous speed of the magnetic field and thus there is always relative motion between the

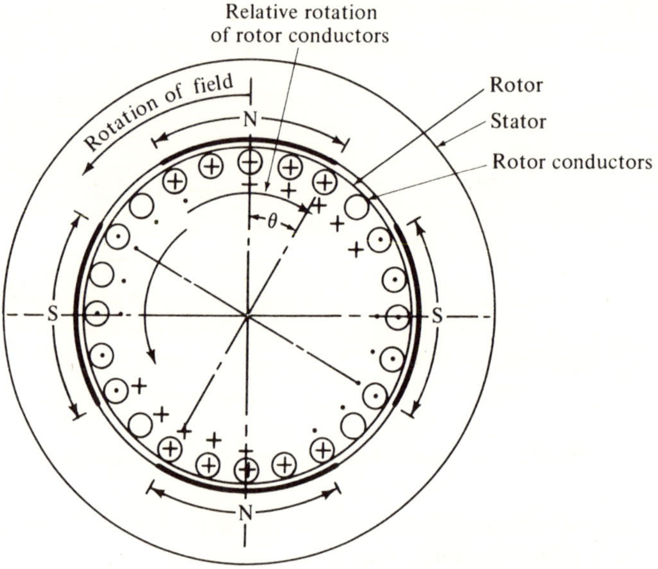

FIGURE 5-5. Squirrel-cage rotor cross section illustrating the principle of an induction motor.

synchronous speed of the rotating field and the rotor conductors. This difference in speeds is known as *slip speed* and it is the speed at which the rotor conductors slip behind the revolving field of the stator. The slip S by definition is given by the formula,

$$S = \frac{\text{rpm}_{\text{syn}} - \text{rpm}_{\text{rotor}}}{\text{rpm}_{\text{syn}}}. \qquad \textbf{(5-1)}$$

In terms of rotor speed the formula may be written as

$$\text{rpm}_{\text{rotor}} = \frac{120f}{P}(1 - S). \qquad \textbf{(5-1a)}$$

Slip is often expressed in percentage slip and the expression becomes

$$\text{percent slip} = \frac{\text{rpm}_{\text{syn}} - \text{rpm}_{\text{rotor}}}{\text{rpm}_{\text{syn}}} \times 100. \qquad \textbf{(5-2)}$$

Example 5-1. A 60-Hz induction motor has an eight-pole stator winding and operates at a speed of 895 rpm at no load and 860 rpm at rated load. (a) Calculate the slip; (b) calculate the percent slip; (c) calculate the speed regulation at rated load; (d) calculate the no-load slip.

Solution:

$$\text{(a) } S = \frac{\text{rpm}_{\text{syn}} - \text{rpm}_{\text{rotor}}}{\text{rpm}_{\text{syn}}} = \frac{900 - 860}{900} = 0.0465$$

$$\left(\text{rpm}_{\text{syn}} = \frac{120f}{P} = 120 \times \frac{60}{8} = 900\right);$$

$$\text{(b) percent slip} = \frac{\text{rpm}_{\text{syn}} - \text{rpm}_{\text{rotor}}}{\text{rpm}_{\text{syn}}} \times 100$$

$$= \frac{900 - 860}{900} \times 100 = 4.65\%;$$

$$\text{(c) speed regulation} = \frac{\text{rpm}_{\text{no load}} - \text{rpm}_{\text{load}}}{\text{rpm}_{\text{load}}} \times 100$$

$$\text{speed regulation} = \frac{895 - 860}{860} \times 100 = 4.07\%;$$

$$\text{(d) percent slip (no load)} = \frac{900 - 895}{900} \times 100 = 0.555\%.$$

Example 5-2. A 50-Hz induction motor operates at full load with a speed regulation of 4.1 percent. The no-load speed is 1494 rpm. (a) Calculate the speed at full load and (b) calculate the full-load percent slip.

Solution:

(a) Solving the speed-regulation formula of part (c) in the previous example, we have

$$\text{rpm}_{\text{load}} = \frac{\text{rpm}_{\text{no load}} \times 100}{100 + \%\text{ regulation}} = \frac{(1494)(100)}{100 + 4.1} = 1435\text{ rpm};$$

(b) $P = \dfrac{120f}{\text{rpm}} = \dfrac{(120)(50)}{1494} = 4.02$. Therefore, the synchronous speed is 1500 rpm for a four-pole motor.

$$\text{percent slip} = \frac{\text{rpm}_{\text{syn}} - \text{rpm}_{\text{load}}}{\text{rpm}_{\text{syn}}} \times 100$$

$$= \frac{1500 - 1435}{1500} \times 100 = 4.33\%.$$

In explaining the principle of induction-motor operation it was assumed that the rotor current is in phase with the induced voltage. In Figure 5-5 the rotor current directions are shown by crosses (+) and dots (·) outside the conductors for a rotor power factor of 0.50 or a 60-deg phase angle. By the analysis the principle still applies with a reduction in the torque because at this phase angle not all of the current-carrying rotor conductors are under the influence of the rotating field. In Figure 5-5 the angle θ is the space angle between the induced voltage in the rotor E_2 and the rotor current I_2. Because the figure is for a four-pole winding, the electrical degrees are twice the space degrees and hence; because the space angle is 30 deg, the electrical degrees are 60 or the phase angle between E_2 and I_2. As will be seen later the rotor frequency is low and hence the phase angle is pretty close to zero degrees.

5-3 QUANTITATIVE ANALYSIS OF CURRENT AND TORQUE[1],[5]–[7]

The revolving magnetic field of the stator is cut by the rotor conductors as the rotor rotates at less than synchronous speed and therefore EMF is induced in these conductors of the rotor winding. The frequency and the magnitude of this EMF depend on the relative speed of the rotor. At standstill or at start the rotor EMF will be maximum and the frequency will be the same as the frequency of the applied EMF to the stator. At a slip of 0.50 (50 percent) or one-half synchronous speed of the revolving magnetic field, the rotor EMF would be one-half of the standstill EMF and the frequency would be one-half the stator frequency. The expression for the *rotor generated voltage* per phase is

$$E_2 = SE_{BR}, \tag{5-3}$$

where

E_2 is the generated voltage per phase in the rotor,
E_{BR} is the rotor generated voltage per phase at standstill (blocked rotor),
S is the slip.

The *rotor frequency* f_2 is given by the expression,

$$f_2 = Sf_1, \tag{5-4}$$

where f_1 is the stator frequency.

Example 5-3. A 440-V, 60-Hz, three-phase, four-pole wound-rotor induction motor whose stator and rotor windings are both connected in wye operates at a speed of 1700 rpm. The rotor has 60 percent as many turns as the stator.

Calculate (a) the slip; (b) the blocked-rotor voltage per phase; (c) the rotor generated voltage per phase; and (d) the frequency of the rotor voltage.

Solution:

(a) Synchronous speed $= \dfrac{120f}{P} = \dfrac{(120)(60)}{4} = 1800$ rpm,

$$S = \frac{\text{rpm}_{\text{syn}} - \text{rpm}_{\text{rotor}}}{\text{rpm}_{\text{syn}}} = \frac{1800 - 1700}{1800} = 0.0555;$$

(b) $\dfrac{E_{BR}}{\text{phase}} = \left(\dfrac{440}{\sqrt{3}}\right)\left(\dfrac{60}{100}\right) = 152.5\ \text{V};$

(c) $\dfrac{E_2}{\text{phase}} = SE_{BR} = (0.0555)(152.5) = 8.45\ \text{V};$

(d) $f_2 = Sf_1 = (0.0555)(60) = 3.33$ Hz.

This example shows that the running voltage and frequency of an induction-motor rotor is quite low. In the above example the turns ratio of the stator to the rotor is known since the rotor has a wound winding. With a squirrel-cage rotor, similar voltage and frequency relationships exist but since the rotor has no definite winding turns, the secondary voltage E_2 is expressed in primary terms. Therefore, in a squirrel-cage rotor the resistance and leakage reactance values are expressed in stator terms. This is similar to the secondary resistance and leakage reactance of a transformer being expressed in primary terms.

In any type of motor the power is always developed in the rotor; unlike dc motors, however, the power is not conducted by conductors to the rotor but is transferred by the mutual flux across the air gap between the stator and rotor. In a transformer the power is transferred to the secondary through mutual flux but without an air gap. A phase-rotor circuit is shown by the successive Figures 5-6(a), 5-6(b), and 5-6(c) where E_2 is the induced voltage

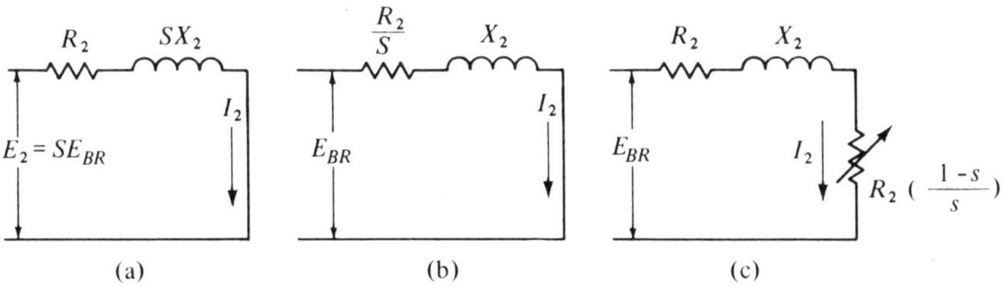

FIGURE 5-6. **(a) Circuit diagram per phase of rotor for blocked rotor. (b) Equivalent rotor circuit of Figure 5-6(a) for Equation (5-5a). (c) Equivalent rotor circuit with resistance of rotor and load for Equation (5-6a).**

per phase of the rotor and, from Equation (5-3), is equal to SE_{BR}. The impedance per phase is made up by R_2, the dc resistance, and the leakage reactance SX_2. The resistance is the dc resistance since the rotor frequency is comparatively low. The leakage reactance varies as the frequency, $X = 2\pi fL$, and therefore it is given by the expression SX_2. The rotor current I_2 per phase is then given by the expression,

$$I_2 = \frac{E_2}{\sqrt{(R_2)^2 + (SX_2)^2}} = \frac{SE_{BR}}{\sqrt{(R_2)^2 + (SX_2)^2}} \text{ A.} \qquad \textbf{(5-5)}$$

Dividing the numerator and denominator of the formula by the slip S, the formula for the rotor current I_2 becomes

$$I_2 = \frac{E_{BR}}{\sqrt{(R_2/S)^2 + (X_2)^2}} \text{ A.} \qquad \textbf{(5-5a)}$$

The current I_2 of the above two formulas is shown in Figures 5-6(a), 5-6(b), and 5-6(c). The power input to the rotor, or the power transferred across the air gap, is given by the expression,

$$RPI = (I_2)^2 \frac{R_2}{S}. \qquad \textbf{(5-6)}$$

By substituting the value of I_2 from Equation (5-5a) in Equation (5-6), the rotor power input RPI becomes

$$RPI = \frac{E_{BR}{}^2}{(R_2/S)^2 + X_2{}^2} \times \frac{R_2}{S} = \frac{S(E_{BR})^2 R_2}{R_2{}^2 + (SX_2)^2}. \qquad \textbf{(5-7)}$$

The value of R_2/S may be divided into two components as shown in Figure 5-6(c) by adding R_2 and subtracting SR_2/S to the expression R_2/S. By combining and factoring, the value R_2/S becomes

$$\frac{R_2}{S} = R_2 + R_2 \frac{1 - S}{S}. \qquad \textbf{(5-8)}$$

Substituting the value of R_2/S in Equation 5-6, it becomes:

$$RPI = I_2{}^2 \left(R_2 + R_2 \frac{1 - S}{S}\right) \text{ W,}$$

$$RPI = I_2{}^2 R_2 + I_2{}^2 R_2 \frac{1 - S}{S} \text{ W.} \qquad \textbf{(5-8a)}$$

As shown in Figure 5-6(c), R_2 is the rotor resistance and $R_2[(1 - S)/S]$ represents the equivalent resistance of the load. The first term of the equation $I_2{}^2R_2$ represents the copper loss of the rotor per phase, and the second term $I_2^2R_2[(1 - S)/S]$ represents the power developed per phase of the rotor. Therefore the rotor copper loss per phase is

$$RCL = I_2^2R_2 = (RPI)S \text{ W}. \tag{5-9}$$

The rotor power developed per phase RPD is

$$RPD = I_2^2R_2 \frac{1 - S}{S} = RPI(1 - S) \text{ W}. \tag{5-10}$$

In Chapter 2, the expression for *torque* is given as

$$T = \frac{7.04 E_g I_a}{S} \text{ lb-ft}.$$

The torque developed for a polyphase induction motor may be written as

$$T = 7.04 \frac{RPD_t}{\text{rpm}_{\text{rotor}}}, \tag{5-11}$$

where RPD_t is the total power developed by the rotor and for a three-phase motor is three times the power of Equation (5-10). If the value of Equation (5-10) is substituted in the last formula, it becomes

$$T = 7.04 \frac{RPI_t(1 - S)}{\text{rpm}_{\text{rotor}}} = 7.04 \frac{RPI_t}{\text{rpm}_{\text{rotor}}/(1 - S)}$$

$$= 7.04 \frac{RPI_t}{\text{rpm}_{\text{syn}}}. \tag{5-11a}$$

Example 5-4. In Example 5-3, if the rotor resistance (dc) per phase is 0.08 ohm and the standstill or blocked-rotor leakage reactance is 0.108 ohm/phase, calculate (a) the power developed in the rotor and (b) the torque developed, both at a speed of 1700 rpm.

Solution:

(a) $$I_2 = \frac{E_{BR}}{\sqrt{(R_2/S)^2 + X_2^2}} = \frac{152.5}{\sqrt{(0.08/0.0555)^2 + (0.108)^2}},$$

$$I_2 = \frac{152.5}{\sqrt{(1.44)^2 + (0.108)^2}} = \frac{152.5}{1.444} = 105.5 \text{ A},$$

$$RPI_{(\text{total})} = 3(I_2)^2 \frac{R_2}{S} = 3(105.5)^2(1.44) = 48{,}100 \text{ W},$$

$$RPD = RPI(1 - S) = 48{,}100(1 - 0.0555) = 45{,}430 \text{ W}$$

$$= \frac{45{,}430}{746} = 60.8 \text{ hp};$$

(b) $$T = 7.04 \frac{RPI}{\text{rpm}_{\text{syn}}} = 7.04 \frac{48{,}100}{1800} = 188 \text{ lb-ft}.$$

In this example only the rotor circuit was considered and the assumption was that there is no drop in the blocked-rotor voltage E_{BR} due to impedance drop. In the induction motor the air gap represents a high reactance and must be considered as shown in the complete equivalent phase circuit diagram depicted in Figure 5-7; X_M shows the reactance of the air gap. A resistance

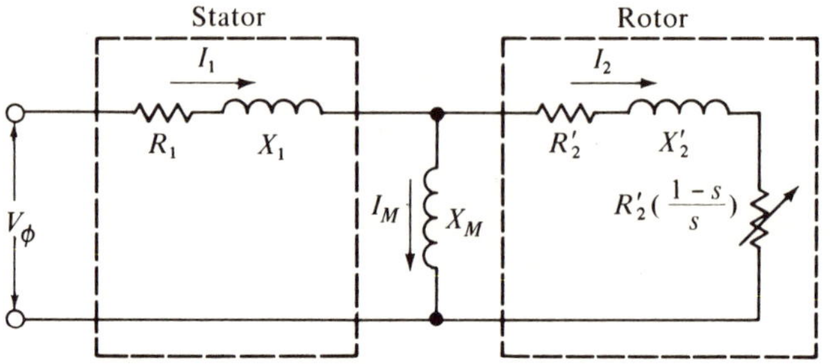

FIGURE 5-7. Equivalent circuit per phase of an induction motor.

representing the core losses (hysteresis and eddy current) should be shown in parallel with X_M. It is omitted since this resistance is so small in comparison to the reactance X_M. In this circuit R_2' represents the rotor resistance in dc ohms per phase in stator (primary) terms; R_2' is equal to a^2R_2 when R_2 is known or can be measured in the case of a wound-rotor motor; X_2' is the rotor reactance per phase at standstill in stator terms; $R_2'[(1 - S)/S]$ represents the resistance of the load; R_1 and X_1 represent the stator resistance (ac) and reactance, respectively. The continuation of Example 5-4 illustrates the computations based on the complete diagram as shown in Figure 5-7.

Example 5-4 (*continued*). $R_1 = 0.25$ ohm; $X_1 = 0.30$ ohm, and $X_M = 18.5$ ohms. Find (a) the power developed in the rotor and (b) the torque developed. Figure 5-8(a) shows the phase diagram for these conditions.

Solution:

Rotor resistance in stator terms:

$$a^2R_2 = \left(\frac{100}{60}\right)^2(0.08) = 0.222 \text{ ohm}.$$

Rotor reactance in stator terms:

$$a^2X_2 = \left(\frac{100}{60}\right)^2(0.108) = 0.30 \text{ ohm},$$

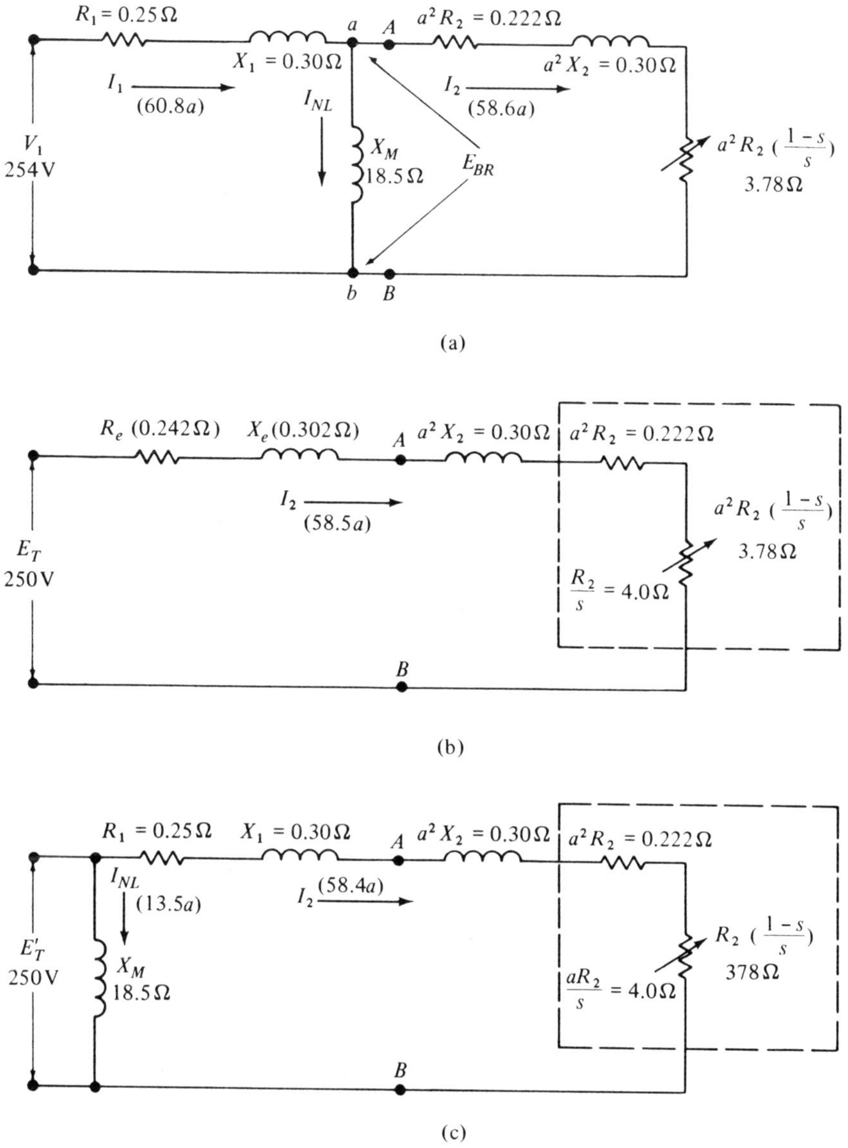

FIGURE 5-8. **(a) Equivalent circuit diagram per phase for Example 5-4. (b) Equivalent Thevenin's circuit per phase for Example 5-4. (c) Approximate equivalent circuit per phase for Example 5-4.**

$$Z_{ab} = \frac{\left(a^2R_2 + a^2R_2\frac{1-S}{S} + ja^2X_2\right)(X_M)}{\left(a^2R_2 + a^2R_2\frac{1-S}{S} + ja^2X_2\right) + X_M}$$

$$= \frac{(0.222 + 3.78 + j0.3)(j18.5)}{(0.222 + 3.78 + j0.3) + (j18.5)}$$

$$= \frac{(4.00 + j0.3)(j18.5)}{4.00 + j18.8} = \frac{-5.55 + j74.0}{19.25\angle 78^\circ} = \frac{74.3\angle 94.3^\circ}{19.25\angle 78^\circ}$$

$$= 3.86\angle 16.3^\circ \text{ ohms}$$

$$= 3.70 + j1.08 \text{ ohms,}$$

$$Z_t = (3.70 + j1.08) + (0.25 + j0.30) = 3.95 + j1.38 \text{ ohms,}$$

$$I_1 = \frac{V}{Z_t} = \frac{254}{3.95 + j1.38} = \frac{254}{4.18} = 60.8 \text{ A,}$$

$$I_2 = \frac{I_1 Z_{ab}}{a^2R_2 + a^2X_2 + a^2R_2[(1-S)/S]} = 60.8\frac{3.86}{4 + j0.30}$$

$$I_2 = 58.6 \text{ A.}$$

(a) Rotor power developed:

$$RPD = I_2^2R_2\left(\frac{1-S}{S}\right) \times 3 = (58.6)^2(3.78)(3) = 39{,}100 \text{ W}$$

$$= 52.3 \text{ hp.}$$

(b) Torque developed:

$$T = 7.04\frac{RPD_t}{\text{rpm}_{\text{rotor}}} = 7.04\frac{39{,}100}{1700} = 162 \text{ lb-ft.}$$

The value of the current I_2 may also be solved by the use of Thevenin's theorem. The equivalent circuit is shown by Figure 5-8(b). The value of E_T is the open-circuit voltage across A-B of Figure 5-8(a) with the rotor impedance and equivalent load resistance removed.

$$E_T = V\frac{X_M}{R_1 + jX_1 + jX_M} = 254\frac{j18.5}{0.25 + j0.30 + j18.5}$$

$$= 254\frac{18.5\angle 90^\circ}{18.81\angle 89.2^\circ} = 250 \text{ V.}$$

The value of series impedance is found by replacing the voltage source by its internal impedance (0) and is equal to

$$R_e + jX_e = \frac{(R_1 + jX_1)(X_M)}{R_1 + jX_1 + X_M} = \frac{(0.25 + j0.30)(j18.5)}{0.25 + j0.30 + j18.5}$$

$$= \frac{-5.55 + j4.625}{0.25 + j18.8} = \frac{7.23\underline{|140.2^\circ}}{18.81\underline{|89.2^\circ}} = 0.384\underline{|51.0^\circ}$$

$$= 0.242 + j0.302 \text{ ohm},$$

or

$$R_e = 0.242 \text{ ohm} \quad \text{and} \quad X_e = j0.302 \text{ ohm},$$

$$I_2 = \frac{E_T}{R_e + jX_e + a^2R_2 + a^2R_2[(1 - S)/S] + a^2X_2}$$

$$= \frac{250}{0.242 + j0.302 + 0.222 + 3.78 + j0.3}$$

$$= \frac{250}{4.242 + j0.602} = \frac{250}{4.28\underline{|8.08^\circ}} = |58.5| \text{ A}.$$

The rotor power developed and the torque are the same as found before. It can be seen that R_e and X_e of Figure 5-8(b) are practically equal to R_1 and X_1 of the circuit of Figure 5-8(a).

Unless the value of X_M is known from the design, it is more convenient to use the no-load value of the motor current. In the foregoing example the no-load current is 13.75 A as shown in Figure 5-8(c). The following formulas show how the I_2 value of current is attained.

$$E'_T = V_1 - I_{NL}(R_1 + jX_1) = 254 - (-j13.75)(0.25 + j0.30),$$
$$E'_T = 254 - (4.12 - j3.44) = 254 - 4.12 + j3.44 = 249.9 + j3.44,$$
$$= 250 \text{ V}$$

$$I_2 = \frac{E'_T}{\sqrt{\left[R_1 + a^2R_2 + a^2R_2\left(\frac{1 - S}{S}\right)\right]^2 + [X_1 + a^2X_2]^2}}$$

$$= \frac{250}{\sqrt{(0.25 + 0.222 + 3.78)^2 + (0.3 + 0.3)^2}},$$

$$I_2 = \frac{250}{\sqrt{(4.25)^2 + (0.6)^2}} = \frac{250}{4.29} = 58.4 \text{ A}.$$

The rotor power developed and the torque is substantially the same as found for the first calculation using Figure 5-8(a). The calculations by use of Figure 5-6(c) are considerably higher. The value of E_{BR} used was 152.5 V or the stator to rotor ratio (turns ratio) and does not account for the drop in the stator (transformer primary). Calculating this drop from Figures 5-8(b) and (c) the value of E_{BR} on the rotor was about 141 V. This represents a 7-percent drop and had this value been used, the value of rotor power developed and torque would have been the same as the later calculations.

As the load increases in the dc motor, the speed decreases, so more current can flow in the armature conductors to increase the increased torque requirements. Similarly, as the load increases in the induction motor, the speed of the rotor decreases, the slip S increases, and by Equation (5-5) the current in the rotor conductors increases. The power transferred across the air gap is increased and the torque increases. The torque reaches a maximum value which is the maximum torque that can be developed. Beyond this point the motor will stall. By the theory of maximum power transfer from stator to rotor in Thevenin's circuit, Figure 5-8(b),

$$a^2R_2 + a^2R_2\frac{1-S}{S} = \frac{a^2R_2}{S} = \sqrt{R_e^2 + (X_e + a^2X_2)^2}.$$

Solving the previous equation for slip S, the slip at maximum torque S_{mt} is

$$S_{mt} = \frac{a^2R_2}{\sqrt{R_e^2 + (X_e + a^2X_2)^2}}. \tag{5-12}$$

For the approximate circuit of Figure 5-8(c), the slip at which maximum torque occurs, becomes

$$S_{mt} = \frac{a^2R_2}{\sqrt{R_1^2 + (X_1 + a^2X_2)^2}}. \tag{5-12a}$$

The following example illustrates the calculation for *maximum torque.*

Example 5-5. For the three-phase induction motor of Examples 5-3 and 5-4, (a) at what speed is maximum torque developed? (b) What is the maximum torque that can be developed?

Solution:

(a) $$S_{mt} = \frac{a^2R_2}{\sqrt{R_e^2 + (X_e + a^2X_2)^2}} = \frac{0.222}{\sqrt{(0.242)^2 + (0.302 + 0.30)^2}}$$

$$= \frac{0.222}{0.649} = 0.342,$$

$$\text{rpm} = \text{rpm}_{\text{syn}}(1 - S) = 1800(1 - 0.342) = 1184 \text{ rpm}.$$

The solution by using the circuit of Figure 5-8(c) is

$$S_{mt} = \frac{a^2R_2}{\sqrt{R_1^2 + (X_1 + a^2X_2)^2}} = \frac{0.222}{\sqrt{(0.25)^2 \times (0.3 + 0.3)^2}} = \frac{0.222}{0.651} = 0.341,$$

$$\text{rpm} = 1800(1 - 0.341) = 1800(0.659) = 1186 \text{ rpm}.$$

$$\text{(b)}\qquad I_2 = \frac{E_T}{\sqrt{(R_e + a^2R_2/S)^2 + (X_e + a^2X_2)^2}}$$

$$= \frac{250}{\sqrt{(0.242 + 0.222/0.342)^2 + (0.302 + 0.30)^2}}$$

$$= \frac{250}{\sqrt{(0.891)^2 + (0.602)^2}} = \frac{250}{1.075} = 232.5 \text{ A},$$

$$RPI = 3I_2^2\,\frac{a^2R_2}{S} = 3(232.5)^2\,\frac{0.222}{0.342} = 105{,}100 \text{ W}.$$

$$T_{\max} = 7.04\,\frac{RPI}{\text{rpm}_{\text{syn}}} = 7.04\,\frac{105{,}100}{1800} = 411 \text{ lb-ft}.$$

Because the slip by both methods is the same and the equivalent resistance and reactance are substantially the same, the value of maximum torque also comes out to 411 lb-ft.

The maximum torque is independent of the rotor resistance. Rotor resistance does effect the slip and hence the speed. In Figure 5-9 there are curves

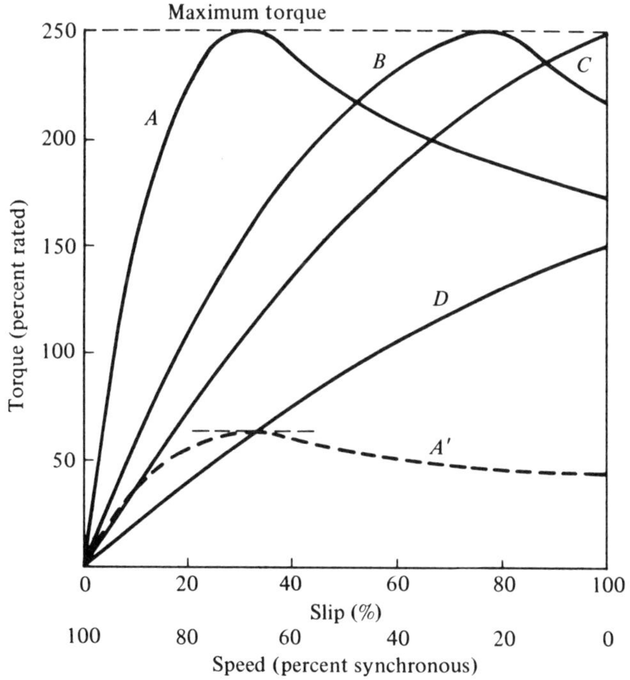

FIGURE 5-9. Torque versus slip and torque versus speed curves for a wound-rotor polyphase induction motor.

of percent rated torque versus slip and speed. Curve A shows the torque versus slip and speed for the motor in the preceding example. Curve B is a curve for R_2 greater than 0.08 ohm/phase. Curve C is for a larger value for R_2 than in curve B. Also in this curve the maximum torque occurs when slip is 100 percent or zero speed. In curve D, R_2 is still greater than for curve C and the starting torque is less than the maximum torque.

Starting torque or blocked-rotor torque occurs when the slip is 100 percent. For a wound-rotor induction motor, resistance can be added in the rotor circuit. Added resistance increases the starting torque to a point where the starting torque is equal to the maximum. Also the speed decreases when resistance is added. An examination of Equations (5-5) and (5-5a) shows that the value of current at start is lowered when resistance is added. The following example illustrates how the starting torque is increased and the starting current is reduced by adding resistance in the rotor of an induction motor.

Example 5-6. In the preceding examples, (a) what is the starting current? (b) What is the starting torque? (c) What value of resistance per phase must be added so that the starting torque is equal to the maximum torque? (d) What is the starting current for the conditions of (c)?

Solution:

Refer to Figure 5-8(a) and Example 5-4.

(a) At start,

$$S = 1 \qquad \text{and} \qquad a^2R_2\left(\frac{1-S}{S}\right) = 0,$$

$$Z_t = R_1 + jX_1 + \frac{X_M(a^2R_2 + ja^2X_2)}{X_M + a^2R_2 + ja^2X_2}$$

$$= 0.25 + j0.30 + \frac{j18.5(0.222 + j0.30)}{j18.5 + 0.222 + j0.30}$$

$$= 0.25 + j0.30 + \frac{-5.55 + j4.11}{0.222 + j18.8} = 0.25 + j0.30 + \frac{6.90\underline{|143.45^\circ}}{18.8\underline{|89.3^\circ}}$$

$$= 0.25 + j0.30 + 0.367\underline{|54.15^\circ} = 0.25 + j0.30 + 0.215 + j0.298$$

$$= 0.465 + j0.598 = 0.759\underline{|52.15^\circ} \text{ ohm},$$

$$I_1(\text{stator}) = \frac{254}{0.759} = 335 \text{ A}.$$

(b) $$I_2 = I_1\frac{X_M}{X_M + a^2R_2 + ja^2X_2} = 335\frac{18.5}{18.8} = 329 \text{ A},$$

$$RPI = 3I_2^2\frac{a^2R_2}{S} = 3(329)^2(0.222) = 72{,}200 \text{ W},$$

$$T = 7.04\frac{RPI}{\text{rpm}_{\text{syn}}} = 7.04\frac{72{,}200}{1800} = 282 \text{ lb-ft}.$$

(c) When $S = 100$ percent,

$$a^2(R_2 + R_a) = \sqrt{R_e^2 + (X_e + a^2X_2)^2}$$

for maximum power transfer from Thevenin's circuit, Figure 5-8(b), and R_a is the resistance per phase added in the rotor.

$$a^2R_2 + a^2R_a = \sqrt{R_e^2 + (X_e + a^2X_2)^2},$$

$$0.222 + \left(\frac{100}{60}\right)^2 R_a = \sqrt{(0.242)^2 + (0.302 + 0.30)^2},$$

$$0.222 + (2.78)R_a = 0.649 \text{ ohm},$$

$$R_a = \frac{0.649 - 0.222}{2.78} = \frac{0.427}{2.78} = 0.1535 \text{ ohm}.$$

(d) Referring to Figure 5-8(a), the resistance $a^2R_2 + a^2R_a = 0.649$ ohm and substituting this value for a^2R_2 in the equation for Z_t under part (a),

$$Z_t = 0.25 + j0.30 + \frac{j18.5(0.649 + j0.30)}{j18.5 + 0.649 + j0.30}$$

$$= 0.25 + j0.30 + \frac{-5.55 + j12.0}{0.649 + j18.8} = 0.25 + j0.30 + \frac{13.23\angle 114.8^\circ}{18.8\angle 88^\circ},$$

$$Z_t = 0.25 + j0.30 + 0.703\angle 26.8^\circ = 0.25 + j0.30 + 0.626 + j0.317,$$

$$Z_t = 0.876 + j0.617 = 1.07\angle 35.2^\circ \text{ ohms},$$

$$I_1(\text{stator}) = \frac{V}{Z_t} = \frac{254}{1.07} = 238 \text{ A}.$$

5-4 QUALITATIVE ANALYSIS OF OPERATION[1],[4],[5],[7]

Efficiency of a polyphase induction motor is calculated in a manner similar to that for any rotating machine or transformer. Smaller induction-motor efficiencies are readily made by direct measurement with a brake or dynamometer and the efficiency is simply the ratio of output to input expressed in percent.

Another method for smaller motors involves three tests: (a) a no-load test, (b) a load test, (c) a stator-resistance test. For a wound rotor, the resistance of the rotor winding is also measured. The no-load and the load tests can be made with the circuit of Figure 5-10. The following example illustrates this method.

Example 5-7. A test on a 10-hp, 220-V, three-phase, 60-Hz, six-pole induction motor was tested as follows:

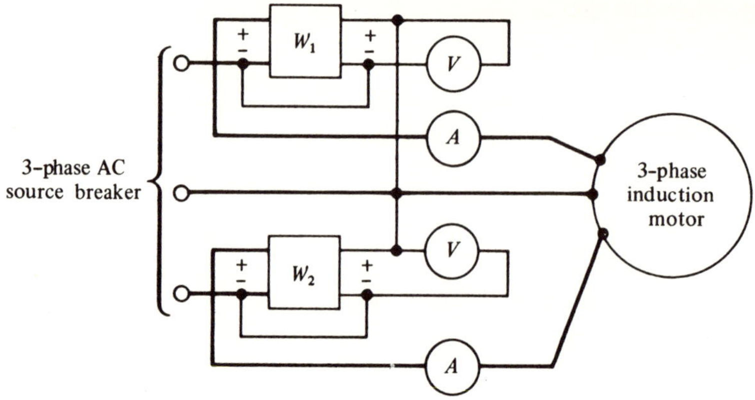

FIGURE 5-10. Connection diagram for no-load, load and blocked rotor tests.

No-load test: $E_T = 220$ V, 60 Hz; $P_1 = 1450$ W; $P_2 = -830$ W; $I_{NL} =$ 9.8 A.

Stator resistance measurement: average resistance between terminals = 0.288 ohm (dc).

Load test: $E_T = 220$ V, 60 Hz; $P_1 = 5900$ W; $P_2 = 2985$ W; $I_L = 27.0$ A; speed = 1135 rpm.

Find (a) the horsepower output, (b) the torque, (c) the efficiency, and (d) the power factor at this load condition. The stator was assumed to be wye connected and the ac resistance is 125 percent of the dc resistance. Also, the tests were made at operating conditions; that is, the motor had been operating for some time and the resistance was at the hot temperature values.

Solution:

Resistance of stator per phase:

$$R_s = \frac{0.288}{2} \times 1.25 = 0.180 \text{ ohm.}$$

Rotational losses (friction, windage, and iron losses) (No-load test):

$$\begin{aligned} W_{\text{rot}} &= P_1 + P_2 - 3I^2R_s \\ &= (1450 - 830) - 3(9.8)^2(0.180) \\ &= 620 - 52 = 568 \text{ W.} \end{aligned}$$

Stator copper loss:

$$W_{sc} = 3I_L{}^2R_s = 3(27)^2(0.180) = 394 \text{ W.}$$

Rotor power input:

$$RPI = (P_1 + P_2) - (W_{\text{rot}} + W_{sc})$$
$$= (5900 + 2985) - (568 + 394) = 7923 \text{ W}.$$

Rotor copper loss:

$$RCL = (RPI)S = \frac{1200 - 1135}{1200}(7923) = 431 \text{ W}.$$

Rotor power developed:

$$RPD = RPI - RCL = 7923 - 431 = 7492 \text{ W}.$$

(a) Horsepower output $= \dfrac{RPD}{746} = 10.05$ hp;

(b) torque $T = 7.04 \dfrac{RPD}{\text{rpm}_{\text{syn}}}$

$$= 7.04 \frac{7923}{1200} = 46.5 \text{ lb-ft};$$

(c) percent efficiency $= \left(1 - \dfrac{\text{losses}}{\text{output} + \text{losses}}\right) \times 100,$

$$\text{percent efficiency} = \left(1 - \frac{568 + 394 + 431}{7492 + 568 + 394 + 431}\right) \times 100$$

$$= \left(1 - \frac{1393}{8885}\right) \times 100 = 84.3\%;$$

(d) power factor $= \dfrac{\text{power input}}{\sqrt{3}EI} = \dfrac{8885}{\sqrt{3}(220)(27)} = 0.861.$

In the foregoing example the rotational losses are assumed to be constant and are charged as stator losses. This is not quite true, however, since the friction and windage losses are actually a rotor loss; but this assumption does not cause any significant error. Also, in practice, the tests are usually made on a cold machine that is at the same temperature as the ambient temperature of the room in which the machine is located. The resistance value should be corrected to the running temperature, which is usually assumed to be 75°C. Other points for a curve can be obtained for other current values. If the rated nameplate current is used, however, the value of efficiency will be very close to rated horsepower output.

For larger machines or when it is not feasible to make load tests, the

American Standard Test Code[1] recommends three tests as follows: (a) No-load test; (b) dc stator-resistance test; (c) blocked-rotor test. With these tests, the values of stator and rotor resistance are found and the leakage reactance of the stator and the air-gap reactance. Calculations are made as in Example 5-4 with the circuit of Figures 5-7 and 5-8(a) or the approximate equivalent circuit of Figure 5-8(c). The procedure and formulas for these tests now follow.

No-Load Test. This test is identical to the no-load test for small motors using the connection diagram of Figure 5-10. From this test and the resistance test, the rotational losses are found by the formula,

$$W_{\text{rot}} = P_1 - P_2 - 3(I_{NL})^2 R_s \tag{5-13}$$

where

- P_1 is the no-load power of wattmeter 1 (the two-wattmeter method is used),
- P_2 is the no-load power of wattmeter 2,
- I_{NL} is the line no-load current,
- R_s is the dc stator-resistance corrected temperature at which the no-load test was made.

(Note that all values are on a phase basis.) The no-load impedance is found by the following formulas,

$$Z_{NL} = \frac{V_{NL}}{\sqrt{3}(I_{NL})}, \tag{5-14}$$

$$R_{NL} = \frac{P_1 + P_2}{3I_{NL}^2}, \tag{5-15}$$

$$X_{NL} = \sqrt{Z_{NL}^2 - R_{NL}^2}. \tag{5-16}$$

Direct-Current Stator-Resistance Test. The resistance across each pair of terminals is measured by the voltmeter-ammeter method or with a bridge. The average resistance per phase at ambient temperature is the average of the measured values divided by two. The reason for the above is that for the three-phase induction motor, it is assumed that the stator is wye connected. The value of the stator resistance R_1 per phase must be converted to the operating temperature for use in the calculations.

Blocked-Rotor Test. This test is made with the rotor clamped so that it cannot rotate. The connection diagram shown in Figure 5-10 is used. It is recommended that this test be made at 25 percent of rated frequency, which

[1] IEEE No. 112A, September 1964.

is 15 Hz for a 60-Hz motor. It should also be run at reduced voltage and care must be taken that the applied voltage is such that no excessive currents flow. Usually a value of impressed voltage (variable from an induction regulator) is used that will yield rated current. This test is similar to the short-circuit test of a transformer. The reduced frequency is recommended to compensate for the low-frequency rotor currents. The following formulas are calculated from this test,

$$Z_{BR} = \frac{E_{BR}}{\sqrt{3}I_{BR}}, \tag{5-17}$$

$$R_{BR} = \frac{P_1 + P_2}{3I_{BR}^2} \quad \text{(from blocked-rotor test)}, \tag{5-18}$$

$$X'_{BR} = X'_1 + X'_2 = \sqrt{Z_{BR}^2 - R_{BR}^2}, \tag{5-19}$$

X'_{BR}, X'_1, and X'_2 are at frequency of blocked-rotor test and

$$X_1 + X_2 = X'_{BR}\frac{f}{f_{BR}}. \tag{5-20}$$

According to the test code, X_1 and X_2 are as follows:

design A and D:	$X_1 = 0.5(X_1 + X_2)$	and	$X_2 = 0.5(X_1 + X_2)$,
design B:	$X_1 = 0.4(X_1 + X_2)$	and	$X_2 = 0.6(X_1 + X_2)$,
design C and F:	$X_1 = 0.3(X_1 + X_2)$	and	$X_2 = 0.7(X_1 + X_2)$,
wound rotor:	$X_1 = 0.5(X_1 + X_2)$	and	$X_2 = 0.5(X_1 + X_2)$.

At no load, the slip is very low, so R_2/S is very high. Therefore in Figures 5-7 and 5-8(a),

$$X_M + X_1 = X_{NL} \quad \text{or} \quad X_M = X_{NL} - X_1. \tag{5-21}$$

From the blocked-rotor test and the stator-resistance test, the rotor resistance R_2 in stator terms is

$$R_2 = R_{BR} - R_1. \tag{5-22}$$

In this expression R_{BR} as found by using Equation (5-18) must be converted to the operating temperature value.

The following example illustrates how these three tests are utilized to find the efficiency, output, torque, and current of a polyphase induction motor.

Example 5-8. The following test data was obtained for a 100-hp, 2300-V, 60-Hz, three-phase, six-pole, design-A induction motor:

No-load test: $I_{NL} = 7.9$ A, power input $(P_1 + P_2) = 2940$ W, $E_L = 2300$ V at 60 Hz, speed = 1196 rpm.

Stator dc resistance: 4.02 ohms (average) between terminals.

Blocked-rotor test: frequency = 15 Hz, E = 199 V (line to line) I = 25 A, power input $(P_1 + P_2)$ = 7420 W.

All data are taken with the motor at 25°C. Calculate (a) the stator current, (b) the output in horsepower, (c) the efficiency at these conditions, and (d) the power factor for a slip of 4.2 percent.

Solution:

Solution of the parameters is as follows,

$$Z_{NL} = \frac{V_{NL}}{\sqrt{3}I_{NL}} = \frac{2300}{\sqrt{3}(7.9)} = 168 \text{ ohms},$$

$$R_{NL} = \frac{P_1 + P_2}{3I_{NL}^2} = \frac{2940}{3(7.9)^2} = 15.7 \text{ ohms},$$

$$X_{NL} = \sqrt{(168)^2 - (15.7)^2} = 167.2 \text{ ohms},$$

$$Z_{BR} = \frac{E_{BR}}{\sqrt{3}I_{BR}} = \frac{199}{3(25)} = 4.60 \text{ ohms},$$

$$R_{BR} = \frac{P_1 + P_2}{3I_{BR}^2} = \frac{7420}{3(25)^2} = 3.96 \text{ ohms} \qquad \text{(at 25°C)},$$

$$X'_{BR} = X'_1 + X'_2 = \sqrt{(4.60)^2 - (3.96)^2} = 2.33 \text{ ohms} \qquad \text{(at 15 Hz)},$$

$$X_1 + X_2 = X'_{BR}\left(\frac{f}{f_{BR}}\right) = 2.33\left(\frac{60}{15}\right) = 9.32 \text{ ohms},$$

$$X_1 = X_2 = 0.5(X_1 + X_2) = 0.5(9.32) = 4.66 \text{ ohms},$$
$$X_M = X_{NL} - X_1 = 167.2 - 4.66 = 162.6 \text{ ohms},$$

$$R_2 = R_{BR} - R_1 = 3.96 - \frac{4.02}{2} = 1.95 \text{ ohms} \qquad \text{(at 25°C)},$$

$$R_2 = R_{2(25)}\frac{(\alpha + T_2)}{(\alpha + T_1)} = 1.95\frac{(234.5 + 75)}{(234.5 + 25)} = 2.32 \text{ ohms (75°C)}.$$

All of the above values are on a per phase basis. Figure 5-11 shows the equivalent circuit diagram per phase for this example.

$$\frac{R_2}{S} = \frac{2.32}{0.042} = 55.2 \text{ ohms},$$

$$Z_{AB}(\text{equivalent}) = \frac{(Z_2)(X_M)}{Z_2 + X_M} = \frac{(55.2 + j4.66)(j162.6)}{(55.2 + j4.66) + (j162.6)}$$

$$= \frac{-758 + j8970}{55.2 + j167.3} = \frac{9100\underline{|94.8^\circ}}{176.5\underline{|71.6^\circ}} = 51.1\underline{|23.2^\circ} \text{ ohms}$$

$$= 46.9 + j20.0 \text{ ohms},$$

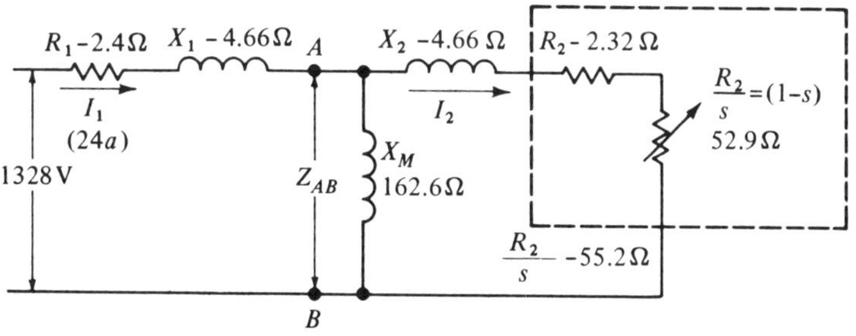

FIGURE 5-11. Equivalent circuit diagram per phase for motor of Example 8.

$$\begin{aligned} Z_t &= Z_1 + Z_{AB} = 2.4 + j4.66 + (46.9 + j20.0) \\ &= 49.3 + j24.66 \\ &= 55.1\underline{|26.5^\circ} \text{ ohms.} \end{aligned}$$

(a) $I_1 = \dfrac{V/\sqrt{3}}{Z_t} = \dfrac{2300}{\sqrt{3}(55.1)} = 24.0$ A (stator current)

(b) $RPI = 3I_1^2R_{AB}^* = 3(24.0)^2(46.9) = 81{,}000$ W

R_{AB}^* is the real component of Z_{AB},

$$\begin{aligned} RPD &= RPI(1 - S) = 81{,}000(1 - 0.042) = 77{,}600 \text{ W} \\ \text{output} &= RPD = 77{,}600/746 = 104 \text{ hp.} \end{aligned}$$

This value for output could have been solved by solving for I_2 with Thevenin's circuit or by the use of the approximate circuit of Figure 5-8(c).

(c) For efficiency, the following are the calculated losses,

Rotor copper loss $= S(RPI) = 0.042(81{,}000)$,	3400 W
Stator copper loss $= 3I_1^2R_1 = 3(24)^2(2.4)$,	4150 W
Rotational losses[2] $= 2940 - 3(7.9)^2(2.4)$,	2490 W
Stray load loss (taken for motors 200 hp or more at 1 percent of output),	0 W
Total losses,	10,040 W

Thus,

$$\begin{aligned} \text{efficiency} &= \left(1 - \frac{\text{losses}}{\text{output} + \text{losses}}\right) \times 100 \\ &= \left(1 - \frac{10{,}040}{77{,}600 + 10{,}040}\right) \times 100 = 88.57\%. \end{aligned}$$

[2] From the no-load test.

(d) Power factor $= \dfrac{\text{watts input}}{\sqrt{3}E_L I_1} = \dfrac{87{,}640}{\sqrt{3}(2300)(24)} = 0.916$ lag.

Note that the power factor for an induction motor is always lagging since the reactances are inductive.

5-5 OPERATING CHARACTERISTICS[1],[4]

The performance characteristics of induction motors are largely dependent on design details. These values are the value of rotor resistance, shape of the slots and teeth of both the stator and the rotor, and the air gap between rotor and stator.

The first of these is *rotor resistance*. Figure 5-12 shows the running charac-

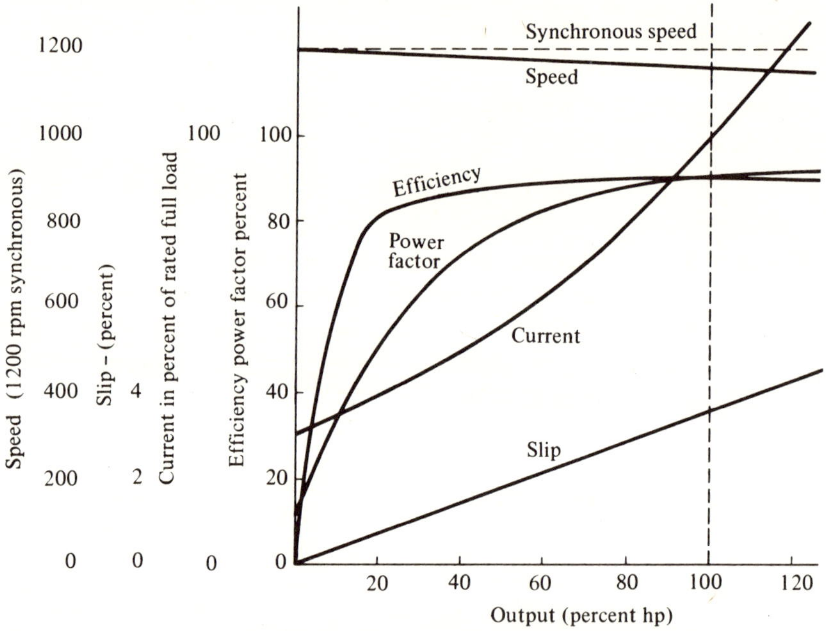

FIGURE 5-12. Typical performance curves for a three-phase squirrel-cage induction motor, 100–200 hp.

teristics of a squirrel-cage induction motor that has comparatively low rotor resistance. The effect of increasing rotor resistance is to increase the slope of the slip–horsepower output curve. In turn speed regulation is increased which causes the speed–horsepower output to have a larger drop. As in dc motors by adding resistance in the armature circuit, which increases the copper loss and lowers the efficiency, the increased rotor resistance causes the efficiency to drop. The power factor is reduced slightly and the running current is increased.

In squirrel-cage motors rotor resistance is increased by the cross-sectional size of the rotor bars and end rings. The wound rotor has a winding of the same number of phases as the stator. The ends of the winding are brought out through slip rings. To operate the motor the rotor winding terminals must be short circuited to operate. Since there is a rotor winding, the rotor resistance on short circuit is inherently higher than the resistance of a squirrel-cage motor. Star-connected rheostats are connected to the rotor terminals. Starting the motor with all resistances in will reduce the starting current, increase the starting torque, and reduce the operating speed.

Referring to Figure 5-9, the effect of added rotor resistance can be seen on the speed, starting torque, and the slip or speed. Added rotor resistance has no effect on the value of maximum torque. The curves A, B, C, and D are for rotor resistances where $R_D > R_C > R_B > R_A$. As the rotor resistance is increased, the slip increases and the speed decreases. As for the starting characteristics, added rotor resistance increases the value of starting torque (when slip is 100 percent) up to a point. The curve D is for the greatest resistance, with torque less than any of the others. As mentioned before and noted from Example 5.6, the starting current decreases as the rotor resistance is increased. The torque at starting when the slip is 100 percent, or for any given value of slip, varies closely as the square of the applied terminal voltage. This can be shown by combining Equation (5-6) into the torque formula of Equation (5-11a), from which T varies directly as the square of the blocked-rotor voltage. Therefore it can be said that the torque varies closely as the square of the applied terminal voltage. The A' curve on Figure 5-9 shows the torque-slip or speed curve for curve A if the motor is operated at 50 percent terminal voltage. Here the maximum torque and starting torque are each 25 percent of the full voltage values.

The shapes of the slots and teeth of the rotor affect the value of motor leakage reactance. When the teeth are wide and the slots are narrow and deep, the rotor reactance is high. The effect of this is to reduce the maximum torque and the starting current. On operating conditions the power factor is decreased slightly.

Induction motors have as small an air gap as is mechanically possible. The shorter the air gap, the lower will be the magnetizing current and, in turn, the higher the value of air-gap reactance X_M. Also, the smaller the magnetizing current, the higher will be the power factor. Due to magnetizing current and its being relatively high compared to that of a transformer, the power factor of an induction motor is always lagging. At light loads it is small with a large phase angle and, conversely, at heavy loads the power factor is large.

5-6 STANDARD TYPES OF SQUIRREL-CAGE INDUCTION MOTORS[2],[7],[11]

Polyphase squirrel-cage integral horsepower induction motors may be one of four designs. The National Electrical Manufacturers Association (NEMA) standard MG1-1967 classifies these as Design A, B, C, and D. These designs by the type of rotor cage, the teeth, and slot dimensions of the rotor and the resistance of the rotor bars determine the starting current, starting torque, breakdown torque (maximum torque), and slip of the motor.

Design A is considered to have normal starting torque, normal starting current, and low slip at full load, with the slip less than 5 percent except for motors with ten or more poles in which case the slip may be slightly above 5 percent. The starting torque varies from 175 to 100 percent of rated full-load torque. Several sizes are over 200 percent with a 1-hp–1800-rpm motor rated at 275 percent to a low for a 500-hp–3600-rpm motor rated at 70 percent. The maximum or breakdown torque averages about 250 percent of normal full-load torque with a 1-hp–1800-rpm motor in excess of 300 percent to a 250-hp or larger motor in excess of 175 percent. The starting current or locked-rotor currents run from over 600 to 800 percent of normal full load. The higher percentage currents are for the smaller-size motor. The rotor cage is a single cage of low resistance and the slots are of medium depth. Figure 5-3(a) is a typical example of the rotor slots and cross section of the rotor bars for this design. Motor sizes to $7\frac{1}{2}$ hp and 200 hp and above may be started across the line.

Design B is considered to have normal starting torque, low starting current, and low slip, with the slip at full load less than 5 percent unless the motor has ten or more poles in which case the slip may be slightly above the 5-percent limit. The starting and breakdown torques are the same as for design A. Starting currents are less than for design A and do not exceed 600 percent of the full-load running torque. The rotor slots are deeper than the previous design and thus motors of this design have higher rotor reactance, so that motors operate at a lower power factor. Figure 5-3(b) shows a typical design for the rotor slots and rotor-bar cross section for design-B polyphase induction motor. All motors of this group can be started across the line without special starting equipment.

Design-C group motors have high starting torque, low starting current, and slip at full load less than 5 percent. The starting torques range from 250 to 200 percent of the normal full-load torque with a 3-hp–1200-rpm motor having a minimum starting torque of 250 percent to the 20-hp–1800-rpm and larger motor having a minimum starting torque of 200 percent. The rotor cages for this design are of the double squirrel-cage type. Typical rotor slot and cross section are shown in Figure 5-3(c). Starting currents are the same as design B. Maximum or breakdown torque values are similar to design-A and design-B motors. Section 5-7 describes the principle of the double squirrel cage.

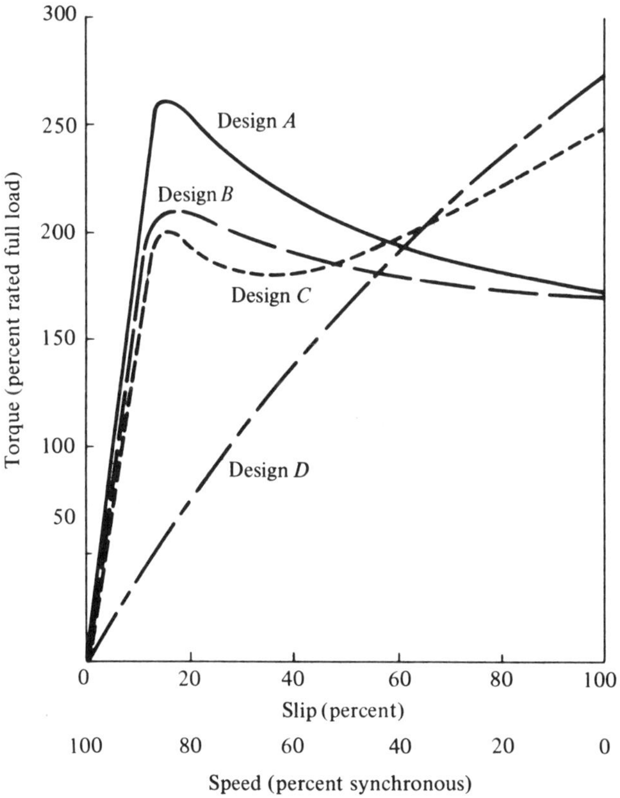

FIGURE 5-13. Torque versus slip or speed curves for small horsepower (5 hp–1200 rpm) polyphase induction motor.

Design-D classification has high starting torque, low starting current, and high slip values. The starting or locked-rotor torque for 150 hp and smaller sizes is a minimum of 275 percent developed torque at rated full-load conditions. The slip at rated load is greater than 5 percent. Consequently the speed regulation is high and efficiency is lower than for designs A, B, and C. Starting currents are the same as for designs B and C. This type is started across the line and is chiefly used for intermittent duty with rather severe starting torque requirements or for high-impact loads such as punch presses and shears.

Figure 5-13 shows typical torque-slip or speed curves for small polyphase induction motors of the four classifications. These curves show a slip of 4 percent or 96-percent synchronous speed at 100 percent torque for design A, a breakdown torque of 260 percent at a slip of 16 percent or 84-percent synchronous speed, and a starting or locked-rotor torque of 175 percent. Design B has a breakdown torque of 210 percent with a starting torque of about 172 percent. Design C has a breakdown torque of 200 percent but a starting torque of 250 percent. Design D has a breakdown torque equal to the starting torque of 275 percent. All percentages are in terms of rated full-load torque at rated conditions.

5-7 DOUBLE SQUIRREL-CAGE INDUCTION MOTOR[1],[4],[8]

It was mentioned in the preceding section that design-C induction motors have a double-cage rotor. In Figure 5-3(c) typical slots for the double cage are shown. Figure 5-14 shows enlargements of the slots of Figure 5-3(c) with the rotor bars shown as the cross-hatched sections. The bars in the upper part of the slots have a high resistance and because they lie near the air gap, they have a low reactance. The bottom bars are larger and have low resistance, but they are deeply imbedded and have a high reactance. At start, the frequency of the rotor is the same as the stator and because the reactance is proportional to the frequency, the reactance ($\omega f_2 L$) of the lower bars is high. The impedance of the lower cage is $R_1 + j(\omega f_2 L_1)$ and because R is very small, the impedance is practically equal to the reactance. The upper cage bars have a low reactance which is given by the expression, $R_u + j(\omega f_2 L_u)$. The optimum starting torque occurs when R_u is equal to the reactance $\omega f_2 L_u$ and the impedance is approximately 1.4 times the resistance of the upper bars. Therefore, at the instant of start the impedance of the upper cage is considerably less than the impedance of the lower cage, so most of the rotor current flows through the upper cage bars. As the rotor accelerates to full

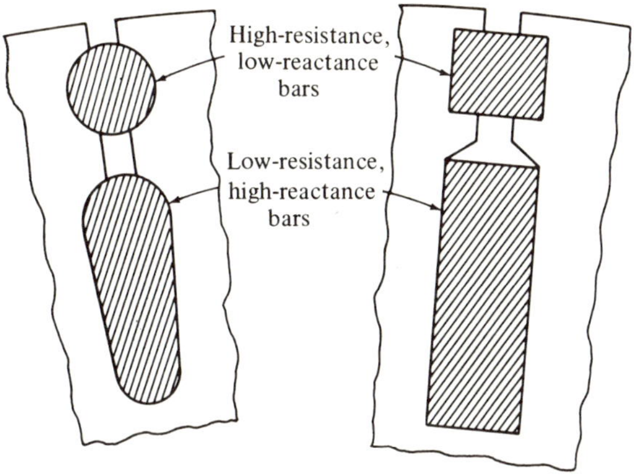

FIGURE 5-14. Enlargement of Figure 5-3(c) showing slots and rotor bars of double squirrel-cage induction motor.

speed, the rotor frequency f_2 becomes equal to the slip times the stator frequency Sf_1. Because the running slip is less than 5 percent, the reactance of the lower cage is less than 5 percent of the starting value. If $R_1 = j(\omega f_2 L_1)$, the impedance $(R_1 + j\omega f_2 L_1)$ is approximately 7 percent of its starting value. Since now the lower cage has a lower impedance than the upper cage, most of the current in the rotor flows through the lower cage. In the starting phase and in normal operation there is no complete absence of current flow in either cage. Also, the mention of "high" and "low" means "relatively high" or "relatively low" and the value of the resistances and reactance must be carefully calculated so that on start most of the current will flow through the upper cage and under normal running conditions most of the current will flow through the lower cage bars.

In smaller motors the difference of cross-sectional area can account for the relative resistance values, but in larger motors the upper and lower bars are of different alloys. The bars shown in Figure 5-14 are for a larger motor with the upper bars of a high-resistivity copper alloy and the lower bars of low-resistivity copper or aluminum.

5-8 WOUND-ROTOR INDUCTION MOTOR[2],[6],[9]

In Section 5-2 the construction of the wound-rotor induction motor was described. The winding must be made to have the same number of poles as the stator. In the preceding section it was shown how the adding of resistance

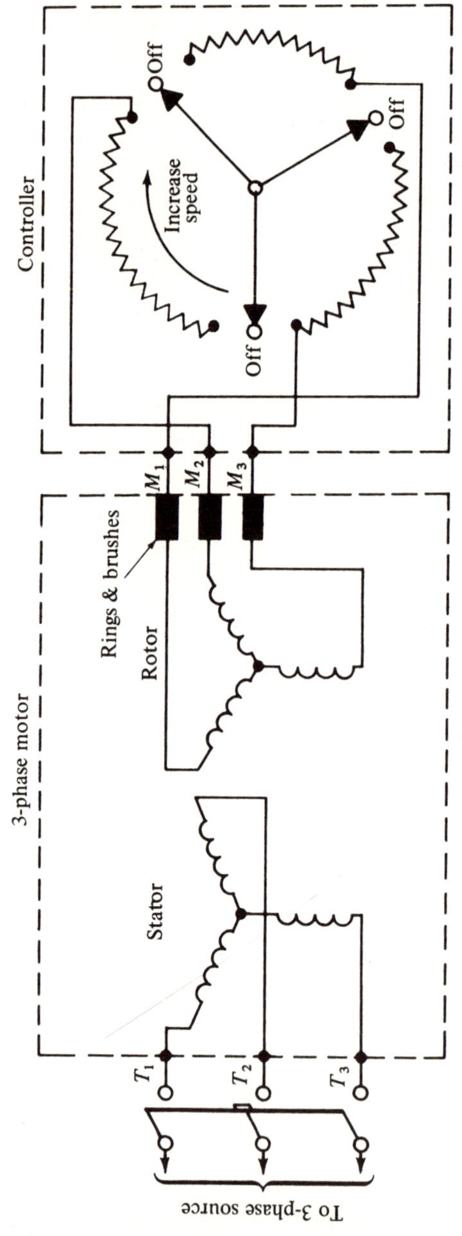

FIGURE 5-15. Schematic diagram showing the adding of resistance for a wound-rotor induction motor.

in the rotor increases the starting torque, decreases the speed (increases the slip), and decreases the starting current. The double squirrel-cage rotor will give one value of starting torque and one value of starting current and no changes can be made after the machine is manufactured. With the wound rotor the value can be varied and any value from short-circuited rotor torque to breakdown torque can be achieved. The curves of Figure 5-9 show how added resistance affects starting torque and speed. Speed may be reduced to below 50 percent of rated full-load speed. The efficiency drops due to the increased I^2R loss in the rotor.

Figure 5-15 shows the circuit diagram of a three-phase wound-rotor induction motor with a controller connected to rotor terminals M_1, M_2, M_3. The three arms slide over three equal resistances as the rotor is turned clockwise from the off position. The arm is hand operated just as the arm of the manual face-plate starter for a dc motor. Because this device controls the speed, it is called a *controller* and the resistances must be of such rating to dissipate the heat and operate continuously at any setting. The controller is designed so that the arms must be brought to the off position before the motor can be started. The value of resistance per phase is usually of such a value to give starting torque equal to or close to the breakdown value of torque. Example 5-6 shows how the value of resistance and the starting currents are calculated.

The NEMA standard MG1-1967 lists the breakdown torque values for *wound-rotor polyphase induction motors* with an average of 250 percent of rated full-load torque. A 2-hp–1800-rpm motor rated at 275 percent, a 2-hp–900-rpm motor at 250 percent, and a 20-hp–200-rpm motor rated at 225 percent. No values are given for starting currents or starting torques because these values depend on the added resistance per phase. With the rotor short circuited, the maximum torque values are similar to designs A, B, and C for squirrel-cage motors. Because the rotor is wound with coils, the short-circuited resistance is slightly higher than the rotor resistance of a squirrel-cage rotor. This means that at rated speed the slip is higher than that of a comparable squirrel-cage motor.

The cost of manufacture of the wound-rotor motor is higher than the cost for the same size squirrel-cage type. Efficiency is lower, especially when operating at a lower speed than rated. Maintenance costs are higher for the wound-rotor motor. Such motors may be used in conveying systems, hoists, elevators, cranes, and pumps. A wound rotor has the advantage over the dc series motor because it can be operated at no load at a fixed speed near its synchronous speed.

5-9 STARTING OF POLYPHASE INDUCTION MOTORS[1],[10],[11]

This section is confined to starting methods for three-phase squirrel-cage induction motors which are the most widely used motors of the polyphase ac variety. The starting of wound-rotor motors was covered in the preceding section and shown by the circuit of Figure 5-15. There are five basic starting methods, four of which are based on the principle of reducing voltage. These methods are (a) across-the-line starters, (b) autotransformer starters, (c) primary resistor or reactor starters, (d) wye-delta starters, and (e) part-winding or split-winding starters.

All of these methods may use manual or automatic operation. Manual operation is more applicable to smaller motors up to $7\frac{1}{2}$ hp since the size of manual equipment becomes too large for large motors.

Across-the-Line Starters. These starters are the simplest and easiest to maintain. Most of the squirrel-cage motors can be started this way, but because of the high starting currents, this method is confined to small motors up to 5 to $7\frac{1}{2}$ hp and to large sizes if the supply is of sufficient capacity. In any event the rules of the power utility should be checked before planning to use an across-the-line starter, especially if any size larger than 5 hp is involved. The normal types are the three-pole snap switches, enclosed knife switches, or the drum-type contactor.

The automatic push-button across-the-line starter is shown by the schematic diagram for a low-voltage squirrel-cage motor in Figure 5-16(a). Figure 5-16(b) shows the control circuit for a high-voltage motor. This simply adds a transformer for supplying the control circuit with low voltage for operational safety. The operation is quite simple, for when the start button is pressed, the magnetic contactor coil M is energized. The three M contacts and the M_1 contact close. The M contacts connect the motor across the line if the main contactor switch has been closed. The M_1 contact seals the start button and keeps the contactor coil M energized after the operator releases the start button. The motor is stopped by pressing the stop button, which is normally closed, or by an overload relay opening one of the normally closed contacts OL_1 or OL_2. The contactor coil is de-energized and the motor is disconnected when all M contacts open. Large-size motors would use relays in the control circuit with the relays operating the larger size contactors required.

Autotransformer Starters. This method reduces the voltage applied to the motor through autotransformers. Three autotransformers connected in wye or two connected in open delta may be used. The open-delta system is most

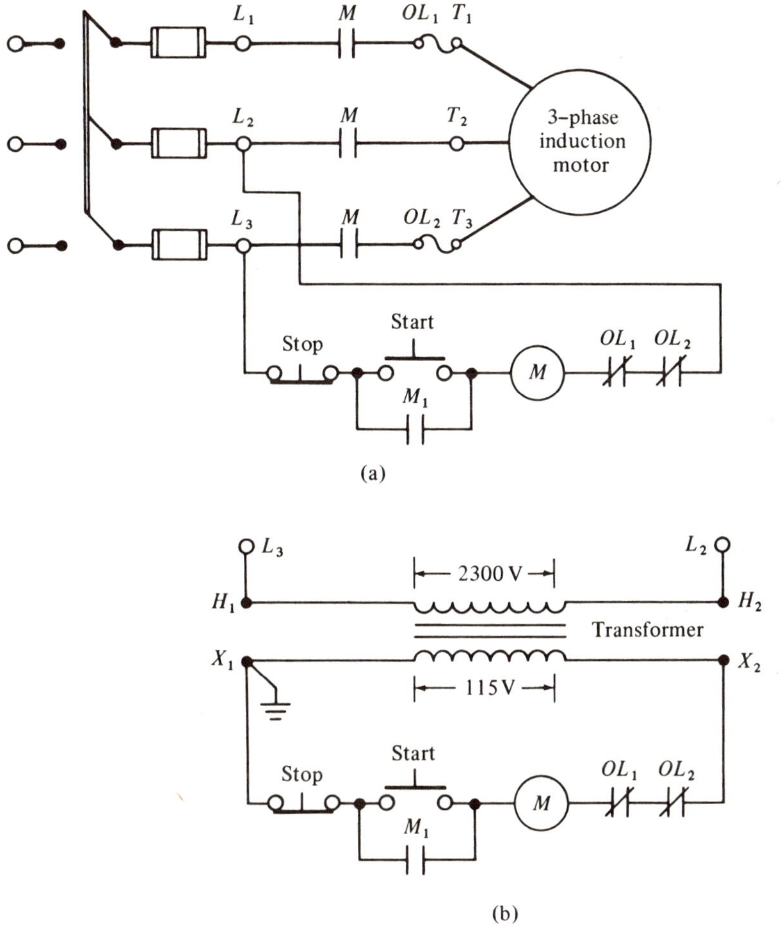

FIGURE 5-16. (a) Push-button across-the-line starter schematic diagram for a three-phase squirrel-cage induction motor. (b) Schematic diagram of the control circuit for high-voltage across-the-line starter of Figure 5-16(a).

commonly used. A manual open-delta schematic connection is shown in Figure 5-17. The *S* contacts are closed when the handle is moved back to the start position. A spring-loaded handle is mounted on the side of the metal case in which the contacts and transformers are mounted. By following the schematic diagrams, the terminals of the motor are connected to S_2, S_3, S_4, and S_5. The other sides of S_2 and S_5 are connected to the 65-percent taps of the transformers. The other sides of contacts S_3 and S_4 are connected to L_2 and the common connection of the transformers, respectively. The bottom sides of S_1 and S_6 are connected to L_1 and L_3, respectively, through overload coils. The upper sides of S_1 and S_6 are connected to the full-winding terminals of the transformers. So when the handle is pushed back, the S contacts are closed connecting the line terminals to the open-delta primary and the motor is connected to the 65-percent taps and the common transformer connection. On this tap the terminal voltage is 65 percent and the starting torque is reduced to about 42 percent of the breakdown or starting torque. The motor current is 65 percent of the starting current value and by transformer action the current drawn from the line is about 42 percent of the starting motor current.

When the speed of the motor is near its normal speed, the handle is quickly pulled to the forward or run position. In this position the S contacts are opened and the R contacts are closed. L_1 is connected through the overload coil and the R_1 contact to T_1 of the motor. L_2 is connected through R_2 contact to T_2 of the motor. L_3 is connected through overload coil and R_3 contact to T_3 of the motor. The motor is now across the line. The spring-loaded handle is held by an undervoltage latch coil *UV*. To stop the motor the stop button is pressed which de-energizes the latch coil and the handle by spring action returns to the off or center position. If the motor is overloaded, the overload coils open the normally closed contacts in the *UV* circuit releasing the latch as if the stop button were pressed. This type of starter does not go through a series of increasing voltages because the tap setting is a permanent connection. The tap settings can be changed for the voltage best suited for the starting application. A disadvantage of this starting compensator is that the motor is open circuited when transferring from the starting S contacts to the running R contacts. This transfer must be done quickly and for a motor of some size this requires a great deal of physical exertion. For large or high-voltage motors, this manual starter is cumbersome for size, manipulation, and safety. The functions can be done with a start button and relays to operate the S and R contacts with a time delay for the closing of the R contacts. Interlock is also necessary, otherwise series trouble results when both contacts are closed at the same time.

If the three line currents or the voltages are too unbalanced for good operation, three transformers in the wye connection may be used. Generally only two transformers in open delta are used since an unbalance of 15 percent or less does not materially alter the starting characteristics.

Primary Resistor or *Reactor Starters.* This type of starter is analogous to the starters employed for dc motors in which resistance is added in the arma-

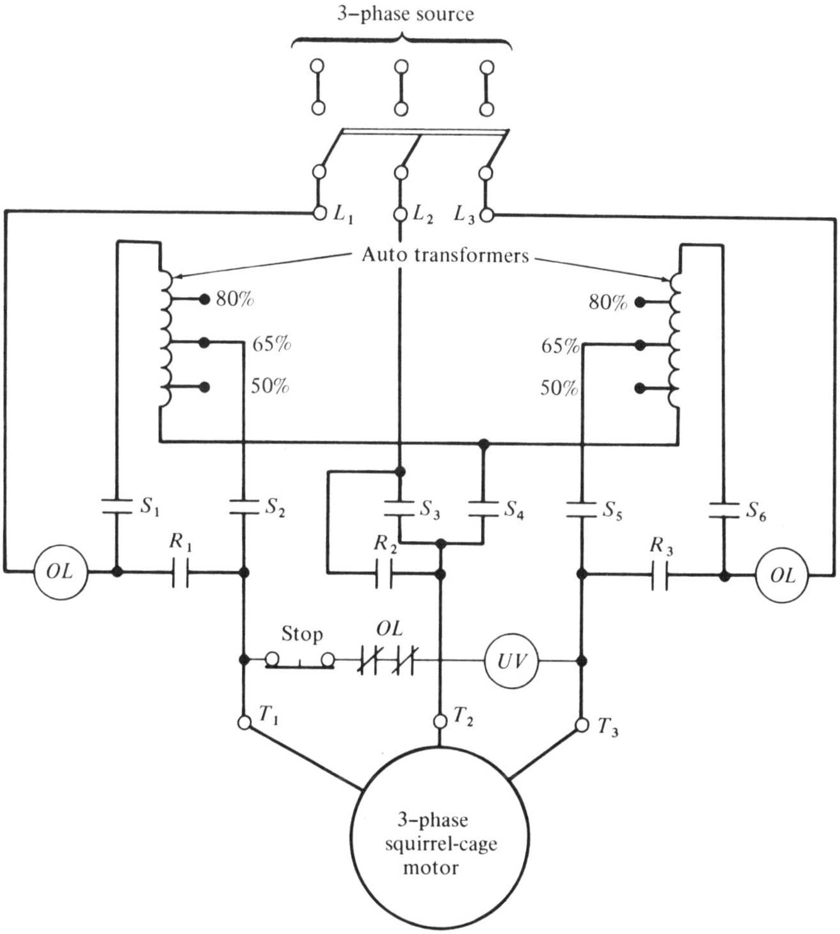

FIGURE 5-17. Circuit diagram for manually-operated compensator (autotransformer) starter for reduced voltage for a three-phase squirrel-cage induction motor.

ture circuit. The drop across the added resistance or reactance reduces the impressed voltage across the motor stator and the current is reduced in the line only proportionally to the reduction in voltage. Either resistance or reactance adds impedance by the relationship, $Z = \sqrt{R^2 + X^2}$. If resistance is used, the starting power factor is improved, but greater power losses are produced. It is somewhat easier to vary resistance, and this method is more commonly used than added reactance. For smaller motors the scheme as

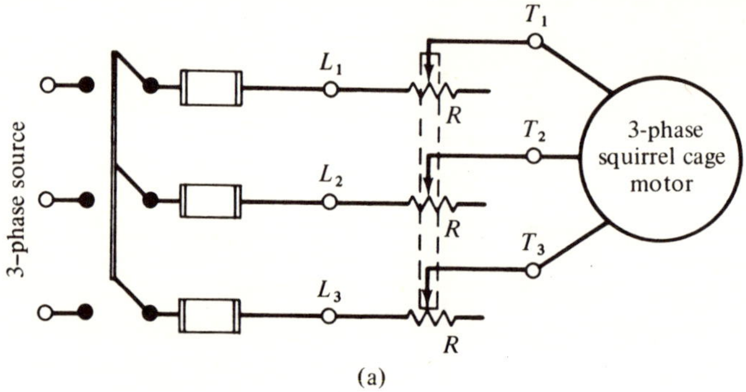

(a)

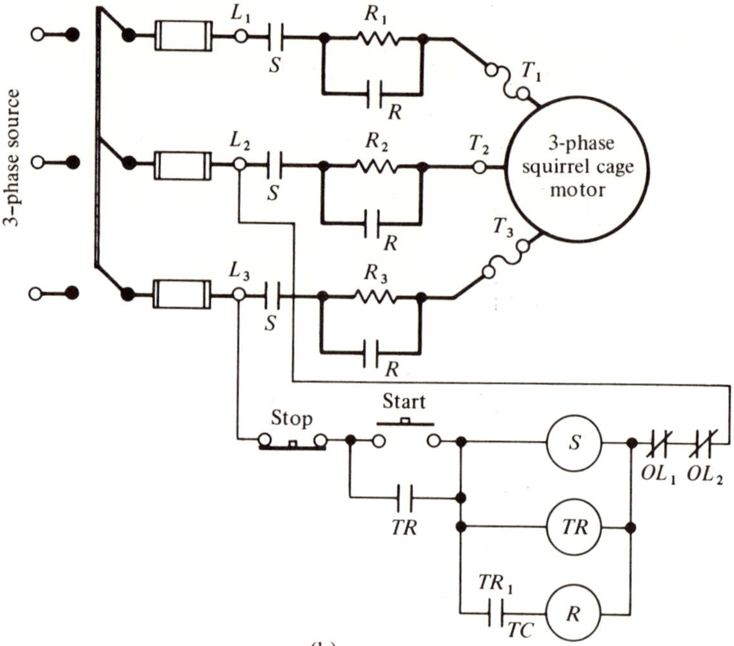

(b)

FIGURE 5-18. (a) Circuit diagram for manually-operated line resistor starter for three-phase squirrel-cage induction motor. (b) Circuit diagram for a one-step automatic push-button starter for three-phase squirrel-cage induction motor.

shown in Figure 5-18(a) is used. The three equal resistances are changed equally. This is usually a manual face-plate device. This method produces an even acceleration as the resistance is cut out and at no time during the acceleration is the circuit open.

For large motors and for automatic operation the schematic diagram of Figure 5-18(b) is used. This scheme provides one-step acceleration. However, steps can be introduced with the use of definite time relays *TR* with steps of resistance cutting out in a set time pattern. The operation of the circuit of Figure 5-18(b) proceeds as follows: After the main switch is closed, the start button is pressed and contactor *S* is energized; the *S* contacts close, energizing the motor at reduced voltage due to line *IR* drops; the timing relay *TR* is energized; the *TR* contact seals the start button; the TR_1, *TC* contact closes after a set time delay. When TR_1 is closed, the *R* contactor coil is energized; and the *R* contacts close shorting out the starting resistances.

Another type of starter, which is semiautomatic, uses graphite disks placed in three insulated cylinders. When the handle is moved to the start position, the motor is connected to the line through the carbon-disk resistances. A solenoid, also energized, moves the handle up and compresses the graphite disks. Compressing the carbon disks reduces the resistance. At the end of travel, the contactor is energized by an *r* contact, the carbon-disk resistances are shorted out by three *R* contacts, and the motor operates across the line. A sealing contact seals the *r* contact and the lever returns to the off-run position. A stop button de-energizes the *R* contactor. Thermal overload relays (normally closed contacts) can also de-energize the *R* contactor. This starter gives a smooth acceleration, high power factor, and increased torque as the motor is accelerated.

Wye-Delta Starter. This type of starter requires that the terminals of each phase of the stator winding must be brought outside the motor frame. NEMA standards[3] give the terminal designation as phase A, T_1 and T_4; phase B, T_2 and T_5; and phase C, T_3 and T_6. The standard states that for start T_4, T_5 and T_6 are connected and L_1 is connected to T_1; and L_2 is connected to T_2 and L_3 to T_3. For running, T_1 and T_6 are connected to L_1; T_2 and T_4 are connected to L_2; and T_3 and T_5 are connected to L_3. Figure 5-19 shows how this is done manually with a three-pole, double-throw switch. As with the autotransformer type, the circuit is opened when the change from start to run position is made. On start, the stator winding is connected in wye; when in the running position, the stator winding is connected in delta. Therefore, on starting the voltage is $1/\sqrt{3}$ or 58 percent of the running

[3] MG1-1967 (MG1-2.62).

voltage, while the starting current per phase is 58 percent of the delta phase current. The delta phase current is 58 percent of the line current and so the wye line current is 33 percent of the full-voltage starting current.

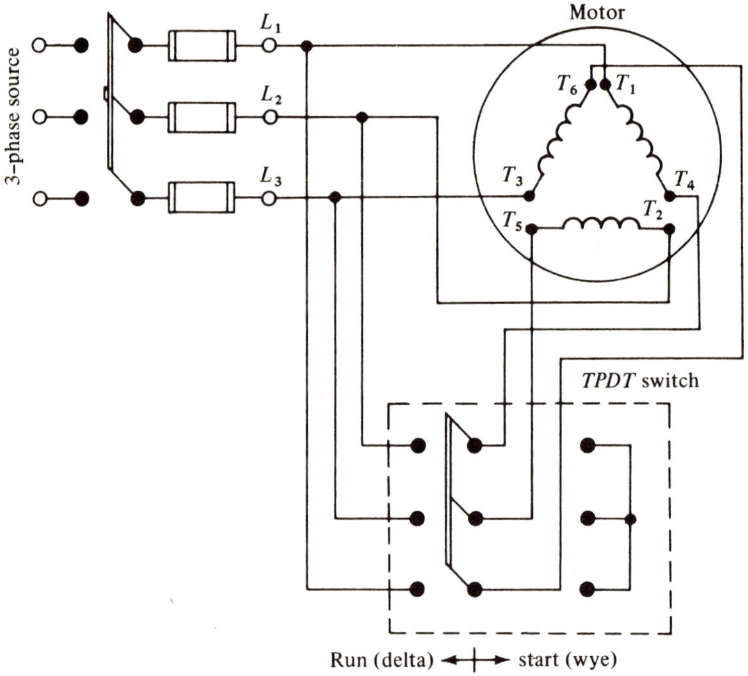

FIGURE 5-19. Circuit diagram for manually-operated wye-delta starter for three-phase squirrel-cage induction motor.

The wye-delta system may be done automatically with a 4-pole relay, a 2-pole time-delay relay of which one contact is normally closed and one normally open, two 3-pole contactors and one 2-pole contactor.

Part-Winding or *Split-Winding Starter.* A wye-connected dual-voltage motor whose terminals (nine brought out) is shown in Figure 5-20(a). This method is considered dual voltage; for if T_1 and T_7, T_2 and T_8, and T_3 and T_9 are each connected and in turn connected to L_1, L_2, and L_3, respectively, of a three-phase source while T_4, T_5, and T_6 are connected together, a parallel wye-connected stator would result. If T_4 is connected to T_7, T_5 connected to T_8, and T_6 is connected to T_9 with T_1, T_2, and T_3 connected to L_1, L_2, and L_3,

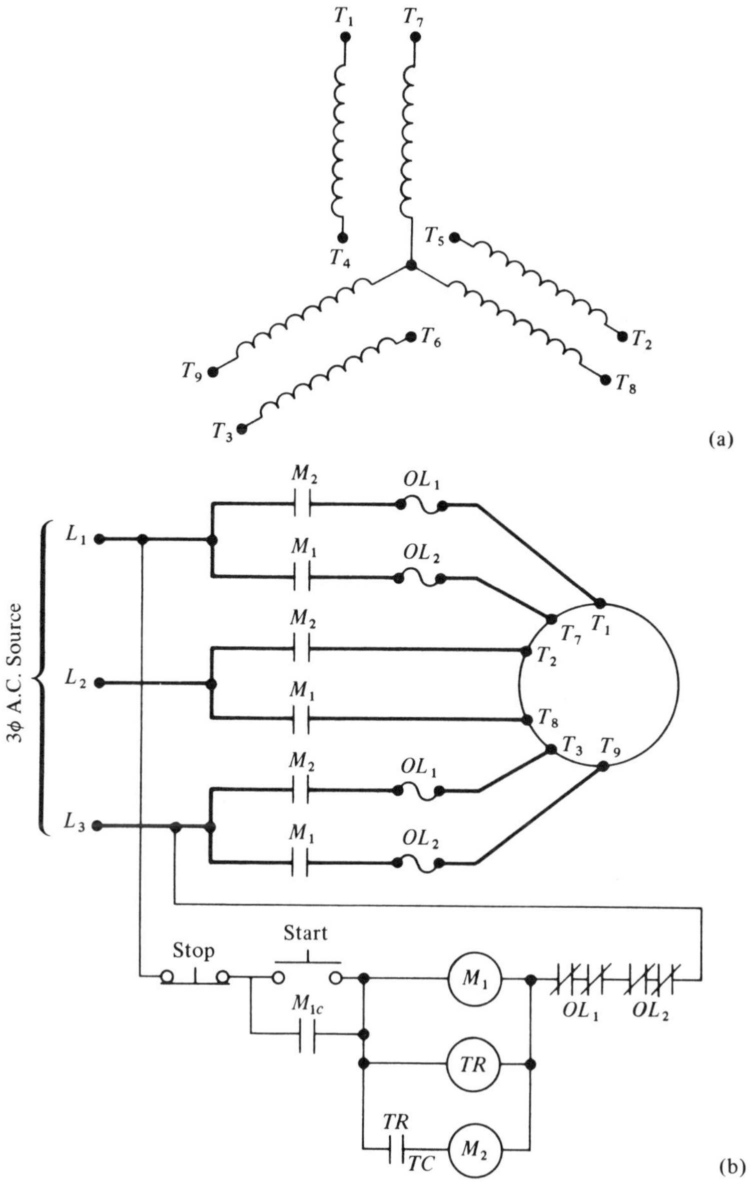

FIGURE 5-20. **(a) Terminal markings for a two-part wye-connected, dual-voltage three-phase squirrel-cage induction motor. (b) Schematic diagram for an automatic push-button starter for a two-part winding motor. Note that T_4, T_5, and T_6 are connected together.**

respectively, a wye with the phases in series would result. For the same coil voltage, the three-phase supply would have to be double the voltage for the parallel connection.

A method of part winding starting is depicted by Figure 5-20(b). When the start button is depressed, the M_1 coil of the three-pole contactor M_1 is energized. The closing of the M_1 contacts connects the source to the half-wye winding of terminals T_7, T_8, and T_9. The M_{1C} contact seals the start button. The timing relay TR is energized at the same time as M_1. After a time delay TR contact closes. The closing of this contact energizes the M_2 contactor. The closing of the M_2 contacts connects T_1, T_2, and T_3 to the same line. T_4, T_5, and T_6 are connected together within the motor so that the second wye is now placed in parallel with the first wye. With this method the inrush current is approximately 60 percent of the inrush value had the parallel wye been connected across the line. The starting torque is slightly less than full-voltage starting torque.

The following example shows how the starting currents, voltages, and torques compare for the various starting methods.

Example 5-9. A 75-hp, 220-V, 60-Hz, three-phase squirrel-cage induction motor draws 185 A at full-load conditions. Starting current at full voltage is 1085 A with 135 percent of normal full-load operating torque. The speed at full load is 1145 rpm.

(1) If the 65-percent taps are used with the open-delta autotransformer compensator for starting, what is the starting motor current, the starting line current, and the starting torque? (Give answers in both quantitative and qualitative values.)

(2) If the motor is started by the primary resistor starting method with the resistances yielding a 65-percent starting voltage, find the same quantities as in (a).

(3) If the motor is started by the wye-delta method, assuming the motor above has a delta-connected stator for rated conditions, find the same quantities as in (a).

Solution:

$$\text{(a) Rated torque, } T_r = \frac{\text{hp}}{0.00019(\text{rpm})} = \frac{75}{0.00019(1145)} = 345 \text{ lb-ft,}$$

$$\text{motor current, } I_m = \frac{E_s/E_R}{I_s} = \frac{220 \times 0.65}{220}(1085)$$

$$= \frac{143}{220}(1085) = 706 \text{ A,}$$

$$\text{percent of full-load current} = \frac{706}{185}\,100 = 382\%,$$

$$\text{line current } I_L = I_m(\text{percent tap}) = (706)(0.65) = 459\text{ A},$$

$$\text{percent of full-load current} = \frac{459}{185}\,100 = 248\%,$$

$$\text{starting torque, } T_S = \left(\frac{E_s}{E_R}\right)^2 (1.35 \times T_r)$$

$$= \left(\frac{143}{220}\right)^2 (1.35 \times 345) = 196.5\text{ lb-ft},$$

$$\text{percent of full-load torque} = \frac{T_S}{T_r} \times 100$$

$$= \frac{196.5}{345}(100) = 57\%.$$

(b) $I_m = \dfrac{E_m}{E_l} I_S = \dfrac{220 \times 0.65}{220}\,1085 = 706\text{ A},$

percent of full-load current (same as (a)) = 382%,
$I_L = I_m = 706$ A,
T_S, [same as (a)] = 196.5 lb-ft = 57%.

(c) $\dfrac{I}{\text{phase}}$ (delta), $\dfrac{1}{\sqrt{3}}(1085) = 626\text{ A},$

$$I_L = I_m = \frac{1}{\sqrt{3}}\,626 = \frac{58}{100}\,626 = 362\text{ A},$$

$$\text{percent, } \frac{362}{185}\,100 = 196\%,$$

$$\text{starting torque, } T_S = \left(\frac{1}{\sqrt{3}}\right)^2 1.35 \times 345$$

$$= (0.58)^2\,465 = 155\text{ lb-ft},$$

$$\text{percent of full-load torque} = \frac{155}{345}\,100 = 45\%.$$

5-10 SPEED CONTROL OF POLYPHASE INDUCTION MOTORS[1],[2],[8],[10]–[12]

The speed of dc motors is varied by changing the field flux or the generated voltage of the armature in the speed relationship,

$$\text{speed} = K\frac{E_g}{\phi}.$$

The *speed equation* for the induction motor is

$$\text{rpm}_{\text{rotor}} = \text{rpm}_{\text{syn}}(1 - S). \tag{5-23}$$

By substituting the value of synchronous speed (rpm_{syn}), this formula may be written as

$$\text{rpm}_{\text{rotor}} = \frac{120f_1}{P}(1 - S) \tag{5-23a}$$

where

f_1 is the stator frequency in hertz,
P is the number of poles for the stator winding,
S is the slip (as a decimal fraction).

This last expression suggests three methods of speed control—namely, (a) control the frequency of the impressed voltage on the stator, (b) control the number of poles of the stator winding, and (c) control the slip of the motor.

The first method by controlling the frequency of the voltage source not only offers a wide variation of speed, it also requires a variable-frequency supply for the motor. This last requirement means considerable auxiliary equipment and from a practical point of view is not feasible for most requirements. However, where the motor or a group of motors have their own prime movers such as reciprocating engines or turbines used for ship propulsion, this is an excellent means of speed control. In turbines, however, the efficiencies of the turbine decrease over a wide range of speed. Therefore the second method of pole changing is usually used in connection with frequency changing. When the frequency changing method is used, the impressed voltage should be directly proportional to the frequency if the flux densities are to remain constant. If this is done, it also means that the torque will remain constant.

The second method of speed control by changing the number of poles is applied to squirrel-cage motors. Because the number of poles of the rotor must be the same as the stator, the poles of a squirrel-cage rotor adjust by induction and the current flow through the bars adjusts to phase values by flow through the end rings. If the winding of the stator is changed to double the number of poles, a wound-rotor winding would also have to be changed; this is not practical for the number of slip rings required. Because of the complications for the wound-rotor pole changes, this method is limited to squirrel-cage motors. The two ways of changing the number of poles is by using multiple windings on the stator or by using the consequent-pole method.

If multiple windings are used, only one winding is used while the others are open and unused. Usually this method is confined to two windings, for then only one-half of the conductors are active. For a three-winding scheme only one-third of the conductors would be active. Motors using this method must have larger circumferences to accommodate the multiple windings. The two windings use alternate slots for alternate coils. One winding for four poles with three coils per group would require (3 × 4 × 3) 36 slots. A second winding for six poles with two coils per group would also require 36 slots. On a stator with 72 slots the two windings could be used. When operating at a given speed on one winding, voltages are generated in the other winding. If the open winding is connected in wye, no currents will flow; but if the winding is connected in delta or parallel connected as it might be if the winding is used for consequent-pole changing, the winding ends must be open to prevent circulating currents.

The *consequent-pole* method of pole changing employs the same windings and by reconnecting the terminals the poles are doubled by reversing currents in each alternate coil group so that each coil group produces like instantaneous magnetic poles. Figure 5-21(a) shows one phase A of a four-pole winding. The solid arrows indicate the current directions for the conventional connection of four poles or the high-speed connection. The dotted arrows indicate the consequent poles with four like north poles within the turns and the opposite or consequent south poles between the coil groups. There are now eight poles or the speed is one-half of the previous speed. For the high-speed winding A to A'' is phase A start and finish, while A to A' is the start and finish for phase A for the consequent poles. Figure 5-21(b) shows the terminal markings as indicated in the NEMA standards[4] for the constant-torque connections. Here, for the high-speed connection the inputs L_1, L_2, and L_3 are connected to T_4, T_5, and T_6, respectively, and T_1, T_2, and T_3 are tied together; for the slow speed the source is connected to T_1, T_2, and T_3 with T_4, T_5, and T_6 open. The first or fast-speed connection has two wyes in parallel. The double-pole or slow-speed connection is a series delta. These connection changes may be accomplished manually with a six-pole double-throw switch or automatically with push buttons. Terminal markings for constant-horsepower and variable-torque connections are specified in the NEMA standards.[5] Schematic diagrams may be found in the references.

Three-speed and four-speed induction motors are possible by the combined use of two individual windings and consequent-pole switching. Terminal markings for such three-phase induction motors are specified in the NEMA standards.[6] Two-phase terminal markings are also specified.

[4] See footnote 3.
[5] See footnote 3.
[6] MG1-1967 (MG1-2.65).

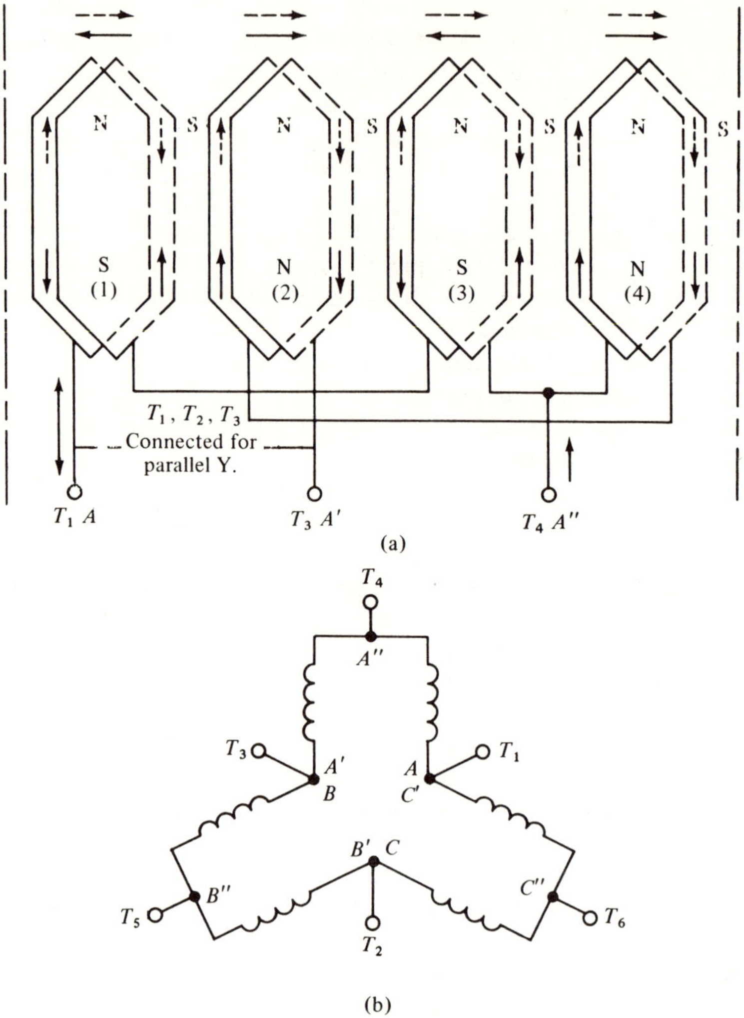

FIGURE 5-21. (a) Phase *A* of a three-phase split winding showing formation of consequent poles. (b) Three-phase stator winding showing terminal and phase connections for winding of Figure 5-21(a).

The third method of speed control by variation of slip applies to wound-rotor motors. In the Section 5-8 discussion of the wound-rotor induction motor it was pointed out that slip and, therefore, the speed of a wound rotor was decreased by the addition of resistance in the rotor. Figure 5-15 shows such a circuit. This method does adjust the speed below the full-load speed but at the expense of efficiency by the copper losses $I_R{}^2R_R$ in the rotor.

Slip may also be increased (speed lowered) by lowering the impressed voltage on the motor. Since torque varies approximately as the square of the stator voltage, however, the torque drops off rapidly. A 25-percent drop of terminal voltage would result in about a 44-percent drop in torque. Obviously this is not satisfactory for most applications and has limited use; the chief use is for small fan motors.

The *concatenation* method of speed control involves the cascading of two induction motors with their shafts coupled together. The first machine must be a wound-rotor motor and the second can be either a squirrel-cage or wound-rotor motor. The wound rotor used for the second machine can give some additional speed adjustment with a variable three-phase wye rheostat connected to its slip rings. Figure 5-22 shows the schematic circuit diagram

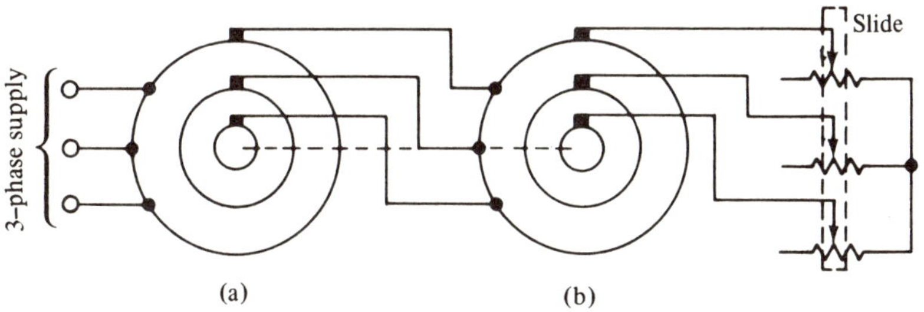

FIGURE 5-22. Two three-phase wound-rotor induction motors connected in concatenation with a three-phase rheostat on the second motor.

for two wound-rotor induction motors connected in concatenation. Here the stator winding of machine (a) is connected to the three-phase supply. The output from the slip rings of machine (b) is connected to the stator of the second machine. The synchronous speed of this combination is given by the formula,

$$\text{rpm}_{\text{syn}} = \frac{120f_1}{P_1 + P_2} \tag{5-24}$$

where

f_1 is the frequency of the stator of machine (a),
P_1 is the number of poles of machine (a),
P_2 is the number of poles of machine (b).

In the above formula the fields of the rotating fields of both machines are in the same direction. If the fields rotate in opposite directions, the denominator becomes the difference of the poles ($P_1 - P_2$). This last connection is not usually used.

Another method involving the variation of slip of a wound-rotor induction motor is the introduction of a voltage in the rotor circuit. To do this, various methods and types of motors have been devised. These are explained briefly in the next section. However, frequency converters or variable-frequency devices supply a voltage at the slip frequency Sf_1 of the rotor. Also, the energy involved with this voltage is recovered since no variable rotor resistance in the rotor circuit is used.

5-11 SPECIAL SYSTEMS AND MOTORS FOR SPEED CONTROL[8],[10]–[12]

The third method of speed control by adding resistance to the slip rings of a wound rotor adjusts the speed to some value lower than the synchronous speed. It was shown that the slip depends on the rotor resistance, and the decreased speed is at the expense of added heat loss in the resistance. The following systems add a foreign voltage to the rotor circuit to vary the rotor induced voltage. The voltage substituted for the I_2R_2 voltage drop has the same effect as adding or subtracting resistance. Adding voltage to aid the

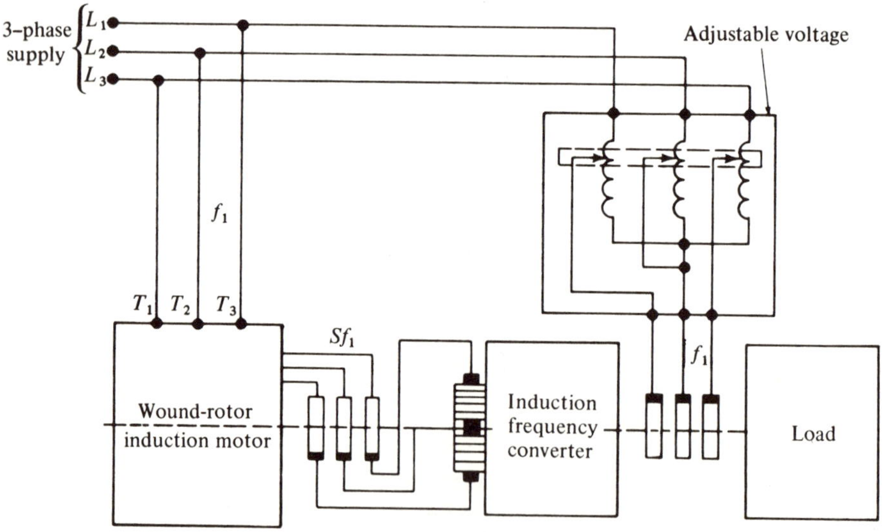

FIGURE 5-23. Schematic diagram for a LeBlanc control system for a three-phase wound-rotor induction motor.

rotor induced voltage increases the speed. The foreign voltage can raise the speed above synchronous speed. This foreign voltage must obviously be at the same frequency as the rotor frequency, $Sf_1 = f_2$.

The *LeBlanc* system for speed control is shown in the circuit of Figure 5-23. A wound-rotor induction motor is connected to a three-phase source. Three adjustable autotransformers connected in wye are connected to the same source. The variable-voltage output of the autotransformers is connected to the slip rings of an induction frequency converter. The slip rings of the motor rotor are connected to the brushes of the commutator brushes that are spaced 120 electrical deg around the commutator. The armature shaft of the frequency converter is coupled to the shaft of the wound-rotor motor. The frequency of the field induced in the iron of the frequency converter is $f_1 - (1 - S)f_1$ or Sf_1. Therefore the frequency of the voltage added to the induction rotor is the same frequency Sf_1. By shifting the three brushes together the value by phase shift can be varied. Therefore the value of voltage can be adjusted by both the autotransformer settings and by the brush position of the frequency converter. The brushes are moved by a mechanical lever or by a worm drive. When the voltage adds to the induced rotor voltage, the speed of the motor is above synchronous and the energy to the rotor is supplied both by the motor stator and the frequency converter. When the foreign voltage subtracts from the induced voltage, the speeds are below synchronous and the energy from the motor stator to the rotor supplies energy back to the line through the autotransformers.

The *Kramer* system of speed control is more flexible and sophisticated. A modified system for constant horsepower is shown in Figure 5-24. In this system a separately excited dc motor is coupled to the shaft of the wound-rotor motor. The brushes of the motor are connected to the dc brushes of a separately excited frequency converter. The excitation of the motor and the frequency converter is usually supplied by a small exciter on the same shaft as the motor The output of the frequency converter is connected to the slip rings of the wound rotor. The motor field excitation controls the speed of the wound-rotor motor, and the frequency of the rotor output controls the dc output of the converter. Decreasing the motor excitation causes it to increase its speed and hence its output adds to the output of the induction motor. By increasing the excitation the speed of the dc motor decreases and it acts as a generator with the energy coming from the induction-motor rotor. The dc motor electrical output causes the rotary converter to feed voltage to the rotor slip rings and increase speed by adding to the rotor voltage. The excitation at the rotary converter regulates the power factor at which the induction motor operates. This system is used for large power units 500 hp

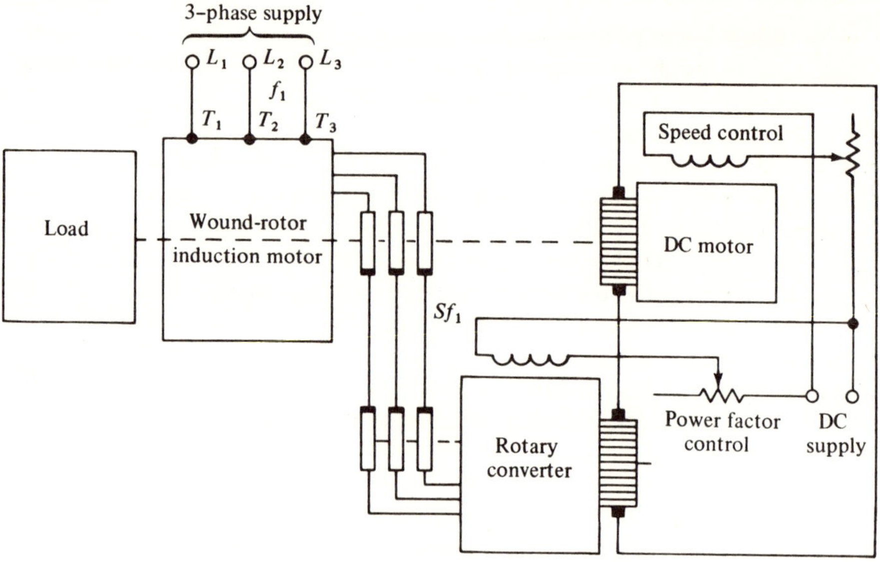

FIGURE 5-24. Schematic diagram for a modified Kramer constant-horsepower control system for a three-phase wound-rotor induction motor.

or over. It uses standard parts and it is efficient. It is used for speed reductions to about 40 to 50 percent. The size of the dc motor and the converter should have 40 to 50 percent of the rating of the main motor. The system can be converted to a constant-torque system if the dc motor is also coupled to an alternator whose terminals are connected to the three-phase source.

The *Scherbius* system is shown in Figure 5-25. In this system no dc motor is used, but a variable-frequency converter is coupled to the wound-rotor shaft. The output of the rotor (slip rings) is connected to the brushes of a regulating-frequency converter and to a three-bank wye autotransformer primary. The secondary of the autotransformers and the brush output of the variable-frequency converter are connected to the stator of the regulating-frequency converter. The slip rings of the variable-frequency converter are connected to the same line as the wound-rotor motor. The regulating frequency converter is coupled to a squirrel-cage motor that is also connected to the three-phase supply. This system overcomes the instability of the Kramer system at and near synchronous speed. It is obvious that at synchronous speed the voltage of the variable-frequency converter at synchronous speed is zero and the foreign voltage at that point should be zero frequency or dc. In the Kramer system shown in Figure 5-24, at synchronous speed the input to the

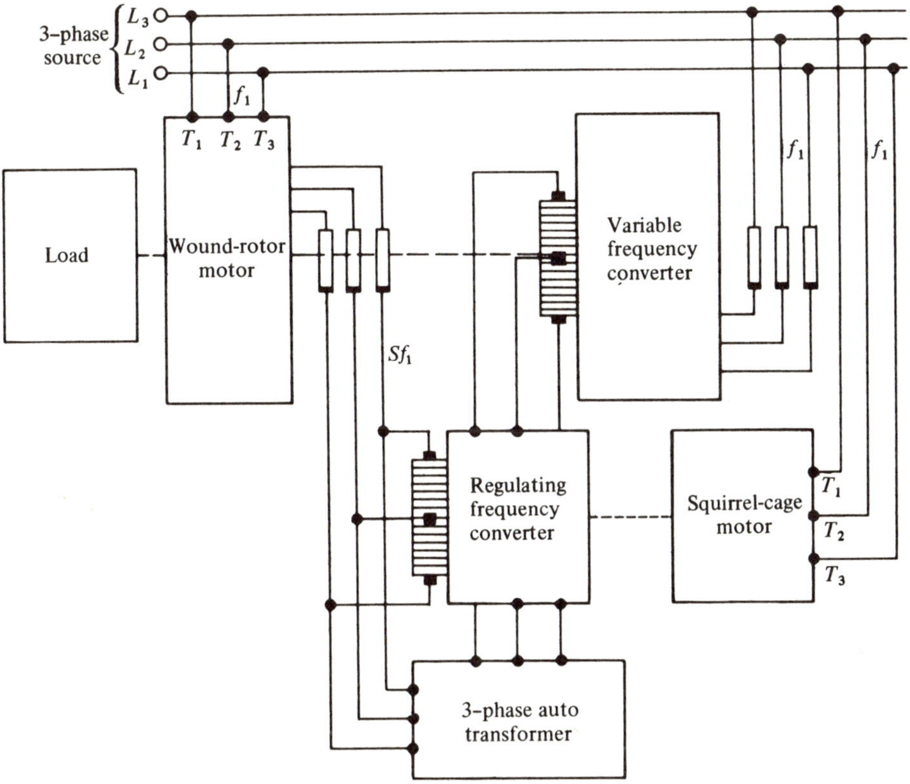

FIGURE 5-25. Scherbius speed control system diagram for a three-phase wound-rotor induction motor.

rotary converter is at zero frequency and there is no output from the commutator.

The *Schrage* motor takes its name from its Swedish inventor, Karl H. Schrage. In the United States it is produced and sold by the General Electric Company as the BTA brush-shifting motor. This motor combines into one unit the wound-rotor induction motor, the three-phase autotransformer unit, and the frequency converter. The primary is on the rotor and the secondary is on the stator in three sections of approximately 120 deg. The ends of the three stator sections terminate on pairs of brushes that ride on a commutator whose winding is not electrically connected to the primary. Voltage control is accomplished by the three pairs of brushes that can be moved together or separated further to decrease or increase the voltage. The brushes are operated by a hand wheel or a lever. The entire brush rigging may also be moved. The moving of the pairs of brushes to change the voltage is analogous to the autotransformer setting of the LeBlanc system, and the moving of the entire rigging is analogous to the movement of the brush rigging of the frequency converter of the LeBlanc system. This type of motor is a smaller sized motor —50 hp or less. Except for instability at synchronous speed, it operates smoothly over a wide range of speeds with good efficiency and speed regulation. It has good starting torque and a high breakdown torque.

The *Rossman drive* is an electromechanical system for a special squirrel-cage motor whose stator also rotates. A controlled dc motor drives the stator. When the stator is driven in the opposite direction from rotating field, the speed is controlled below the stationary stator speed. When the stator is rotated in the same direction as the stator field, speeds are at synchronous and above. The drive does provide smooth control over a wide range of speeds. This drive is not widely used because of its expense. The dc motor needs a reversing controller and the range of speeds is dependent on the speeds of the dc motor.

Other electromagnetic, mechanical, and hydraulic drives are used to provide adjustable speeds for squirrel-cage induction motors. *Eddy current coupling*, which is an electromagnetic slip clutch device, can be used for motors up to 500 hp.

5-12 BRAKING OF INDUCTION MOTORS[1],[8]

Just as dc motors in many applications need to be stopped quickly and smoothly, so must ac induction motors. The magnetic-mechanical brake is not as smooth and quiet as electrical braking methods. The same methods used for stopping dc motors also apply to induction motors. These methods are dynamic braking, regenerative braking, and plugging.

Dynamic braking for induction motors, both polyphase and single phase, is accomplished by providing dc through two terminals. If a dc voltage is applied to two terminals of the stator, the effect is that the rotating field produced by the ac has now stopped rotating and is stationary. The rotor by its inertia induces currents in the rotor conductors that are in opposite directions when running at normal speed and when the rotor stops quickly. Figure 5-26 shows the control circuit for an across-the-line starter of Figure 5-16(a) with the addition of dynamic braking. A time relay *TR* with three contacts, a dynamic-brake contactor (*DB*) with four normally open contacts

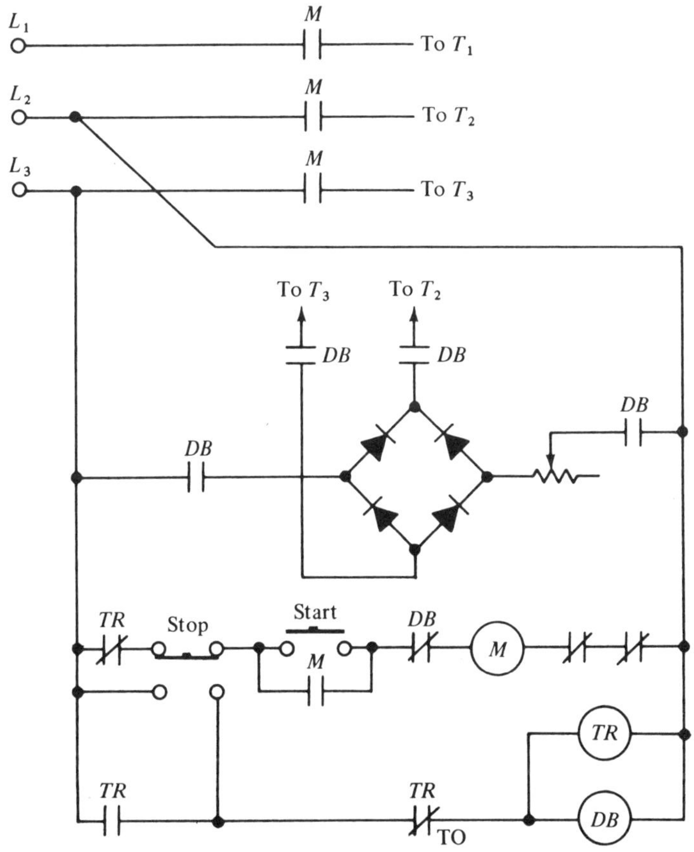

FIGURE 5-26. Schematic diagram of control circuit for dynamic braking of a three-phase induction motor.

and one normally closed contact, a bridge circuit of four diodes, and a variable resistor have been added. The stop button is replaced by one with the bottom contacts normally open. The motor is stopped in the conventional way by pressing the stop button. Contactor *M* is de-energized, relay *TR* and contactor *DB* are energized through the bottom contacts of the stop button and normally closed *TR* contact. Normally closed contact *TR* opens with a time delay when the *TR* coil is energized. Normally open *TR* contact seals the lower contacts of the stop button. The four normally open *DB* contacts close immediately when the stop button is pressed. With the *DB* contacts closed, full-wave dc is placed across two terminals, T_2 and T_3, of the motor.

This causes a stationary field on the stator, and the rotor by generator action stops quickly. The normally closed *TR* contact is timed so that after the motor has stopped, the contact opens, the *TR* coil and the *DB* coil are de-energized, and the motor is ready for start again.

Plugging is braking a motor by reversing the electrical connections so that a counter torque is developed. It is disconnected just before it rotates in the opposite direction. The schematic diagram of Figure 5-16(a) is revised as shown in Figure 5-27 by adding a contactor *R* with three contacts and a nor-

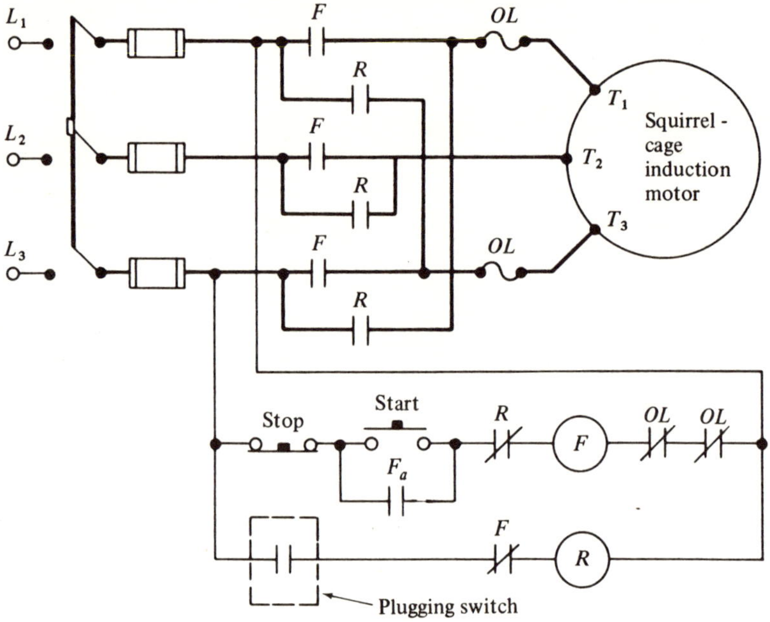

FIGURE 5-27. Schematic diagram of a plugging control for quick stopping of a three-phase squirrel-cage induction motor for operation in one direction.

mally closed auxiliary contact. A plugging switch is attached to the shaft of motor. This is the switch that closes when the shaft rotates and opens when rotation stops. In the schematic diagram the *M* contactor of Figure 5-16(a) is now labeled *F*. A normally closed auxiliary contact *F* is added that is in series with the plugging contact and the *R* contactor coil. When the stop button is pressed, the *F* contactor is de-energized and the motor, instead of coasting to a stop, has its terminals reconnected to the source so that the motor will rotate in a reverse direction. This is done by the additional normally closed contactor *F* closing when the contactor is de-energized. Contactor *R* is energized through the plugging switch and the normally closed contact *F*. When *R* is energized, L_1 connects to T_3, L_3 connects to T_1, and L_2 is again connected to T_2. When the rotation has stopped, the plugging switch opens disconnecting the terminals so that the motor does not start to rotate in a reverse direction. The action of changing two leads to the motor is to reverse the direction of rotation, and by making this connection while the motor is rotating in the opposite direction causes rotation to stop very quickly. The auxiliary normally closed contacts *F* and *R* are interlocks, so the *F* contactor cannot be energized until *R* is de-energized. This means that the motor cannot be started with the start button until the motor is stopped. The *F* contact is an interlock so that when the motor is running, the plugging relay cannot energize the *R* contactor.

Regenerative braking is not as practical a method for induction motors as for dc motors because the rotor of an induction motor must rotate at above synchronous speed for generator action. On hoists, elevators, and lifting equipment it may be used but because of the speed requirement, it is not practical and therefore is not used often.

Another term connected with controlling motor operation is *jogging*, which is the quick repeated energizing of a motor so that only small movements occur in a driven machine. Figure 5-28 shows the schematic diagram of the control circuit for an *across-the-line starter* with a jog button. A jog relay is

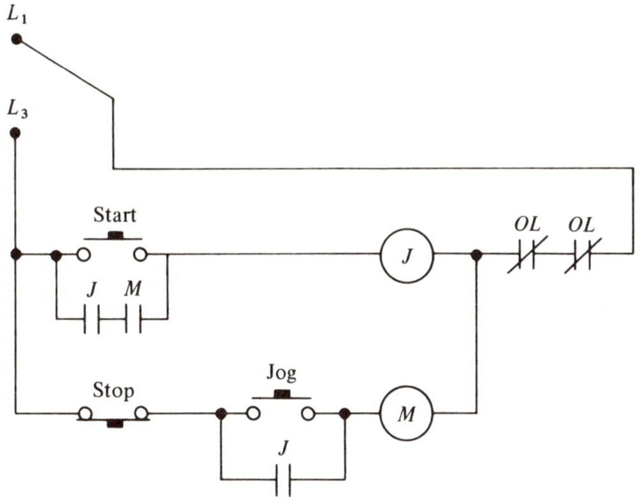

FIGURE 5-28. Schematic diagram for control circuit with jog button for an across-the-line starter (Figure 5-16(a)).

also added. To jog the motor the jog button is pressed quickly by taps that energize the contactor *M*. Pressing the jog button with the start button when the motor has been started has no effect, just as pressing the start button has no effect when a motor is running.

5-13 SINGLE-PHASE AC MOTORS[1],[2],[8],[13]

The single-phase ac motor is the basic electric motor found in homes, offices, farms, stores, industries, schools, and, in fact, anyplace where electric motors are used. This group covers power ratings from 1 mhp to 25 hp. They may be classified as fractional horsepower or integral horsepower motors. A *fractional horsepower* motor is defined by NEMA standards as a motor built in a frame smaller than the frame of a 1 hp (continuous rating) of open construction and a speed rating of 1700 to 1800 rpm. An *integral horsepower* motor isone built in a frame having an open continuous rating of 1 hp at 1700 to 1800 rpm or larger, but no larger than 25 hp. Thus a $\frac{3}{4}$-hp motor at a speed of 1800 rpm is a fractional horsepower motor but a $\frac{3}{4}$-hp motor at a speed of 900 rpm would be an integral horsepower motor because at 1800 rpm its power rating would be $1\frac{1}{2}$ hp. A $1\frac{1}{2}$-hp motor at 3600 rpm is classified as a fractional horsepower motor.

Fractional horsepower motors may be design N or design O, both to withstand full starting voltage. The locked-rotor current is limited to values shown in the NEMA standard, MG1-12.32. Integral horsepower motors may be designs L and M and should develop breakdown torque as shown in NEMA standard, MG1-10.33, and locked-rotor current not to exceed values in NEMA standard MG1-12.33.

Aside from the design or size classifications mentioned, a single-phase motor may be any one of the following types by principle of operation: (a) squirrel-cage induction; (b) wound rotor; (c) Series (dc or ac); or (d) Synchronous.

The rotor of the single-phase induction motor is identical to that of a polyphase induction motor. It has been shown before that a stator properly wound for a polyphase source has a revolving field at synchronous speed. A single-phase stator winding does not produce a revolving field. This is necessary to start the squirrel-cage rotor to turn. After the rotor is turning, a single-phase stator winding with the rotor current carrying conductors produces a rotating magnetic field. Figure 5-29 shows a two-pole wound stator with the rotor at standstill. With the currents in the stator as marked and

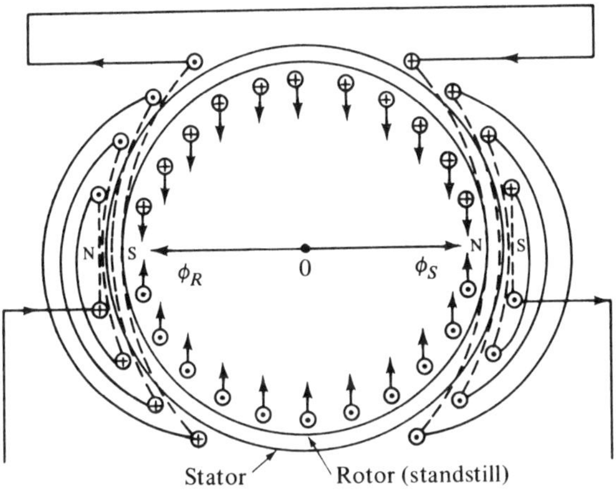

FIGURE 5-29. Cross-section diagram showing rotor and stator flux for a single-phase induction motor at standstill showing no rotational torque.

increasing sinusoidally, by Lenz's law and neglecting any reactance the current flow in the short-circuited rotor bars is as shown. With the application of the left-hand motor rule, a force is produced in each conductor as shown by the small arrows. No net torque is produced since the torque of each conductor is canceled out. The effect is that all the conductors tend to compress the rotor. The other half cycle is similar with polarity of the field and all current directions and forces reversed. In Figure 5-29 the stator coils are radial windings as illustrated in Figure 4-5(a) and 4-5(b) of the preceding chapter.

If in some way the rotor is made to turn, a rotating flux is produced. In Figure 5-30(a) the rotor from Figure 5-29 is rotating in a counterclockwise direction. The polarities of the induced voltage in the rotor bars are indicated outside the bars. The induced rotor voltage lags the stator flux by approximately 90 deg. When the stator flux ϕ_s is maximum, the induced EMF is zero and when ϕ_s is zero, the rotor flux ϕ_R is maximum but approximately 90 deg from ϕ_s. The stator flux ϕ_s varies sinusoidally on the horizontal axis and the rotor flux ϕ_R also varies sinusoidally on the vertical axis but lags by approximately 90 deg or 1/240 sec for a 60-Hz frequency. Figure 5-30(b) shows the relationship of the stator and rotor flux. The net flux is shown by the rotating phasor. If both flux values were equal, the net rotating flux would be constant. Figure 5-30(c) shows the total flux ϕ_T in five positions for a

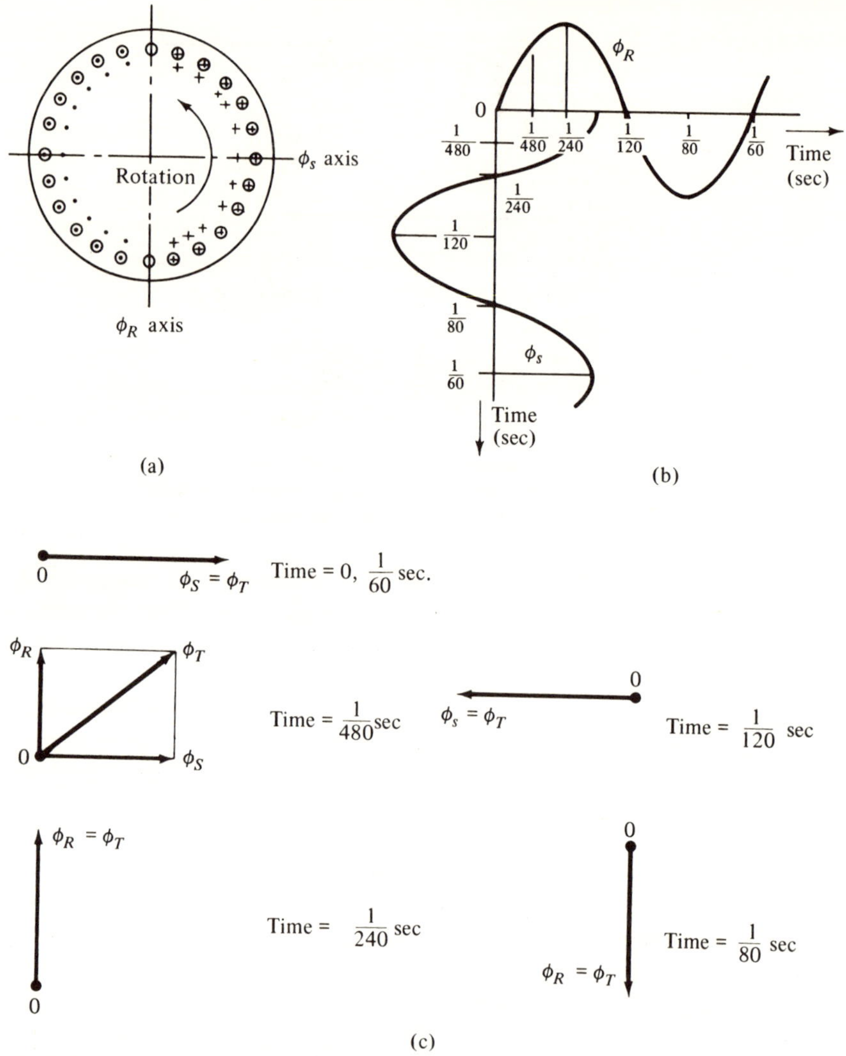

FIGURE 5-30. (a) Rotor of Figure 5-29 rotating, showing the rotor and stator magnetic axes. (b) Showing the sine-wave magnetic-flux variation on the rotor axis and stator axis. (c) Phasor-flux relationships for Figure 5-30(b) showing how the total flux phasor rotates when the rotor is rotating.

60-Hz frequency. The times are 0, 1/480, 1/240, 1/120, 1/80, and 1/60 (same as 0) sec. The total flux ϕ_T has made one complete revolution in a counterclockwise direction.

The squirrel-cage type is broken down into the following types: (a) split phase, which is further divided as resistance start, capacitor start, one-value capacitor, and the two-value capacitor; (b) shaded pole; and (c) reluctance start.

The split-phase type has two windings that are spaced 90 electrical deg apart. In the resistance-start type,[7] the main winding is connected in parallel with the auxiliary winding with a centrifugal switch. The switch is a mechanical device that simply opens when the rotating speed gets up to about 75 percent of the synchronous speed. The auxiliary winding is used only during the start, so that the motor runs on the main winding in normal operation. Figure 5-31(a) shows the schematic diagram and Figure 5-31(b) shows the

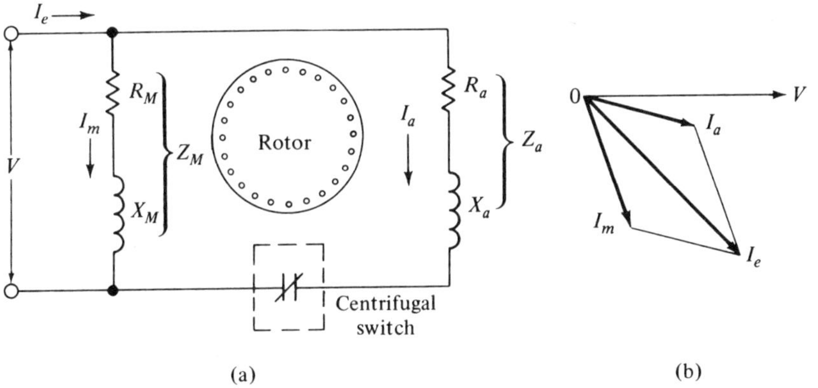

FIGURE 5-31. (a) Schematic diagram of a single-phase standard split-phase induction motor. (b) Vector diagram showing phase relationships of the currents in a standard split-phase induction motor.

phasor diagram. The main winding has low resistance and relatively high reactance, while the auxiliary winding has a relatively high resistance. The currents and flux of each winding are out of phase by about 70 deg as shown in Figure 5-31(b).

The *capacitor-start* motor has a capacitor added in the auxiliary winding; with the auxiliary winding similar to the main winding. With the proper amount of capacitance the value of I_a can be made to lead the impressed

[7] Commonly called the *standard split-phase motor.*

voltage V; the angle between I_a and I_m is nearly 90 deg. The net effect is a true and uniform rotating flux.

The *one-value capacitor* or *permanent split-phase* type uses a capacitor of a fixed value and no centrifugal switch. In the capacitor-start motor the capacitor must be relatively large in rating compared to its needs for running. A $\frac{1}{2}$-hp motor would need a capacitor of about 250 μF. For ordinary full-load running the value might be about 20 μF. Therefore the one-value type motor would use a capacitor of about 15 to 35 μF. Also, since the starting capacitor is used a very short time, the less costly electrolytic type can be used; however, for continuous use an oil type, which is more expensive, must be used.

The *two-value split-phase* motor uses a centrifugal switch and by placing this switch in series with an electrolytic-type capacitor and this in parallel with a low-value oil-filled type capacitor, the dual or two-value capacitance is attained. For the $\frac{1}{2}$-hp motor the small-value capacitor could be 25 μF, while the high-value electrolytic one could be 225 μF. In an autotransformer across a low-value capacitor and with a double-contact centrifugal switch, the starting capacitance is increased for starting and the effective value of the capacitance is increased by the square of the transformer voltage ratio.

Reversing the direction of rotation of any of the split-phase types above is accomplished by the reversing of the auxiliary winding for starting. The direction of rotation in starting is always from the auxiliary winding to the main winding. A switching arrangement is needed; the winding ends must be brought out from the motor and this is easily done with a double-pole double-throw switch.

The shaded-pole motor works on the principle of rotating flux. The poles of the motor are of the salient-pole type with a small portion slotted and a shading coil, usually a band of copper, around this part, called the *shaded part*. Figure 5-32 shows a pole face. The three views show the flux distribution of flux as the flux follows the sinusoidal current of the pole winding. For a 60-Hz frequency, when the current is increasing from zero to 1/360 sec, the change of flux is quite steep and by Lenz's law, current will flow through the shading coil in such a direction as to oppose flux being set up within the shaded part. The flux is concentrated in the unshaded part as shown in Figure 5-32(a). From (b) to (c), or from 1/360 sec to 1/180 sec, the flux change is slight and by Lenz's law the change of flux in the shaded part is nil so that no current flows in the shading coil. The flux opposition is nil and the flux distributes itself uniformly in the pole as shown in view (b) of Figure 5-32. From (c) to (d), or from 1/180 sec to 1/120 sec, the flux decreases and the change is negative. Current flows in the shading coil to oppose the decrease in flux. Now most of the flux is concentrated inside the shaded coil as shown in view (c) of Figure 5-32. The net result, therefore, is to cause the flux to sweep across the pole face. The sweep is not a "jerking" motion as might be indicated from the three views but a smooth sweep, although not of uniform flux value. Motors using pole construction are of the fractional

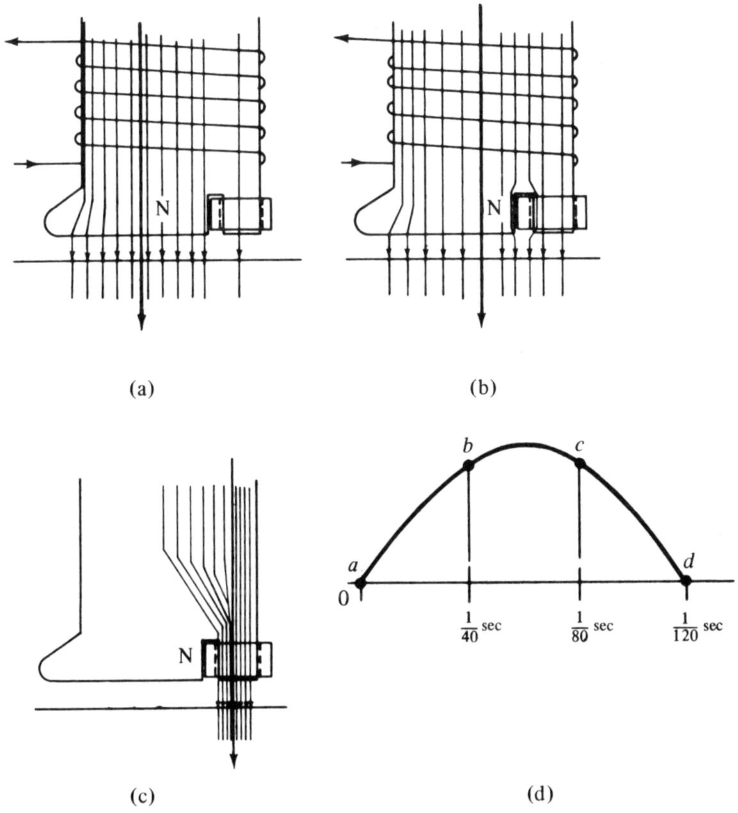

FIGURE 5-32. Magnetic air-gap flux distribution for a shaded pole for one alteration of a sine-wave current.

horsepower type, usually less than $\frac{1}{3}$ hp. Motors of this type have low starting torque and low efficiencies. Speed can be varied by voltage adjustment. They are used for small fans and blowers, phonograph motors, clocks, and other appliances that require low starting torque. Such poles are also used in small synchronous types such as clocks. The direction of rotation cannot be readily changed, with rotation always from the unshaded part to the shaded part of the pole. Some pole construction shades both ends of the pole face; with a double-throw switch either end is shaded and hence direction of rotation changed.

The *reluctance-start induction* motor also employs a salient-pole construction. About 30 to 40 percent of the pole face has a larger air gap. This larger air-gap part has a lower inductance than the small air-gap portion and this causes the magnetic flux to lag the current and crowd the flux to the larger air-gap portion of the pole face. The net effect is that the flux sweeps across the pole face from the larger air gap to the smaller air-gap portion. Starting torque and running torque are smaller than for a comparable shaded-pole motor. Speed can be varied by adjusting the impressed voltage, usually with a tapped autotransformer. The direction of rotation cannot be changed.

The *single-phase wound-rotor* group consists of the repulsion motor, the repulsion-start induction motor, and the repulsion-induction motor. The *repulsion motor* has a stator similar to the split-phase motor without the auxiliary starting winding. The rotor is constructed like the armature of a dc motor with the armature coil ends connected to a commutator. The brushes are short circuited and provided with a movable brush rigging. When the brushes are set on the neutral axis, no torque is developed since the armature field is directly opposed to the main field. When the brush axis is moved by turning the brush rigging, the armature field has a repelling force with the main field and since the armature can rotate, torque is produced. The angle of movement for the brush rigging is from 15 to 30 deg and somewhere between these values maximum torque is produced. The speed is dependent on the angle and the direction of rotation is in the same direction as the movement of the brush axis from the neutral axis of the main field. The direction of rotation is accomplished by the setting of the brush rigging. For clockwise rotation the brush axis is moved clockwise from the neutral axis of the field. The motor itself has high starting torque and may be as high as 350 percent of the full-load torque. Starting currents run about 350 percent of rated full-load current. This type of motor has been largely replaced by the split-phase capacitance-motor group. For special operations when normal speed adjustment is necessary with a mechanical lever adjustment, this type of motor is used. The speed regulation is similar to the dc series motor.

The *repulsion-start* motor uses the repulsion-motor principle for starting and when speed is about 75 percent of normal, the commutator is short circuited by a mechanical device and the motor then runs as a wound-rotor induction motor with a speed regulation of around 5 percent. This type of motor has the same starting advantages as the repulsion motor and the advantage of good induction-motor speed regulation. This type motor is costly to manufacture. They require periodical maintenance. They also are noisy in operation, and they have considerable arcing of the commutator that causes radio and television interference. The capacitor split-phase motor has largely replaced this type of motor except for those applications where lower starting currents and longer starting periods are needed to bring the motor up to speed. They are usually of the fractional horsepower rating.

The *repulsion-induction* motor has a wound rotor with a commutator and a squirrel-cage rotor combined. The short-circuiting device for the commutator also is not used. In starting, the characteristics are those of the repulsion motor. The speed characteristic is somewhat similar to the induction motor but with a larger regulation. This type of motor is used for machine-tool drives where polyphase voltage supplies are not available. In the starting process the reactance is high and the squirrel cage acts similarly to the high-resistance cage of the double-cage polyphase induction motor. Unlike the repulsion-start type, most motors are of the integral horsepower size. A feature of this motor is its ability to drive suddenly applied loads without stalling.

The third group, the *series-motor* group, consists of the universal motor and the large integral horsepower motor. The *universal* motor is a series motor that will operate on either dc or single-phase ac of no greater frequency than 60 Hz at the same speed and voltage. They are motors with a commutator whose armature and field are connected in series. Because of the speed characteristic of a series motor, the size is limited because at light load or no load the speed is high and mechanically integral horsepower motors could not bear up to the centrifugal forces. The size range is from 1 mhp to a maximum of $\frac{3}{4}$ hp. Another problem for integral horsepower motors is the sparking commutator with ac. Theoretically the field and armature current relationship is always such as to produce torque in the same direction for the negative half of the cycle; the reactance causes some phase shift, which causes sparking unless the brush axis is set back. In fractional horsepower sizes this factor is not too severe. For this reason the field turns are made fewer in universal motors. Most household appliances use the universal-type motor for mixers, sweepers, polishers, hair dryers, and carving knives. Hand tools such as drills, saws, sanders, planers also use the universal motor. Because of the high speed at light load, a speed-reducing gear train is used. This is especially true for drills and mixers.

The *large integral horsepower series* motor needs many design modifications, such as well-laminated field pole structures, few series field turns, more armature conductors, special commutating pole windings, and special interpole windings. This type of motor is rarely used for 60-Hz frequencies. The chief use is for railway systems at 25-Hz frequencies.

The fourth group is the *synchronous* group which is made up of hysteresis motors, reluctance motors, and synchronous-induction motors. All of these operate at a constant synchronous speed.

The *hysteresis* motor has its rotor made up of specially shaped, hardened steel disks. By the design of these disks permanent magnets are induced in the rotor and the rotor will follow the rotating field of the stator at constant synchronous speed. The field construction usually uses the shaded-pole construction. Motors of this type are of low rating and are used in electric clock movements.

The *reluctance* motor has a special squirrel-cage rotor on which some teeth are ground down. The stator must be of the split-phase or shaded-pole type to provide a revolving magnetic field. On start the rotor comes up to its speed just as an induction motor, but now the protruding teeth become magnetized and these protruding poles follow the rotating held of the stator just as the poles of the hysteresis motor or the three-phase synchronous motor. If the protruding sections are the same number as the stator poles, the rotor will rotate at the synchronous speed of the stator field. If, however, the number of the protrusions is some integral multiple of the stator poles, the speed is a submultiple of the stator synchronous speed. Such motors are called *sub-synchronous reluctance* motors. The regular synchronous motor will stall or stop at synchronous speed; but if the rotor torque requirement is too great, the rotor of the reluctance motor will fall out of step and rotate at a speed below synchronous just as will an induction motor.

The *synchronous-induction* motor is made to fit in between the reluctance motor, usually of fractional horsepower size, and the regular, doubly fed three-phase synchronous motor, usually in larger sizes above 50 hp. It is similar to the design of the polyphase squirrel-cage induction motor. The stator is identical but the rotor laminations are specially designed for magnetizing the protruding parts. The number of notches (and protrusions) must be the same as the number of poles of the stator winding. In operation the motor starts as an induction motor. When the rotor becomes magnetized with the formation of magnetic poles on the protrusions, the magnetized rotor will follow the rotating field of the stator and, like the regular synchronous motor, at synchronous speed. It will not stop at a given load but will continue to operate as an induction motor until breakdown is reached.

5-14 SUMMARY

The ac induction motor is the most widely used motor for converting electric energy to mechanical energy. This motor requires a primary, the stator, and a secondary, the rotor. Power is developed in the rotor by induced currents in the rotor conductors. They may be operated from polyphase (mostly three-phase) or single-phase ac supplies. A revolving field is essential for operation. A properly designed polyphase winding produces a revolving in the stator. For single phase a revolving field is also necessary, for starting at least. For single-phase induction motors this is usually provided by split-phase windings. The rotor or secondary may be either of the squirrel-cage winding or the conventional windings of coils similar to the stator winding. *Primary* and *secondary* are terms relating to the transformer. The primary is the winding that receives energy and the secondary, the winding that delivers energy. The stator receives the electric energy and the rotor or secondary delivers energy in the form of mechanical power to a mechanical load. The induction motor has an air gap in its magnetic circuit and the secondary winding is on a rotor. Unlike the transformer the secondary frequency is not the same as in the primary.

The magnetic field produced by the stator winding rotates at synchronous speed and the rotor rotates at a slower speed. Slip is the ratio of the difference between the synchronous and the actual speed to the synchronous speed. The speed of the rotor is given by Equation (5-1a),

$$\text{rpm}_{\text{rotor}} = \frac{120 f_1}{P}(1 - S).$$

Unlike a transformer the frequency of the secondary induced voltage is less than the primary frequency. This frequency of the secondary (rotor) is given by Equation (5-4),

$$f_2 = S f_1 \text{ Hz}.$$

The secondary induced voltage is also dependent on the slip and is given by Equation (5-3),

$$E_2 = S E_{BR} \text{ V}.$$

The power is developed in the rotor secondary and the power transferred across the air gap is also the rotor power input. The formula for the rotor power input RPI is given in Equation (5-7) as

$$RPI = \frac{S E_{BR}^2 R_2}{R_2{}^2 + S X_2{}^2} \text{ W}.$$

The power developed in the rotor RPI is given by Equation (5-10),

$$RPD = RPI(1 - S) \text{ W}.$$

The mechanical power is found by dividing the watts by 746, since 746 W is equal to 1 hp.

In the solution of problems for finding power output, the equivalent circuit is used in finding E_2 or E_{BR}. Figures 5-6(a), 5-6(b), and 5-6(c) show the equivalent rotor circuits. The rotor resistance R_2 and the equivalent load resistance is shown by the expression $R_2[(1 - S)/S]$. For the wound-rotor machine the voltage values and the rotor resistance may be found, but for squirrel-cage rotors the equivalent circuit shown in Figure 5-7 is applicable. Figures 5-8(a), 5-8(b), and 5-8(c) show equivalent circuits, Thevenin's, and an approximate equivalent circuit.

Maximum torque or the breakdown torque at which the motor will stall may be calculated by finding the value of slip at which maximum torque occurs. The formula for slip at maximum torque is given by Equation (5-12a),

$$S_{mt} = \frac{a^2 R_2}{R_1^2 + (X_1 + a^2 X_2)^2}.$$

The formula for the torque developed is given by Equations (5-11) and (5-11a),

$$T = 7.04 \frac{RPD_t}{\text{rpm}_{\text{rotor}}} = 7.04 \frac{RPI_t}{\text{rpm}_{\text{syn}}} \text{ lb-ft.}$$

Maximum torque is independent of the value of rotor resistance. Increasing rotor resistance increases the value of slip at which maximum torque occurs (see Figure 5-9). Increasing resistance also increases the starting torque.

Efficiency of induction motors is determined by tests. Small motors may be tested by direct measurement of input and output with a brake or dynamometer. Without a brake or dynamometer, three tests are needed as follows: (a) no-load test; (b) load test; and (c) stator-resistance measurement test. Large motors also require three tests as follows: (a) no-load test; (b) dc stator-resistance test; and (c) blocked-rotor test at reduced voltage. The losses are determined. Example 5-8 illustrates the calculations from the three tests.

The operating characteristics of speed, efficiency, power factor, input current, and slip versus power output are shown for a typical motor in Figure 5-12. Adding resistance in the rotor circuit changes the characteristics by increasing the slope of the speed and slip curves, and by lowering the efficiency curve and the power factor. Increased rotor resistance increases starting torque and decreases starting current values. The design of the slots changes the leakage reactance. Increasing leakage reactance reduces maximum torque and the starting current. The size of the air gap affects the air-gap reactance X_m. The shorter the air gap, the larger the air-gap reactance, which in turn reduces the magnetizing current and improves the load power factor.

Polyphase squirrel-cage motors are designated as design A, design B, design C, and design D. Design A is the most common squirrel-cage motor having a slip of less than 5 percent, a starting torque from 100 percent to 175 percent of rated torque, a breakdown torque of around 275 percent, and a starting current of 600 percent to 800 percent of normal full-load current. The torque-speed characteristics of typical designs are shown in Figure 5-13.

The double squirrel-cage motor is used for design C. The changes from design A are higher starting torque to about 250 percent of normal full-load torque, lower maximum torque to about 200 percent, and a reduction in starting current to less than 600 percent. In this design the high resistance bars are in the top of the slots while the low resistance bars are in the bottom of the slots. The effect is that on starting, due to the high reactance of the bottom bars, the rotor currents use the top bars because their impedance is much lower than the bottom bars. When running at normal speed, the reactance of the bottom bars is much less than at start because of the slip frequency of the rotor. The effect is that the impedance of the bottom cage is now considerably less than the upper high-resistance cage and most of the rotor current passes through the bars of the bottom cage.

In the wound-rotor motor, the rotor winding must be wound for the same number of poles as for the stator. By generally using the wye connection for this winding and bringing out the ends on three slip rings, variable resistances can be added in wye. The adding of resistance for starting increases the starting torque and reduces the starting current drawn. By cutting out resistance the speed is increased and when all resistances are cut out, the slip at full load should be about 5 percent. By the adding of resistance the speed is reduced and this method of speed control is used for hoists and elevators. Speeds at synchronous and above can be attained by more sophisticated schemes that introduce a foreign voltage into the rotor windings. The voltages must be at the same frequency as the rotor frequency. The LeBlanc system, the Kramer systems, and the Scherbius systems are schemes that will give a wide range of speeds above and below synchronous speed. Diagrams for these systems are shown in Figures 5-23, 5-24, and 5-25. The Schrage motor of Sweden, which is manufactured in the United States by the General Electric Company as the BTA brush-shifting motor uses the principle of the LeBlanc system. In this motor the primary winding is on the rotor and the secondary is on the stator. Three sets of brushes can be rotated or each set can be adjusted simultaneously to vary the voltage.

The concatenation method of speed control involves two induction motors with their shafts coupled. The motor connected to the source must be a wound-rotor motor with its rotor connected to the input of the second motor which can be either a squirrel-cage or a wound-rotor motor. The speed is less than the synchronous speed, which is given by Equation (5-24),

$$\text{rpm}_{\text{syn}} = \frac{120 f_1}{P_1 + P_2}.$$

Many polyphase induction motors can be started by direct connection to the supply. Such starters are called *across-the-line* starters. Figures 5-16(a) and 5-16(b) show schematic diagrams for such starters. The manual-autotransformer type for reduced voltage starting using autotransformers either in wye or open delta is shown by the schematic diagram of Figure 5-17 with two transformers in open delta. The primary resistor or reactor starters reduce the motor voltage by dropping voltage across the resistance or reactance. A manual type of resistance is shown in the schematic diagram of Figure 5-18(a). A push-button, one-step resistance type is shown in Figure 5-18(b). A wye-delta type is shown by the diagram of Figure 5-19. A part-winding type of starter is depicted by Figures 5-20(a) and (b).

The speed of an induction motor can be controlled by (a) controlling the frequency of the source; (b) controlling the number of poles; and (c) controlling the slip. The first method requires a controlled source such as used on a ship. The second method by changing the number of poles may be done by multiple windings or the consequent-pole method. In both schemes the winding ends must be brought out and the connections made either manually or automatically. The consequent-pole method is shown in Figures 5-21(a) and (b) for doubling the number of poles from four to eight which will cut the speed almost one-half. The third method by controlling the slip may be done by adding resistance in the rotor circuit, lowering the impressed voltage, and the introduction of a foreign voltage in the rotor circuit. The methods of adding resistance in the rotor circuit and of adding a foreign voltage were discussed under the wound-rotor induction motor.

Braking is accomplished by applying dc current to two terminals when the motor is disconnected from the source. This produces a stationary field and is known as *dynamic braking*. Figure 5-26 shows the schematic control circuit for doing this for a smaller motor. Larger motors would not operate directly with contactors in the control circuit. Plugging is another method for stopping a motor quickly. This method requires a plugging switch and a contactor to reverse two leads to the motor as shown in Figure 5-27. Regenerative braking is a method whereby the induction motor acts as an induction generator.

This method is little used because speeds above synchronous are necessary. Figure 5-28 is a schematic diagram for jogging a motor, which is the term used for having the motor turn very small amounts.

Single-phase ac motors are manufactured in sizes up to 25 hp. One classification uses the terms *fractional horsepower* or *integral horsepower*. This is a frame classification with a fractional horsepower motor being one whose frame is smaller than the frame of a 1-hp motor at a speed of 1700 to 1800 rpm and an integral horsepower motor as one with a frame larger by 1 hp at the same speeds. Fractional horsepower motors may be classified as design N or design O.

Aside from the above classification, they can be classified as squirrel-cage induction motors, which as the name suggests, have squirrel-cage rotors. Under this group are the resistance start or commonly called *standard split-phase motor*, the capacitor-start split-phase motor, the one-value capacitor split-phase motor, the two-value capacitor split-phase motor, the shaded-pole motor and the reluctance-start motor. A single-phase winding does not produce a rotating field, and thus a method, at least for starting, must be provided to produce a rotating field. The split-phase types have an auxiliary winding displaced 90 electrical deg from the main winding; by properly designed values of resistance and reactance the currents in the two windings connected in parallel are out of phase with the optimum being 90 deg. The windings are spiral windings placed in the stator slots for the nonsalient-pole type of stator. The resistance start has a comparatively high-resistance auxiliary winding. The capacitor start has capacitance in the auxiliary winding so that its current is more nearly 90 deg than in the resistance-start type. In these two types a centrifugal switch operating on a speed basis disconnects the auxiliary winding for running conditions. A rotating field is produced by the combined action of the stator and rotor flux for a single-phase stator winding. The shaded-pole motor and the reluctance-start motor use salient-pole stators. The shaded pole causes the flux to sweep across the pole face as the current varies sinusoidally in the pole winding. Figure 5-32 shows how this is accomplished according to Lenz's law by the shading coil. The reluctance-start motor has two varied air gaps that cause the flux to sweep from the larger air gap to the smaller air gap.

The single-phase ac wound-rotor type motor consists of the repulsion motor, repulsion-start motor, and repulsion-induction motor. The rotor has a winding with a commutator similar to the dc motor on the rotor. The repulsion motor uses short-circuited brushes with the brush axis displaced from the neutral axis of the field by 15 to 30 deg. The instantaneous field of the rotor reacts with the instantaneous field of the stator with a repelling action similar to bar magnets placed together with one free to rotate—the one free to rotate is repelled by the other because like poles repel each other. The repulsion-start motor uses the same type of winding except that for normal running the short-circuited brushes are lifted and the commutator bars are short circuited by a centrifugal lever device and the motor runs as an induction motor. The repulsion-induction motor has two rotor windings in the rotor, one of which is a wound winding similar to that of the repulsion-start motor and the other a squirrel-cage winding. On starting, the repulsion winding is in command and in running, both windings operate together as if a repulsion motor and a squirrel-cage motor were coupled together. All three types of this group have excellent starting torque and low starting currents.

The third group is the series-motor group which consists of the universal motor and the larger integral horsepower series motor. This group does not operate on the principle of induction; in this group the rotors receive their currents by conduction. They are identical in principle to the dc series motor and, in fact, the universal motor is defined as a motor that operates on ac or dc with the same characteristics. Due to the high speeds at no loads, the sizes are of the fractional horsepower rating. They find their greatest use for household appliances and small tools. The larger series motor is used at low frequencies to a maximum of 25 Hz. They find some use in electric railroad locomotives. They need careful design as well as compensating winding to reduce the sparking of the commutator.

The fourth group is the synchronous group consisting of the hysteresis, reluctance, and synchronous-induction motors. The hysteresis motor is used for clocks and timing devices. Its stator is most always of the shaded-pole type. The rotor is made of hardened steel disks, whose geometrical shape causes them to become permanently magnetized. The reluctance motor is a squirrel-cage rotor with some teeth lowered. This causes the raised parts of low reluctance to become magnetized. The poles then follow the opposite poles of the stator field at synchronous speed. The stator is of the split-phase type. They are made mostly in the fractional horsepower size. The synchronous-induction motor is similar to the reluctance motor, but the stator is wound for polyphase operation. The rotor, like the reluctance motor, is a squirrel-cage type; unlike the reluctance motor, however, the shaping of the rotor laminations are specially designed for the pole protrusions. This is necessary because this type motor is a fillin between the reluctance motor and the regular doubly fed synchronous motor. The power sizes run up to 50 hp and hence the rotor poles must be made up by higher flux densities than those of the reluctance motor.

Problems

1. The rated speed of a 50-hp, three-phase induction motor is 1145 rpm. The no-load speed is 1196 rpm. (a) What is the percent slip and (b) what is the speed regulation?
2. The slip of a $\frac{1}{2}$-hp, 230-V single-phase split-phase induction motor is 4.17 percent at rated speed for 60 Hz of 3450 rpm. (a) What is the synchronous speed? (b) How many poles does the motor have? (c) What is the no-load speed if the speed regulation is 4.06 percent? (d) What is the slip at no load?
3. A ten-pole, three-phase, 60-Hz induction motor has a slip at full load of 4.86 percent. (a) What is the speed of the rotating field with respect to the stator? (b) What is the speed of the rotor relative to the revolving field? What is the speed of the rotor with respect to the stator?
4. A 50-Hz standard split-phase motor has a no-load speed of 745 rpm and rated-load speed regulation of 4.2 percent. (a) What is the full-load speed? (b) What is the slip at full load? (c) How many poles does the stator have? (d) What is the speed of the rotor with respect to the rotating field?
5. A 230-V, 60-Hz, three-phase wound-rotor induction motor has eight poles and operates at a speed of 856 rpm. Both the stator and the rotor are wye connected. The rotor has 50 percent the number of stator conductors. What is the blocked-rotor voltage per phase? (b) What is the operating generated voltage? (Neglect primary and secondary impedance drops.) (c) What is the operating rotor frequency?
6. A three-phase, six-pole, 60-Hz, 460-V wound-rotor induction motor has a delta-connected stator and a wye-connected rotor. The rotor has 65 percent as many rotor conductors as the stator. (a) What is the blocked-rotor voltage per phase and the rotor frequency? (b) What is the running-rotor voltage per phase if the full-load speed is 1140 rpm? (c) What is the rotor voltage per phase if the motor is driven in a direction opposite that of the stator field at a speed of 750 rpm? (Assume that there are no impedance drops.)
7. In Problem 6 the rotor resistance per phase is 0.12 ohm and the rotor reactance is 0.50 ohm per phase. (a) Calculate the value of starting current. (b) Calculate the current when operating at 1140 rpm. (c) Calculate the starting torque. (d) Calculate the power developed at 1140 rpm. (e) Calculate the torque developed at operating conditions.
8. In the preceding problems (6 and 7), a resistance of 0.38 ohm/phase is

added to the rotor circuit. (a) Calculate the value of starting current. (b) Calculate the current when operating at 1140 rpm. (c) Calculate the starting torque. (d) Calculate the power developed at 1140 rpm. (e) Calculate the torque developed at load conditions.

9. A three-phase, 230-V, 60-Hz, four-pole induction motor is rated at $7\frac{1}{2}$ hp. The following values per phase are given for the wye-wye motor: $R_1 = 0.420$ ohm; $R_2 = 0.140$ ohm; $X_1 = 0.526$ ohm; $X_2 = 0.223$ ohm; $X_M = 14.5$ ohms; stator to rotor ratio is 100/65. For a speed of 1720 rpm calculate the following: (a) starting current, (b) starting torque, (c) the operating current, (d) power developed, (e) torque at operating speed of 1720 rpm.
10. In Problem 9, calculate items (c), (d), and (e) for an operating speed of 1710 rpm.
11. By use of the approximate circuit (Figure 5-8c) solve Problem 9 for the same values at the speed of 1720 rpm.
12. For the motor of Problem 9 at what speed is maximum torque developed and what is the value of the maximum torque?
13. In Problems 9 and 12 what value of resistance must be added per phase in the rotor circuit so that the starting torque is equal to the maximum torque? Also, what is the value of starting current?
14. A series of tests were made on a 230-V, 60-Hz, three-phase squirrel-cage induction motor with a dynamometer whose constant, $(K) = (33{,}000)/(2\pi d) = 5000$ where $d = 12.6$ in. The tare for the dynamometer was compensated and was zero. The input was measured with the two-wattmeter method. The following is the data that was recorded:

Reading	E_{line} (volts)	I_{line} (amperes)	P_1 (watts)	P_2 (watts)	Force (pounds)	Speed (rpm)
1	230	15.0	2020	3440	18.25	1700
2	230	12.5	1500	2880	14.55	1720
3	230	10.1	950	2310	10.80	1740
4	230	7.8	510	1750	7.40	1755
5	230	6.1	65	1245	3.65	1780
6	230	5.5	−260	940	1.20	1792
7	230	5.4	−435	795	0.00	1795

For each reading calculate (a) the horsepower output, (b) the power factor, (c) the efficiency, (d) the torque, (e) the slip, (f) the speed regulation for rated output, and (g) the plot characteristic curves.

15. The above machine was tested at no load (disconnected from dynamometer) as follows: line voltage, 230 V; I_{line}, 5.3 A; P_1, -310 W; P_2, -550 W. Use reading 2 of the above problem without dynamometer readings for the load test. The average dc resistance test yielded a value of 0.96 ohm. Calculate from these three tests (a) the horsepower output, (b) the efficiency, (c) the torque, (d) the slip, and (e) the power factor at load value.
16. A 10-hp, 230-V, 60-Hz, eight-pole squirrel-cage induction motor was tested as follows:

    no-load test: $E_T = 230$ V, $I_{NL} = 17.9$ A, $P_1 + P_2 = 782$ W.
    resistance: Average dc resistance between terminals was 0.287 ohm.
    load test: $E_T = 230$ V, $I = 28.5$ A; $P_1 + P_2 = 8760$ W; speed $= 870$ rpm (all tests made with motor that had attained its temperature rise).

    Find (a) the horsepower output, (b) the efficiency, (c) the torque, and (d) the efficiency.
17. A blocked-rotor test was made on the motor of Problem 16 at reduced voltage at 15 Hz as follows: $E_{BR} = 21$ V, $I_{BR} = 28$ A, $P_1 + P_2 = 876$ W. By the locked-rotor method find (a) the horsepower output, (b) the load torque, (c) the efficiency, and (d) the power factor. (Make calculations at the same speed as in Problem 16, 870 rpm.) (Use design B for the reactance calculation.)
18. A 50-hp, 440-V, 60-Hz, three-phase squirrel-cage induction motor was tested as follows:

    no-load test: $E_T = 440$ V, 60 Nz, $P_1 + P_2 = 2190$ W, $I_{NL} = 27.4$ A; speed $= 1198$ rpm.
    resistance test: Average resistance (dc) between terminals 0.240 ohm.
    blocked-rotor test: $E_{BR} = 32.6$ V at 15 Nz, $I_{BR} = 65$ A, total power $(P_1 + P_2) = 3020$ W.

    All tests were made on motor at ambient temperature of 25°C. Calculate resistances at 75°C. Use a rated speed of 1160 rpm. Find (a) the power output, (b) the torque at rated speed, (c) the efficiency at rated speed, and (d) the power factor at rated speed. (Use design A.)
19. For Problem 17, calculate (a) the starting current, (b) the starting torque, (c) the speed at which maximum torque occurs, and (d) the maximum torque.
20. For Problem 18, calculate (a) the starting current, (b) the starting torque, (c) the speed at maximum torque, and (d) the maximum torque.
21. A $1\frac{1}{2}$-hp, 230-V, 60-Hz, single-phase capacitor-start motor was tested

with a dynamometer whose constant $K = 6000$, ($d = 10.5$ in.) as follows:

rated load: $I = 8.4$ A, $V = 230$ V, $F = 5.25$ lb, $P = 1120$ W, speed = 1170 rpm.
breakdown: $I = 15.9$ A, $V = 230$ V, $F = 12.60$ lb, $P = 2015$ W, speed = 1470 rpm.
blocked rotor: $I = 8.7$ A, $V = 40$ V, $P = 165$ W, $F = 17.8$ lb.

Find (a) the rated load conditions: horsepower output, efficiency, torque, and power factor; (b) the maximum torque conditions: horsepower output, efficiency, torque and percent of rated, and power factor; (c) the locked rotor: starting current, percent starting current, starting torque, percent starting torque, and power factor.

22. A 60-hp, 220-V, 60-Hz, three-phase, 875-rpm squirrel-cage induction motor is started with an open-delta autotransformer starting compensator that has 80 percent, 65 percent, and 50 percent taps. Current at full load and rated voltage is 155 A. Locked-rotor current is 870 A. If the motor is started at rated voltage, the starting torque is 125 percent of full-load torque. (a) For the 80-percent tap connection calculate the motor current, line current, and torque for starting conditions. Also give values in percentages of rated. (b) Repeat part (a) for the 65 percent tap connection. (c) Repeat part (a) for the 50 percent tap connection.
23. Design a line-resistor starter for the motor of Problem 22 for (a) 80-percent voltage, (b) 65-percent voltage, and (c) 50-percent voltage. Assume that the blocked-rotor power factor at rated frequency is 0.45 lag. Resistor power ratings to be 10 percent of steady state rating.
24. Design reactors for a line-reactor starter for the motor of Problem 22 for (a) 80-percent motor voltage, (b) 65-percent motor voltage, and (c) 50-percent motor voltage. (Neglect the reactor resistance.)
25. If the preceding motor is designed for a delta-connected stator for rated operation, and proper leads are brought out for the wye-delta method of starting, calculate (a) the starting current and percentage of rated load current; (b) the starting torque in pound-feet and percentage of full-load torque; and (c) show schematic diagram for this starter for push-button control.
26. A three-phase, ten-pole, 60-Hz, wound-rotor induction motor and a four-pole, three-phase squirrel-cage induction motor are to be connected in concatenation. The electrical connection between rotor of machine 1 and stator of machine 2 provides for reversing two leads. Each motor is

also able to be driven independently. (a) What synchronous speeds are possible? (b) Show schematic diagram for reversing the squirrel-cage motor stator.

27. The across-the-line starter circuit of Figure 5-27 is to be modified to provide three start-stop stations so that all three start buttons must be pressed to start the motor and any stop button can stop the motor. Show a schematic diagram for accomplishing this objective.
28. The schematic diagram as shown in Figure 5-18(b) often needs dc contactors for the *S* and *R* contactors shown. Show a schematic diagram using ac relays and dc contactors with diodes for providing the dc current.
29. Design a schematic diagram to operate a compensator with push-button operation for the manual compensator as shown in Figure 5-17.
30. Design an automatic across-the-line starter with dynamic braking for a single-phase motor to be operated from either of two push-button stations.

References

[1] SISKIND, C. S., *Electrical Machines—Direct and Alternating Current*, 2nd ed. New York: McGraw-Hill, 1959.

[2] *Motors and Generators Standards*, MG1-1967. New York: National Electrical Manufacturers Association.

[3] DAWES, C. L., *Electrical Engineering*, Vol. II, *Alternating Currents*, 4th ed. New York: McGraw-Hill, 1947.

[4] CARR, C. C., *Electrical Machinery*. New York: Wiley, 1958.

[5] FITZGERALD, A. E. and C. KINGSLEY, JR., *Electrical Machinery*, 2nd ed. New York: McGraw-Hill, 1961.

[6] FITZGERALD, A. E. and D. E. HIGGINBOTHAM, *Basic Electrical Engineering*, 2nd ed. New York: McGraw-Hill, 1957.

[7] *Test Procedures for Polyphase Induction Motors and Generators*, IEEE No. 112A. New York: Institute of Electrical and Electronics Engineers, September. 1964.

[8] KOSOW, I. L., *Electrical Machinery and Control*. Englewood Cliffs, N.J.: Prentice-Hall, 1964.

[9] ROSENBLATT, J. and M. H. FRIEDMAN, *Direct and Alternating Current Machinery*. New York: McGraw-Hill, 1959.

[10] HARWOOD, P. B., *Control of Electric Motors*, 3rd ed. New York: Wiley, 1952.

[11] SISKIND, C. S., *Electrical Control Systems in Industry*. New York: McGraw-Hill, 1963.

[12] JAMES, H. D. and L. E. MARKLE, *Controllers for Electric Motors*, 2nd ed. New York: McGraw-Hill, 1952.

[13] SISKIND, C. S., *Induction Motors, Single-Phase and Polyphase*. New York: McGraw-Hill, 1958.

Direct-Current Machines as Amplifiers

6

The generated voltage of a dc generator varies directly as the field magnetic flux per pole and the speed at which the armature turns. If the speed is kept constant and the field flux is varied, the generated voltage is proportional to the flux per pole. Special generators designed on this principle are manufactured to control motors and generators, both ac and dc. Such machines may be called *rotating amplifiers*, *rotary power amplifiers*, *metadynes*, or *control generators*. The types that are studied in this chapter are Rototrol (Westinghouse), Regulex (Allis-Chalmers), and Amplidyne (General Electric). Other rotating amplifiers are VSA drive (Reliance), Magnavolt (English Electric), and Magnicon (McFarlane Engineering Company, Scotland).

6-1 DIRECT-CURRENT GENERATOR AMPLIFIERS[1]–[4]

The self-excited and the separately excited dc generators can be thought of as amplifiers. When a self-excited generator is used as an exciter for an alternator field, or a separately excited generator is used in the Ward Leonard speed control as was shown in Figure 3-12, each acts as an amplifier. In each case the alternator exciter or the control generator is a 250-V, 2.5-kW generator whose shunt field resistance is 475 ohms. If with a field excitation of 0.45 A the output to the alternator field or the motor armature is 9 A at 225 V, the output power is (9 $\times$ 225) 2025 W while the input to the field is 96.0 W. If the exciter current is raised to 10 A with a terminal voltage of 250 V, the power output of the generator is 2500 W and the field input power is 118.7 W. This represents a power amplification of (2500 − 2025)/(118.7 − 96.0) or 16.5. In each case the dc generator acts as a control rotary amplifier—in the case of the alternator field to control the ac voltage of the alternator output; in the case of the Ward Leonard system, to control the speed of the motor.

In both of the above cases there is no control to keep the alternator voltage at its set value or the speed of the motor at its set speed. To do this and other functions such as control of power output, torque, current, power factor, position, and many others, the basic generator can be made to control the functions at precise values, such as keeping the alternator voltage at 2300 V or the speed of the motor at 1800 rpm. In doing this the control generator must have higher power amplification than the generator discussed above. In the majority of cases the control is accomplished by feedback and therefore the feedback signal must be of low power value. Another requirement is that the response time for the corrective action must be quick. These dc generator amplifiers all operate by adding separately excited fields to the poles that are operated by feedback and control voltages and, as mentioned, they must consume very little power.

These special dc amplifiers or *rotary amplifiers* are of two types. One type may be called the *self-excited* type with control and feedback windings on the same structure as the self-energized windings. Except for the extra windings, this generator is quite similar to the conventional dc generator. The Rototrol and the Regulex generators, which are discussed in the next section, are of this group. The other type might be called a *separately excited* type since the primary excitation is a separately excited winding and is referred to as the control winding. The output makes use of the cross magnetism or the armature flux for its output. The Amplidyne and the Metadyne generators are of this armature-excited type. This type has two sets of brushes that are spaced 90 electrical deg from each other. One set is short circuited with the load taken off the quadrature set. The Amplidyne is explained in a later section.

6-2 ROTOTROL AND REGULEX GENERATORS[1],[2],[4],[5]

Rototrol and Regulex generators are self-excited and this field is referred to as the *self-energized* field (SEF). The separately excited fields are referred to as the *control* fields. Figures 6-1(a) and (b) show the schematic diagram and the saturation curve, respectively, for a typical Rototrol generator with a series self-energized field. Figures 6-2(a) and (b) show the schematic diagram and saturation curve for a typical Regulex or Rototrol generator with a shunt self-energized field. The field rheostats are in series with the field and they are referred to as *tuning rheostats* (R_T). In either generator the tuning rheostat is adjusted or tuned to a value of resistance, so that the slope of the field-circuit

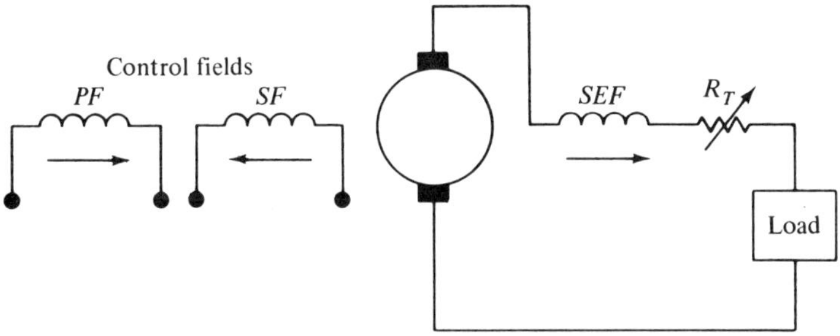

(a)

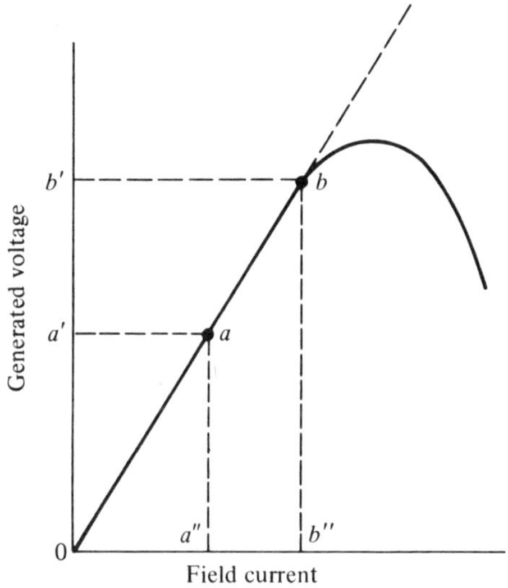

(b)

FIGURE 6-1. **(a) Schematic diagram of a Rototrol dc control generator. (b) Saturation curve for a Rototrol dc control generator.**

Terms: ***PF*—pattern field or reference field.**

***SF*—signal field or pilot field or differential winding or feed-back winding.**

***SEF*—self-energizing field or self-excited winding.**

R_T—tuning resistor.

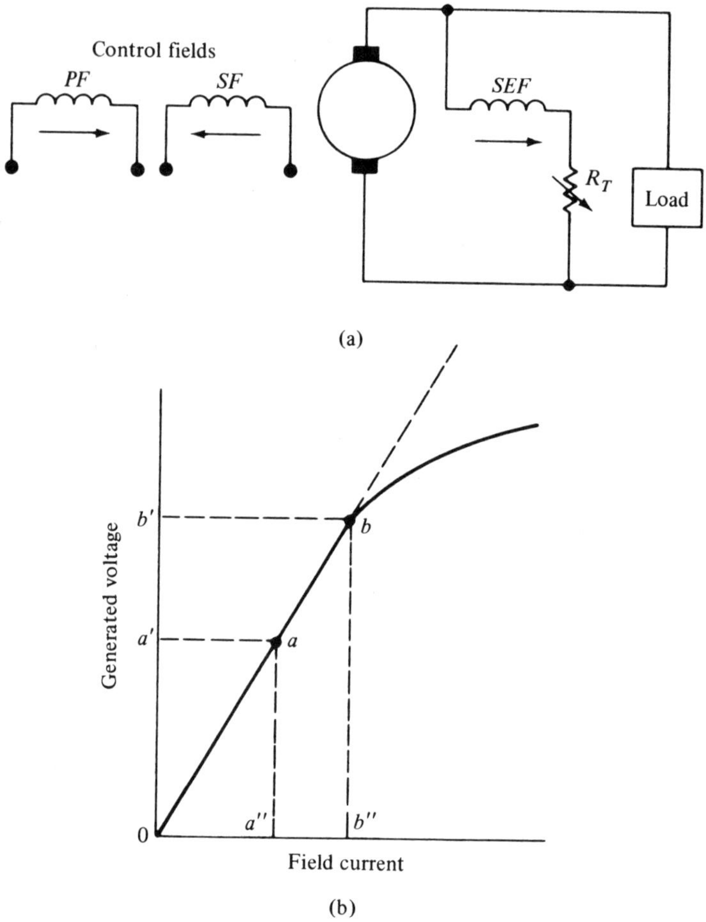

FIGURE 6-2. (a) Schematic diagram of a Regulex dc control generator. (b) Saturation curve for a Regulex dc control generator.

resistance line falls on the straight-line portion of the saturation curve or the air-gap line. The control fields, at least two in number, are adjusted so that the generated voltage falls in the upper half section of the curve, between *a* and *b*, on the air-gap line.

These two types of generators have been almost entirely supplanted by magnetic amplifiers and solid-state operational amplifier feedback control systems using thyristors where the level of power require it.

Figure 6-3 shows a schematic diagram that uses a Rototrol exciter to control the output voltage of a dc compound generator. One of the control fields, referred to as the *pilot* field or *signal* field (SF), is connected across the output terminals of the generator. The feedback from the generator must be small, a requirement for good control practice. By adjusting the voltage

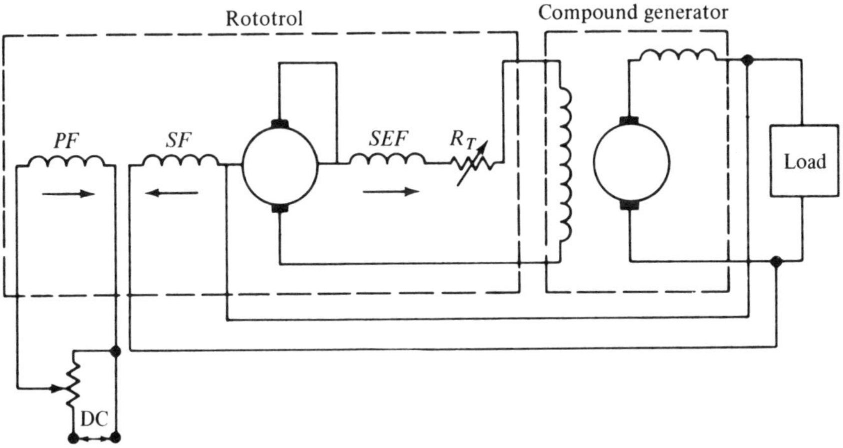

FIGURE 6-3. Schematic diagram for using a Rototrol generator for controlling the output voltage for a dc compound generator.

from a constant voltage dc supply on the other control field, which is called the reference field or pattern field (PF), the value of the desired load voltage is set to the desired value. In the diagram the directions of the magnetomotive forces are indicated. Those of the pattern field and the self-energized field are in the same direction, while that of the signal field is opposite. If the terminal voltage of the generator increases, the signal-field flux will increase due to the increased potential across the field winding. This flux subtracts from the combined pattern field and self-energized field flux. The net is decreased and the potential across the shunt field of the generator which is separately excited by the Rototrol is decreased and the output voltage is reduced. In turn the signal-field flux reduces so that the balance between the two control fields is again established. For a drop in terminal voltage the signal-field flux drops, which creates an increase of the excitation and voltage increase. This may be explained by use of the saturation curve. The generator is operating at some load that requires a voltage on the field at point *a*. The signal-field voltage drops due to increased load on the generator and requires a greater generated voltage and hence more excitation or a generated voltage from the Rototrol equal to *Ob′*. Temporarily the unbalance of the pattern field and the signal field means that the pattern field provides the added voltage required from the Rototrol exciter or the value $a'b'$. When the signal-winding flux again balances the pattern-field flux, the added voltage requirement is again supplied by the self-energized field.

The use of a Regulex generator for speed control is shown by Figure 6-4.

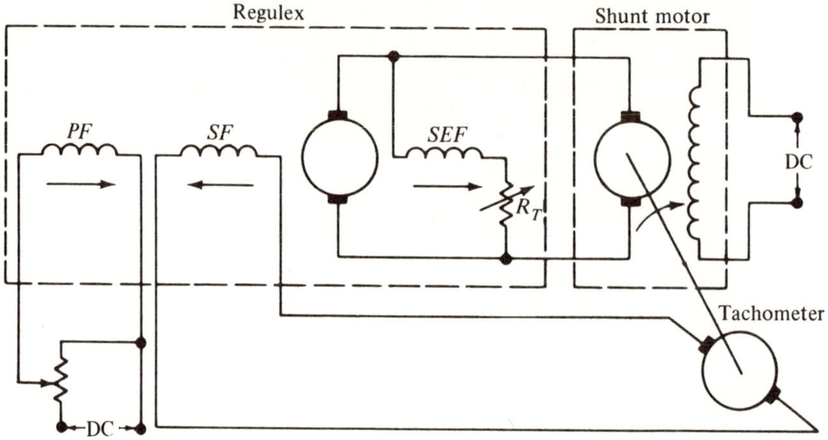

FIGURE 6-4. Schematic diagram for a Ward Leonard control system for a shunt motor using a Regulex generator for the control generator.

In the diagram the Regulex is the control generator for the controlled shunt motor. In high-powered motor applications the control generator controls the field of the supply generator as shown in Figure 6-6 for an Amplidyne application. Because the speed must be transduced into a voltage, the tachometer generator is attached to the motor axis and the voltage generated is proportional to the speed. When there is a decrease in speed, there is a decrease of the voltage on the signal field causing a drop in its flux. The net flux of the pattern field and the self-energized field is increased and the output voltage is increased. The increased voltage on the motor armature causes the speed to increase. After the speed is satisfied, the tachometer speed, also increased, again provides the increase necessary for the signal-field flux to balance the pattern-field flux with the self-energized field supplying the added flux. For an increase in speed there is a decrease in the net flux, a decrease in the armature voltage, and a drop in speed. The pattern field sets the value of speed to be maintained.

The power amplification of the Rototrol and the Regulex amplifiers is considerably higher than that of the straight generator with power-amplification factors in the 100 to 1000 range. The response time is medium quick. Both in response time and power amplification they are not as ideal as the Amplidyne. In construction, however, they are less expensive and in operation are less likely to hunting than the Amplidyne. The following example illustrates the amplification of the Rototrol and the Regulex.

Example 6-1. A Rototrol is used to control the output voltage of a 250-kVA alternator. The alternator field has a resistance of 25 ohms. The reference field of the Rototrol has a resistance of 200 ohms and with 200 mA through it, and the voltage across the alternator field is 250 V. (a) What is the power amplification? (b) What is the voltage amplification?

Solution:

(a) power output for above conditions: $\frac{V^2}{R} = \frac{(250)^2}{25} = 2500$ W,

power input to reference field: $I^2R = (0.20)^2(200) = 8$ W,

$$\text{power amplification} = \frac{\text{power out (rototrol)}}{\text{power-input reference field}} = \frac{2500}{8} = 312.5.$$

(b) voltage output = 250 V,
control-field voltage: $IR = (0.200) \times 200 = 40$ V,

$$\text{voltage amplification: } \frac{\text{voltage out (rototrol)}}{\text{voltage of reference field}} = \frac{250}{40} = 6.25.$$

For the parallel-field connection the calculations are the same.

6-3 THE AMPLIDYNE[1],[5]–[7]

The Amplidyne is a dc generator that has two sets of brushes in quadrature position electrically from each other. One set, which is on the neutral axis of the control field, is short circuited. The load is taken from the brushes that are on the quadrature axis or on the neutral axis of the armature flux or the cross-magnetizing flux. The Amplidyne is a two-stage amplifier unlike the Rototrol which is a one-stage amplifier. A two-machine counterpart would have the output of the first generator feed into the field of the second generator with the excitation of the first machine controlling the output of the second machine. In this amplifier the same armature is used while the excitation for the second stage is the cross-magnetization of the first. In smaller sizes up to $1\frac{1}{2}$ kW the driving motor is a part of the machine. In larger sizes the Amplidyne requires a separate motor drive. The amplification

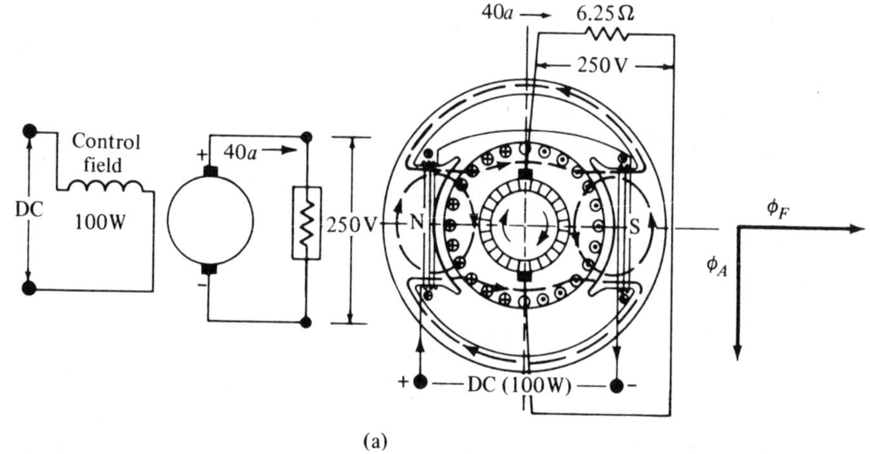

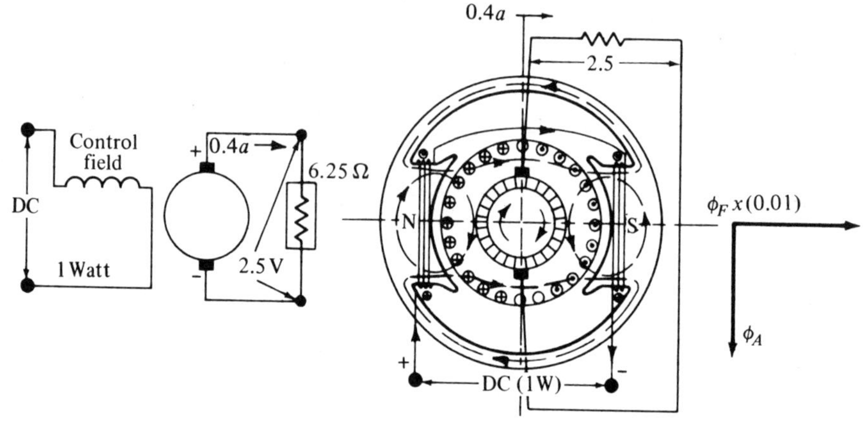

FIGURE 6-5. **(a) Schematic diagram, cross-sectional view and phasor diagram of a separately-excited shunt generator with full-load output. (b) Schematic diagram, cross-sectional view and phasor diagram of the generator of Figure 6-5(a) but with 1/100 of excitation (1 watt). (c) Schematic diagram, cross-sectional view and phasor diagram of generator of Figure 6-5(b) with 1 watt of excitation and brushes short circuited and quadrature set of brushes added. (d) Schematic diagram, cross-sectional view and phasor diagram of generator of Figure 6-5(c). Compensating winding and load added to quadrature brushes. The complete Amplidyne dc generator.**

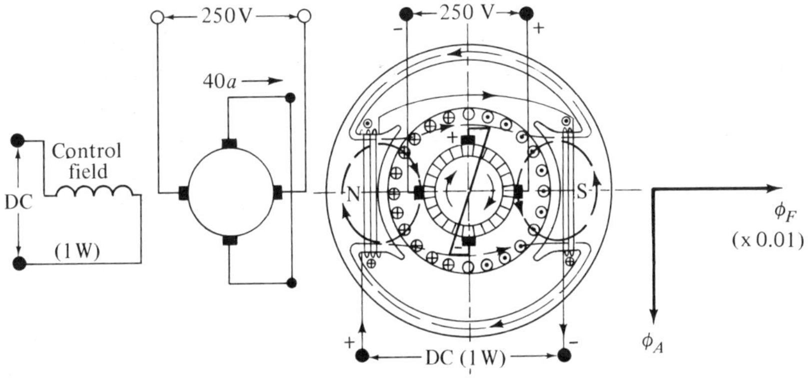

(c)

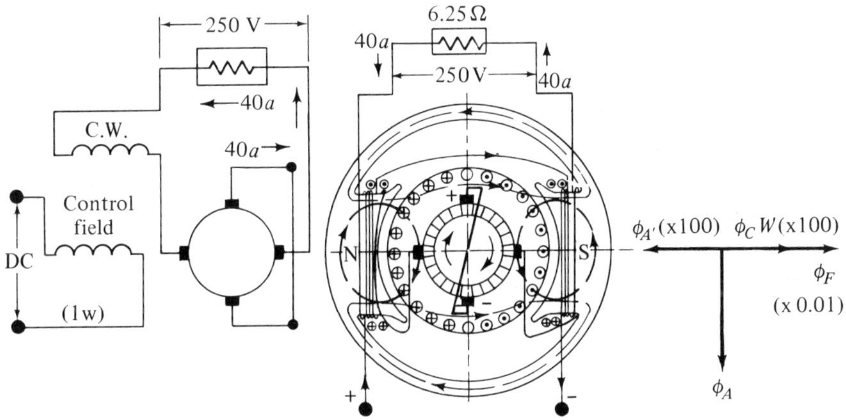

(d)

FIGURE 6-5. (*Continued*).

of power is high, ranging from 5000 to 20,000 and above in the large sizes. Response is also quick with adjustment time around 1/10 sec or less.

The principle of operation is best shown by a series of illustrations. Each illustration shows the schematic diagram, the cross-section view, and the phasor diagrams of the magnetic flux. In each illustration the speed at which the armature is driven is constant. Figure 6-5(a) shows the schematic diagram of a separately excited shunt generator. Flux and current directions are shown in the cross-sectional view while the phasor diagram of the flux is shown. With an input of 100 W to the control field, the output is 10 kW or 40 A at

250 V for a load of 6.25 ohms. If this generator is operating on the air-gap line of the saturation curve, the amplification is 10,000/100 or 100/1.

In the next illustration, Figure 6-5(b), the power input to the control field has been reduced to 1 W by cutting down the turns of the coil and the current or the ampere turns by 1/100. The load current has been reduced cutting the size of the phasor diagram by 1/100. The diagram is shown ten times actual size.

In the next illustration, Figure 6-5(c), the load has been removed and assuming the armature resistance is 0.0625 ohm or 1/100 of the load resistance, 40 A flow through the short circuit and the armature conductors; the armature flux ϕ_A has now been restored to its former value as given in Figure 6-5(a). A pair of brushes has been added 90 deg (two-pole generator) from the short-circuited set of brushes. As noted—and by design—the voltage generated across this pair of brushes is 250 V.

In Figure 6-5(d) the 6.25-ohm resistance load is now placed across the new brushes and the current flow is the same as in Figure 6-5(a), or 40 A at 250 V for a 10,000-W output.

If the current is doubled in the control winding, the field flux ϕ_F is doubled. In turn the short-circuit current is doubled and the armature flux ϕ_A doubles the load voltage to 500 V. The load current is now (500/6.25) 80 A. The new power output is 40,000 W. The input power I^2R is increased from 1 to 4 W. The power amplification is (40,000 − 10,000)/(4 − 1) or 10,000/1. All of this discussion assumed straight-line saturation and no drops in the armature.

In the cross-sectional view the current flow in the armature conductors due to the load in Figure 6-5(d) is indicated outside the conductors. This new armature current causes cross-magnetization flux opposite to the flux of the control winding. This new cross-magnetizing flux is indicated on the phasor diagram by $\phi_{A'}$. For this reason compensating windings (CW) have been added as indicated on the schematic and the cross-sectional view. The flux compensates for the new cross-magnetizing flux. This is necessary; for unless compensated, the net flux ϕ_F could be reduced to zero or even reversed. The schematic diagram of Figure 6-5(d) shows only one control field. As with the Rototrol and Regulex, more control fields are added.

An application using two control windings for controlling the speed of a Ward Leonard system is shown by the schematic diagram in Figure 6-6. In this schematic the Amplidyne is used to control the generator field of the supply generator. Two control fields are utilized. The reference field (RF) sets the speed and the signal field (SF) is connected across a tachometer whose voltage is proportional to the motor speed. The signal field MMF and the reference field MMF are equal but opposing each other. Assuming the speed of the motor drops, the MMF of signal or feed-back winding drops. The net MMF of the reference field increases the net flux and the generator field current by the increased terminal voltage. The increased generator field excitation increases the output voltage to the motor armature. The motor

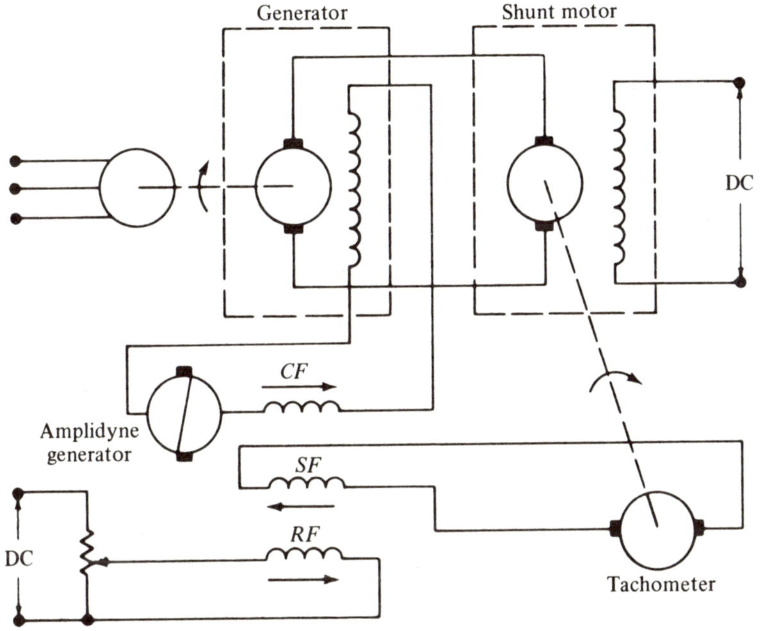

FIGURE 6-6. The Ward Leonard control system for a large shunt motor using an Amplidyne generator for controlling speed by controlling supply-generator field.

speed increases. The tachometer voltage now increases and brings the signal field MMF in balance with the reference field MMF. For an increase in speed the reverse process takes place.

The preceding discussion is but one application of the Amplidyne. It can be used for voltage control and voltage limiting, current control and current limiting, speed control and speed limiting, positioning, power-factor control, and many other functions. These functions are applied to an almost infinite number of drives in the metal industries, power generation, transportation industries, and many more.

The following example illustrates typical amplification factors for an Amplidyne.

Example 6-2. An Amplidyne is used instead of the Rototrol generator of Example 6-1. The reference-field resistance is 100 ohms with 60 mA through the winding; the output voltage across the alternator field is 250 V. (a) What is the power amplification? (b) What is the voltage amplification? (Assume a straight-line saturation curve and no armature drops.)

Solution:

(a) power output at above conditions: $\frac{(250)^2}{25} = 2500$ W,

power input at 250 V output: $(0.060)^2(100) = 0.36$ W,

power amplification: $\frac{2500}{0.36} = \frac{6950}{1}$.

(b) voltage input reference field: $0.060 \times 100 = 6$ V,

voltage amplification: $\frac{250}{6} = \frac{41.7}{1}$.

The Rosenberg generator, although not considered a rotary amplifier, is mentioned here because of its similarity to the Amplidyne. It is used chiefly to provide constant voltage for train lighting. The generator is driven by wheel axles and consequently it is not a constant-speed generator. It is similar to the Amplidyne, but it does not have a compensating winding. At high speeds the reaction of the armature flux subtracts from the main field flux in such a manner as to keep the output voltage nearly constant. At low speeds batteries must provide the source for the lighting. As in the Amplidyne the poles are designed with large shoes to provide low reluctance paths for the cross-magnetizing flux.

6-4 SUMMARY

The dc generator as used in the ordinary Ward Leonard system for speed control is a rotary amplifier. However, for high-power amplification, which is a basic requirement for feed-back control systems, the rotary amplifier for high-power gains has been developed. Because of the fact that speed control of dc motors is more easily accomplished with dc than with ac, the dc rotary amplifier and the silicon-controlled rectifier are advancing the use of dc motors.

The two general types of rotary amplifiers are (a) the one-stage self-energized field type such as the Rototrol, Regulex, and Reliance VSA drive; and (b) the separately excited cross-field or armature-excited generator such as the Amplidyne, Metadyne, and the Magnicon. In the self-energized field type, extra windings or control windings are added to the poles of the generator. With operation on the straight-line portion of the saturation curve, small amounts of power control large power outputs. The Rototrol and Regulex are quite similar, the difference being in the connection of the self-energizing field, with the series connection used by the Rototrol and the shunt connection used by the Regulex. These are one-stage amplifiers and hence have a lower power amplification. They have a slower response time and are used on smaller power applications up to, possibly, $2\frac{1}{2}$ kW.

The Amplidyne is a cross-field magnetized generator with a set of brushes at 90 electrical deg from the main field of the neutral axis. The basic set of brushes are short circuited and the current flow in the armature conductors creates the cross-field flux. The Amplidyne is a two-stage machine that utilizes the same armature windings for both stages. The power amplification is quite high—from 5000 to 20,000 plus.

The response time is short—1/10 sec or less. However, this can lead to overcorrection or a hunting procedure similar to that of the synchronous motor when a sudden change of load takes place. To avoid this antihunt, fields may be added. This is a field across the output with a capacitor or transformer to block out the dc. The oscillations set up MMFs opposite to the control field. This action is quite similar to the action of the damper windings of the synchronous motor when hunting occurs due to a sudden change in load.

Problems

1. Sketch a schematic diagram using a Rototrol or a Regulex generator to control the voltage of an alternator. Explain the operation.
2. Sketch a schematic diagram using Rototrol control generators for controlling a three-phase electric arc furnace. Show for one phase. Explain the operation.
3. Sketch the schematic diagram for a Regulex hoist control. Explain the operation.
4. Sketch an Amplidyne schematic diagram for maintaining constant impressed voltage, limiting motor armature current and speed for a dc shunt motor. Explain the operation.

5. Show a schematic diagram using an Amplidyne control generator for armature current control for a windup reel operation. Explain operation of diagram.
6. Show a schematic diagram using an Amplidyne generator for maintaining a fixed relationship between two motors. Explain operation of the circuit.
7. A Regulex is used for voltage control in a circuit similar to that shown in Figure 6-3. The resistance of the pattern field is 160 ohms and with 125 mA through the winding; the voltage across the generator field of 60 ohms is 250 V. (a) What is the power-amplification ratio? (b) What is the voltage-amplification ratio?
8. An Amplidyne is substituted for the Regulex generator in Problem 7. The pattern or control-field winding has a resistance of 125 ohms and with 40 mA the voltage across the generator field is 250 V. (a) What is the power-amplification ratio? (b) What is the voltage-amplification ratio? Assume straight-line saturation-curve operation and no armature losses.
9. A Rototrol generator is used to control the speed of a shunt motor similar to that in the schematic diagram of Figure 6-6. At full load of 15 hp with an armature voltage of 240 V and a current of 51 A the speed is 1200 rpm. The generator field of 60 ohms and 4 A will provide the full-load conditions. The control field has 500 turns per pole and a resistance of 182 ohms and with 110 mA the generator field current of 4 A is produced. The tachometer has a rating of 100 V per 1000 rpm. (a) What is the power-amplification ratio? (b) If the signal field has a resistance of 820 ohms, how many turns per pole must the signal-field winding have?
10. In Problem 9 an Amplidyne is substituted for the Rototrol. The control field has 1000 turns per pole and a resistance of 75 ohms and with 40 mA provides the 4 A to the generator field of 60-ohms resistance. (a) What is the power-amplification ratio? (b) If the signal field has a resistance of 1500 ohms, how many turns per pole must the signal-field winding have?

References

[1] Ku, Y. H., *Electrical Energy Conversion*. New York: Ronald Press, 1959.
[2] Rosenblatt, J. and M. H. Friedman, *Direct and Alternating Current Machinery*. New York: McGraw-Hill, 1963.
[3] Fitzgerald, A. E. and C. Kingsley, Jr., *Electric Machinery*, 2nd ed. New York: McGraw-Hill, 1961.
[4] Carr, C. C., *Electric Machinery*. New York: Wiley, 1958.

[5] Kosow, I. L., *Electric Machinery and Control*. Englewood Cliffs, N.J.: Prentice-Hall, 1964.

[6] Booklet GEA-7111, *G. E. Amplidyne*. General Electric Company, Direct-Current Motor and Generator Department, Erie, Pa.

[7] Siskind, C. S., *Electrical Control Systems in Industry*. New York: McGraw-Hill, 1963.

Electric Motor Applications

7

Electric motors are rotating machines in which energy conversion takes place. Energy is converted from electrical form to mechanical form. For this conversion process the correct type of motor for the particular application must be used, whether it be the very small battery watch motor or the 10,700-hp dc motor for powering a flight acceleration simulator in the NASA Space Center at Houston. In the case of the watch application the power converted is but a fraction of a watt, but the main factor is to control the speed at a fixed value. For the flight acceleration simulator the speed acceleration must be great to produce accelerating forces up to 30 G. Motors are used in vast and diversified industries, in homes, in schools, in offices, in transportation facilities, and any other place in modern life. Motors are even used in computers. In this section motor applications for industrial use are considered since all types of application of motors may be found here.

7-1 INDUSTRIAL MACHINERY CHARACTERISTICS[1],[2]

Industries are so numerous and the processes of manufacture are so varied that it is not possible to mention all of them here. Some of these industries are steel production and fabrication, nonferrous metals, textiles, oil, food processing, farm equipment, automobiles and trucks, airplanes, electrical appliances and equipment, electric-power generation and distribution, paper, chemicals, rubber, lumber and wood working, business machines and computers, mining, cement, ceramics, printing, pharmaceuticals, and so on and on. Each may be divided and subdivided further.

The industrial machines may be classified as those peculiar to a specific industry such as the rolling mill for rolling structural steel shapes or the rotary kiln for a cement plant, and those typical machines that may be used by a wide variety of industries, such as for fans, blowers, compressors, pumps, and so on.

In the first group of special machines the motor must have special characteristics to fit the characteristics of the machine. In the structural steel shape mill the motor must be reversible, operate at variable speeds, and have high-peak torque. For this application the dc motor is best adapted. The rotory kiln for a cement plant is a huge cylinder and must be rotated at a fairly constant speed. The kiln is not started and stopped frequently. Its peculiar property is the inertia of the kiln. A design D three-phase ac induction motor could be used, or for better efficiency a wound-rotor induction motor might be used. In any case a speed reducer or a gear motor would have to be used.

In the second group are such industrial machines as pumps, fans, blowers, compressors, machine tools, hoists, cranes, elevators, and conveyors, which are not peculiar to any industry. Some machines may, of course, have to be specially designed such as pumps for the chemical industry. Pumps can be of several types—centrifugal, rotary, or reciprocating. Whether the pump is started with load or without load determines whether a high starting torque motor is required. The speed should be constant. The frequency of starts depends upon the application. Compressors are closely allied with pumps; in fact a compressor is a pump but its fluid is usually a gas. The characteristics for compressor loads are the same as for pumps. There is one additional characteristic related to reciprocating pumps and compressors—the pulsating torque. This pulsating torque produces current pulsations in the driving motor. A compressor factor is a number that relates to the requirements of the load to keep the current pulsation to a certain limit such as 66 percent, which is the maximum limit. These factors for various types are listed in Section 21.87 of NEMA standard MG1-1967. Machine tools of the ordinary and smaller types have no special starting torque requirements. Automated machine tools may require braking and reversals and may need high starting torque. Unless the operation is one of starts, stops, and reversals, the speed should be constant. Hoists and cranes require high starting torques and adjustable speeds. Elevators also require high starting torque but do not need adjustable speeds; however, the driving motor must be quiet. Conveyors must operate at constant speed and for starting the torque requirement is of the medium starting torque type.

In addition to the above requirements, the environment of the machine must be known. Temperatures, dust, dirt, moisture, fumes, both corrosive or explosive, and fire hazards are factors that determine the type of motor enclosure, motor ventilation, and insulation system the motor must have.

Space requirements also need to be known, but generally the machine designer provides enough space for the motor. Geographic location and elevation need to be known since this has some bearing on the insulation system. Other mechanical facts such as the drive—direct, coupled, or V-belt pulley—shock load, use of mechanical brakes and clutches, and provision for space for starting and control equipment must also be considered.

7-2 ELECTRIC MOTOR CRITERIA[1]–[7]

When the load characteristics have been determined, the best motor for the machine is selected. With the knowledge gained from the study of rotating machines and the use of the National Electrical Manufacturers Association (NEMA) standards MG1-1967 and MG2-1968, the National Electrical Code (NEC) or local codes, American National Standards Institute (ANSI) C50 series standards, and manufacturers informational literature, the proper type of motor together with the control equipment are selected. Manufacturer's literature, and NEMA and ANSI (ASA) standards give such items as starting torque, maximum torque, speeds, currents, voltage ratings and tolerances, temperature rise, and service factors. The selection is made from the three-phase squirrel-cage induction, wound-rotor induction, synchronous, dc shunt, series or compound, single-phase ac or special type motor.

The induction-motor types account for the majority of the motor applications. Manufacturer's literature and the NEMA and the ANSI (ASA) standards list speeds, locked-rotor torques, maximum torques, normal currents, locked-rotor currents, frame numbers, and so on. The nameplate of a motor specifies these parameters for the motor. The standards specify that the minimum information for and induction motor nameplate be as follows: (1) Manufacturers type and frame designation; (2) horsepower output; (3) time rating; (4) temperature rise; (5) rpm at full load; (6) frequency; (7) number of phases; (8) voltage; (9) full-load amperes; (10) code; (11) design letter for integral hp motors; and (12) motors equipped with thermal protection, the words *thermally protected*, and for integral hp motors a type number. Figure 7-1 shows a squirrel-cage induction motor. The manufacturer is the General Electric Company, the model number is 5K184BK201, and the frame number is 184. It is rated at 5 hp and the time rating is continuous. The speed is 1745 rpm. The frequency is 60 Hz. The supply is three-phase, 208-220/440 V. The full-load current is 14/7 A. The temperature is given as 40°C maximum for class-B insulation. The service factor is 1. The temperature rise for this rating is 70°C. The code is H, which means that the locked-

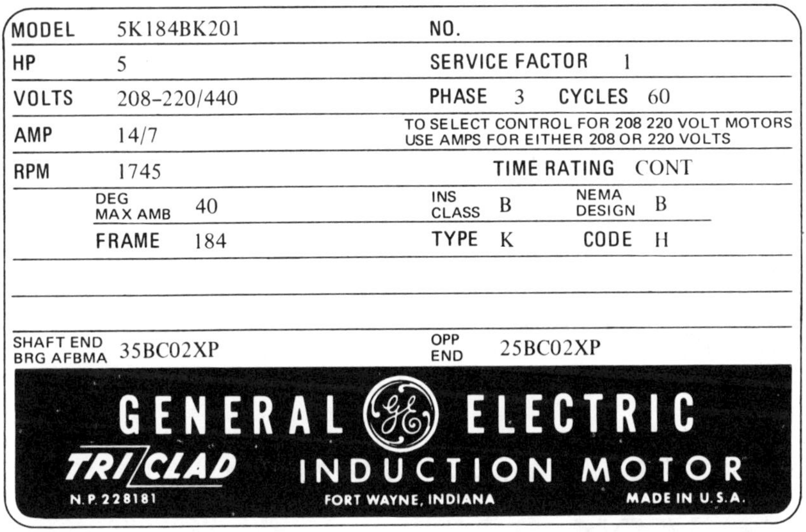

FIGURE 7-1. Sample nameplate for an induction motor. (General Electric Co.)

rotor kVA per hp is 6.3 to 7.09. This is taken from Appendix Table 4-A. For the 5-hp rating the locked-rotor current range is from 82.6 to 93 A. Manufacturer's literature lists the locked-rotor current at 90 A. The design letter is B, which means the motor has a locked-rotor current of about 600 percent normal current and a locked-rotor torque of about 200 percent normal torque. The same literature lists a locked-rotor torque of 185 percent. No thermal protection is listed, but the manufacturer's type is K.

Polyphase wound-rotor motors have no code letter and the locked-rotor current and the locked-rotor torque depend on the resistance that is connected to the rotor. The greater the amount of resistance, the lower the starting current and the higher the starting torque. The nameplate minimum items are the first nine items for the squirrel-cage motor plus item (10), secondary amperes at full load, and item (11), secondary voltage. The suggested NEMA standard MG1-10.39 lists for (4) maximum ambient temperature and adds item (5) insulation system designation. Items (10) and (11) then become (11) and (12). This system was used in the nameplate of Figure 7-1 for a squirrel-cage motor. Also for the suggested standard, the service factor if other than 1 is also added.

The synchronous motor is generally used for drives from 500 hp and above. For slow-speed applications such as reciprocating pumps and compressors, the synchronous motor is better adapted because of its large diameter rotor which acts as a flywheel adding moment of inertia to pulsating loads. In intermediate sizes these motors are often used for steady load drives such as fans, blowers, motor-generator sets, and so on, for the added benefit for power-factor improvement. One disadvantage in their use may be the need for an exciter to provide the necessary dc excitation. The brushless exciter that uses solid-state rectifiers has eliminated the brushes and slip rings of the synchronous motor and really makes the exciter an integral part of the motor.

The following items are on a synchronous motor nameplate: The same ten items as listed for the induction-motor nameplate plus (11) rated field current, (12) rated exciter voltage, and (13) rated power factor. These are the minimum items that must be on the nameplate. Figure 7-2 shows a

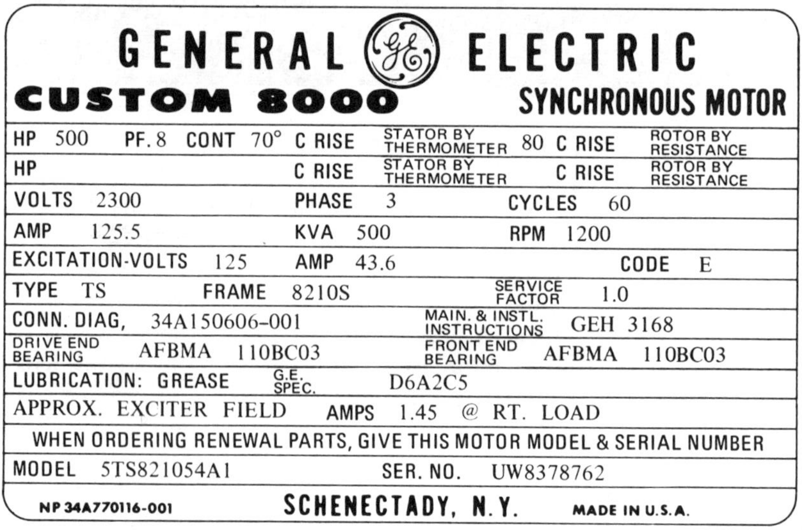

FIGURE 7-2. Sample nameplate for a synchronous motor. (General Electric Co.)

nameplate for a 500-hp, 2300-V, three-phase synchronous motor. Here the exciter information is given as well as the bearing specifications, connection diagram number, and lubrication specification. Because a synchronous motor cannot be started as a synchronous motor, the code letter refers to the starting windings, which start the motor as a squirrel-cage induction motor. This nameplate indicates code E which from Appendix Table 4-A indicates locked rotor kVA per hp of 4.5 to 4.99. This represents a locked-rotor current of

565 to 625 A. This is 450 to 500 percent of the full-load current of 125.5 A at rated voltage of 2300 V. Another aspect of the synchronous motor is the pull-in torque, which is the torque requirement when the dc field excitation is applied and the motor pulls into synchronous speed. NEMA standard MG1-21.86 lists values of locked rotor, pull-in and pull-out torques as well as ratios of load moment of inertia (wk^2) to the normal moment of inertia of the load for 56 specific loads. The selection from manufacturer's listings can then be made. The inertia factor and excessive heating is an important consideration for multiple starts. This aspect is covered in the NEMA MG1-21.43 and ANSI (ASA) C50.11-1965 standards. The pulsating-current limitation for slow-speed applications such as reciprocating pumps or compressors is also covered in these standards. As mentioned in the previous section compressor factors are listed in NEMA MG1-21.87 for ammonia or freon, air, and carbon dioxide compressors.

Direct-current motors are used primarily for speed control and therefore in the selection of a dc motor the control equipment needs to be considered at the same time. It was found that speed could be controlled by three ways: (a) weakening of the field, (b) adding armature resistance, and (c) armature voltage control. The ANSI (ASA) standard C50.4-1965 and NEMA standard MG1-1967 list the allowable speeds, voltages, and base speeds for industrial motors to 800 hp, large industrial motors (larger than 0.75 hp per rpm) and metal rolling mill motors (class N) from 250 hp to 8000 hp, metal rolling mill motors (class S) from 250 hp to 1000 hp, and reversing hot mill motors for base speeds 60 to 200 rpm. The minimum nameplate data for dc motors includes the following:

(1) manufacturer's name and frame designation;
(2) horsepower output at base speed;
(3) time rating at base speed;
(4) temperature rise at base speed;
(5) rpm at rated-load base speed;
(6) maximum safe speed for all integral-horsepower series-wound motors and for compound motors whose no load exceeds the full-load speed by 35 percent or more;
(7) voltage;
(8) field voltage when different from armature voltage;
(9) rated current at base speed horsepower;
(10) winding shunt, stabilized shunt, compound or series;
(11) for fractional horsepower motors, the words *thermally protected* if the motor is equipped with thermal protection.

Items (6), (8), and (11) are not listed on nameplates of large-sized motors. No starting torque is mentioned since the starter or controller limits the starting armature current. Torque is given by the expression $T = K_t\phi I_a$. From this expression, if the armature current is limited to 175 percent of full-load value, the starting torque is 175 percent of rated load torque. For a series motor this would approach 300 percent. Starting torque is not a major consideration in the selection of a dc motor.

Very often dc is not available in an industrial plant for a dc motor. The Ward Leonard system of speed control overcomes this problem but has the high cost objection. Adjustable-speed drives using either the motor-generator set, the silicon-controlled rectifier and solid-state rectifier for low power rating to 15 hp, and eddy current coupled drives all use ac supply. All except the eddy current coupling operate from rectified ac. These drives are available up to 750 hp. In these rectifier units a conventional dc motor is used and the rectified supply is filtered so that the pulse number is 6 or greater. Direct-current motors will not operate satisfactorily from unfiltered supply.

7-3 MOTOR ENCLOSURES[2]–[4],[6],[8],[9]

The motor enclosure must be correct for the environment in which the motor is used. The two types of enclosures are the open machine (NEMA MG1-1.25) and the totally enclosed machine (NEMA MG1-1.26). An open machine has ventilating openings or is a large machine without any ventilation restrictions allowing external air to pass over and around the windings to cool them. A totally enclosed machine is one that is enclosed so as to prevent a free exchange of outside cooling air.

Open Machines. The types of open machines are as follows:

(1) Drip proof: Provided with ventilating openings in such a manner that any liquid or object will not enter if striking the machine at an angle 15 deg or less downward from the vertical.

(2) Splash proof: With ventilating openings as in the previous type except that the angle is 100 deg downward from the vertical.

(3) Semiguarded: With part of the machine, usually the top half, provided with ventilating openings so guarded that $\frac{3}{4}$-in. round rod or larger cannot enter through the openings.

(4) Guarded: With all ventilating openings guarded as in the semiguarded machine.

(5) Drip-proof fully guarded: A combination of the drip-proof and the guarded machines.

(6) Open externally-ventilated: Any of the previous types that are ventilated by an external blower.

(7) Open pipe-ventilated: Any of the previous types (1) through (5) with openings for pipes or ducts for forced ventilation.

(8) Weather protected: An open machine constructed so as to minimize the entrance of rain, snow, or air-borne particles and with openings guarded as in the guarded machine.

Enclosed Machines. These machines are enclosed so that rain, snow, or air-borne particles are not blown into electrical parts by storms or high wind. The totally enclosed machine-enclosure types are as follows:

(1) Totally-enclosed nonventilated (TENV): Totally enclosed with no external means of ventilating closed parts.

(2) Totally enclosed fan-cooled (TEFC): Totally enclosed with exterior cooling by means of a fan to outside of enclosure.

(3) Explosion proof: Totally enclosed to withstand an explosion within the enclosure, and not ignite outside surrounding atmospheric gases, vapor, or dust.

(4) Dust-ignition proof: Totally enclosed but not allowing sufficient amounts of dust to enter machine to cause ignition by sparks, arcs, or heat to ignite exterior dust suspended in atmosphere.

(5) Water proof: Totally enclosed so that a water stream will not enter except around the shaft with provision to drain any water from a reservoir provided to gather water seeping in at the shaft.

(6) Totally enclosed pipe ventilated: Totally enclosed with piped ventilation either internally through ducts or separately ventilated.

(7) Totally enclosed water cooled: Totally enclosed with water piped to circulate cooling water around internal parts.

(8) Totally enclosed water-air cooled: Totally enclosed with air cooling inside, the air in turn cooled by water through a heat exchanger.

(9) Totally enclosed air-to-air cooled: Same as for previous type, but with internal air cooled by external air in an air-to-air heat exchanger.

(10) Totally enclosed fan-cooled guarded: Similar to the above air-to-air cooled, but with openings to fan for external air guarded for a $\frac{3}{4}$-in. diameter or larger rod.

Two types of open machines are manufactured with encapsulated or sealed windings to protect the working parts from moisture and corrosive fumes and vapors. The one type has the windings filled with an insulating resin to form a protective coating for moisture and more severe atmospheric conditions. The sealed-winding type has the winding sealed by materials such as epoxy to form a solid covering for the windings. The metal parts are also protected with protective coatings to resist corrosion by atmospheric vapors.

The National Electrical Code (NEC) sponsored by the National Fire Protection Association (NFPA) defines hazardous locations. The locations are classified as Class I, division 1, and Class I, division 2, and so on. Class I indicates atmospheric mixtures of a list of gases and vapors that are hazardous. Division 1 indicates that the hazardous atmosphere is surrounding the application, while in division 2 these hazardous atmospheres are contained and not present around the application.

Class II indicates the presence of combustible dust and division 1 indicates that the dust is suspended in the atmosphere around the application, while in division 2 the dust is not ordinarily suspended in the air.

Class III refers to ignitible fibers or flyings with division 1 having these fibers or flyings being used around the application and division 2 having the fibers and flyings stored.

In selecting the enclosure the type of material making the atmosphere hazardous should be known. The type should be checked with the code and with manufacturer's literature.

7-4 INSULATION SYSTEMS[2]–[4],[6]

The insulation systems are based on the types of insulation used. The type determines the highest temperature that the insulation may attain. The systems are as follows:

Class A: Cotton, paper, and so on, or class-105 insulation with a maximum temperature of 105°C.

Class B: Mica, glass fiber, and so on, or class-130 insulation with a maximum temperature of 130°C.

Class F: Mica, asbestos, and so on, or class-155 insulation with a maximum temperature of 105°C.

Class H: Mica, glass fibers, and so on, or class-180 insulation with a maximum temperature of 180°C.

Classes A and B are the usual systems used. For insulation system A on an open machine for continuous time rating, the rise from a maximum ambient temperature of 40°C which allows 15°C for the hot spot to surface with a 10°C service factor allowance allows a temperature rise of 40°C. Without service factor, the temperature rise is 50°C. For the B insulation system for a maximum ambient temperature of 40°C, the rise is 70°C without a service factor, since at the higher operating temperature the differential between hot-spot temperature and surface temperature is 20°C. An enclosed machine adds 5°C to the temperature rise because the differential between surface and hot-spot temperature is 5°C less. The ratings for any machine are listed in the NEMA and ANSI (ASA) standards. Manufacturer's literature also gives the rise or the maximum ambient temperature and the insulation letter. All temperatures are in degrees centigrade. The 40°C maximum for ambient temperature is equal to 104°F. On the nameplate of Figure 7-1 no rise is given. The maximum ambient temperature is given as 40°C and the insulation as class B. The service factor is given as 1, so the rise would be 70°C. The temperature-rise rating is important in the study of the environment in which the motor is used. If the environment temperature is more than 40°C, the proper insulation should be used so that the maximum temperature of the insulation class is not exceeded.

The following illustrates a rather simple selection for a motor. A motor is required to drive a 750-gal/min rotary pump with a head of 120 ft. It is calculated that a 25-hp motor at 1150 rpm will fill the need. The pump will run at intervals of an average of four 10-min intervals/hr. The pump must start under load when started and requires a starting torque of 200 lb-ft which is about 175 percent of the full-load torque. The pump and motor are in an enclosed pump building that is well ventilated. The pump is located in a temperate climate where the temperature record shows that the temperature never goes over 100°F. Remembering that the speed of such a pump must be nearly constant, an induction motor will do the job, especially since 230-V three-phase ac is available. The only other problem is the starting torque. Again, design C, which has about 200 percent starting torque looks all right. Manufacturer's literature specifies 200 percent for starting torque. Also, the starting current should be about 600 percent for across-the-line starting. The starting current is listed as 365 A for a full-load current of 65.2 A. Since the motor is inside a building, an open drip-proof enclosure should be used. So a 220-V, three-phase, 60-Hz, design-C, code-letter F, maximum temperature 40°C, Class-A insulation motor could be chosen (Class B would probably be a better choice). Its speed is 1150 rpm, time rating continuous. Service factor is 1 or 1.15, whichever is standard. Starting is with across-the-line with a pressure controlled contactor.

References

[1] Libby, C. C., *Motor Selection and Application*. New York: McGraw-Hill, 1960.

[2] *Motor and Generator Standards*, MG1-1967. New York: National Electrical Manufacturers Association.

[3] *American Standard, Alternating-Current Induction Motors, Induction Machines in General, and Universal Motors*, ASA C50.2-1955. New York: American National Standards Institute.

[4] *American Standard, Direct-Current Generators, Direct-Current Motors, and Direct-Current Commutating Machines in General*, ASA C50.4-1965. New York: American National Standards Institute.

[5] *American Standard Requirements for Synchronous Motors*, ASA C50.11-1965. New York: American National Standards Institute.

[6] Bulletin HBK-304, July 6, 1964, *Induction Motors*—Frames 143–149, *Selection and Application*. Schenectady, New York: General Electric Company.

[7] Booklet GEA-8095, *A New Generation of Motors, Tri-Clad 700 Line*. Schenectady, New York: General Electric Company.

[8] Bulletin HBK-321, July 6, 1964, *Induction Motors—Three-Phase*, Frames 143–449, *Drip-Proof, Totally Enclosed, Explosion-Proof*. Schenectady, New York: General Electric Company.

[9] Abbott, A. L. and F. Stetka, *National Electrical Code Handbook*. New York: McGraw-Hill, 1960.

Appendix

Appendix

TABLE 1-A.[1] Full-Load Currents, in Amperes, Direct-Current Motors (NEC 430-147)

Horsepower	120 V	240 V
$\frac{1}{4}$	2.9	1.5
$\frac{1}{3}$	3.6	1.8
$\frac{1}{2}$	5.2	2.6
$\frac{3}{4}$	7.4	3.7
1	9.4	4.7
$1\frac{1}{2}$	13.2	6.6
2	17	8.5
3	25	12.2
5	40	20
$7\frac{1}{2}$	58	29
10	76	38
15		55
20		72
25		89
30		106
40		140
50		173
60		206
75		255
100		341
125		425
150		506
200		675

The values of full-load currents are for motors running at base speed.

[1] All tables in the Appendix are reproduced by permission from the *National Electrical Code*, 1968 edition, copyright National Fire Protection Association, Boston, Mass. 02110.

TABLE 2-A. Full-Load Currents, in Amperes, Single-Phase ac Motors (NEC 430-148)

Horsepower	115 V	230 V	440 V
$\frac{1}{6}$	4.4	2.2	
$\frac{1}{4}$	5.8	2.9	
$\frac{1}{3}$	7.2	3.6	
$\frac{1}{2}$	9.8	4.9	
$\frac{3}{4}$	13.8	6.9	
1	16	8	
$1\frac{1}{2}$	20	10	
2	24	12	
3	34	17	
5	56	28	
$7\frac{1}{2}$	80	40	21
10	100	50	26

These values of full-load currents are for motors running at usual speeds and motors with normal torque characteristics. Motors built for especially low speeds or high torques may have higher full-load currents, in which case the nameplate current ratings should be used.

To obtain full-load currents of 208- and 200-V motors, increase corresponding 230-V motor full-load currents by 10 and 15 percent, respectively.

The nominal power system voltages are 120, 240, and 480 V.

TABLE 3-A. Full-Load Currents, in Amperes, Three-Phase ac Motors (NEC 430-150)

	Induction-Type, Squirrel-Cage and Wound-Rotor					Synchronous Type, Unity Power Factor*			
Horse-power	110 V	220 V	440 V	550 V	2300 V	220 V	440 V	550 V	2300 V
½	4	2	1	.8					
¾	5.6	2.8	1.4	1.1					
1	7	3.5	1.8	1.4					
1½	10	5	2.5	2.0					
2	13	6.5	3.3	2.6					
3		9	4.5	4					
5		15	7.5	6					
7½		22	11	9					
10		27	14	11					
15		40	20	16					
20		52	26	21					
25		64	32	26	7	54	27	22	5.4
30		78	39	31	8.5	65	33	26	6.5
40		104	52	41	10.5	86	43	35	8
50		125	63	50	13	108	54	44	10
60		150	75	60	16	128	64	51	12
75		185	93	74	19	161	81	65	15
100		246	123	98	25	211	106	85	20
125		310	155	124	31	264	132	106	25
150		360	180	144	37		158	127	30
200		480	240	192	48		210	168	40

For full-load currents of 208- and 200-V motors, increase the corresponding 220-V-motor full-load current by 6 and 10 percent, respectively.

These values of full-load current are for motors running at speeds usual for belted motors and motors with normal torque characteristics. Motors built for especially low speeds or high torques may require more running current, in which case the nameplate current rating should be used.

* For 90 and 80 percent pf the above figures should be multiplied by 1.1 and 1.25, respectively.

The nominal power system voltages are 120, 240, 480, 600 and 2400 V.

TABLE 4-A. Locked-Rotor Indicating Code Letters (NEC 430-(7b))

Code Letter	Kilovolt-Amperes per Horsepower with Locked Rotor
A	0 – 3.14
B	3.15– 3.54
C	3.55– 3.99
D	4.0 – 4.49
E	4.5 – 4.99
F	5.0 – 5.59
G	5.6 – 6.29
H	6.3 – 7.09
J	7.1 – 7.99
K	8.0 – 8.99
L	9.0 – 9.99
M	10.0 –11.19
N	11.2 –12.49
P	12.5 –13.99
R	14.0 –15.99
S	16.0 –17.99
T	18.0 –19.99
U	20.0 –22.39
V	22.4 –and up

The above table is an adopted standard of the National Electrical Manufacturers Association.

The code letter indicating motor input with locked rotor must be in an individual block on the nameplate, properly designated. This code letter is to be used for determining branch-circuit overcurrent protection by reference to Table 5-A.

TABLE 5-A. Maximum Rating or Setting of Motor-Branch-Circuit Protective Devices for Motors Marked with a Code Letter Indicating Locked Rotor kVA (NEC 430-152)

	Percent of Full-Load Current		
Type of Motor	Fuse Rating	Circuit-Breaker Instantaneous Type	Setting-Time-Limit Type
All ac single-phase and polyphase squirrel-cage and synchronous motors with full-voltage, resistor or reactor starting:			
Code letter A	150	...	150
Code letter B to E	250	...	200
Code letter F to V	300	...	250
All ac squirrel-cage and synchronous motors with autotransformer starting:			
Code letter A	150	...	150
Code letter B to E	200	...	200
Code letter F to V	250	...	200

For certain exceptions to the values specified see NEC Secs. 430-52 and 430-54. The values given in the last column also cover the ratings of nonadjustable, time-limit types of circuit breakers which may also be modified as in NEC Sec. 430-52.

Synchronous motors of the low-torque, low-speed type (usually 450 rpm or lower), such as are used to drive reciprocating compressors, pumps, etc., which start up unloaded, do not require a fuse rating or circuit-breaker setting in excess of 200 percent of full-load current.

For motors not marked with a Code Letter, see Table 6-A.

TABLE 6-A. Maximum Rating or Setting of Motor-Branch-Circuit Protective Devices for Motors not Marked with a Code Letter Indicating Locked-Rotor kVA (NEC 430-153)

	Percent of Full-Load Current		
Type of Motor	Fuse Rating	Circuit-Breaker Instantaneous Type	Setting-Time-Limit Type
Single phase, all types	300		250
Squirrel-cage and synchronous (full-voltage, resistor and reactor starting)	300		250
Squirrel-cage and synchronous (autotransformer starting)			
Not more than 30 A	250		200
More than 30 A	200		200
High-reactance squirrel cage			
Not more than 30 A	250		250
More than 30 A	200		200
Wound rotor	150		150
Direct current			
Not more than 50 hp	150	250	150
More than 50 hp	150	175	150
Sealed (hermetic type)			
Refrigeration compressor*			
400 kVA locked rotor or less	175†		†175

For certain exceptions to the values specified see Sections 430-52, and 430-59 (NEC). The values given in the last column also cover the ratings of nonadjustable, time-limit types of circuit breakers which may also be modified as in Section 430-52 (NEC).

Synchronous motors of the low-torque, low-speed type (usually 450 rpm or lower) such as are used to drive reciprocating compressors, pumps, etc., which start up unloaded, do not require a fuse rating or circuit-breaker setting in excess of 200 percent of full-load current.

For motors marked with a code letter, see Table 5-A.

*The locked-rotor kVA is the product of the motor voltage and the motor locked-rotor current (LRA), given on the motor nameplate, divided by 1000 for single-phase motors, or divided by 580 for 3-phase motors.

†This value may be increased to 225 percent if necessary to permit starting.

Answers to Problems

Answers to Problems

CHAPTER 1

1. 1.68 V
2. 1.05 V
3. (a) 0.02 Wb; (b) 2×10^6 Mx
5. 1.25 V
6. (a) 4.25 V; (b) 6.0 V; (c) 4.6 V
7. 1.27 V
8. (a) 0 V; (b) 4.8 V; (c) 3.39 V; (d) −4.8 V; (e) −3.39 V
10. 231 V
11. 900 rpm
12. 1035 rpm
13. (a) 2.99 lb; (b) 2.59 lb
14. (a) 9×10^5 dyn; (b) 9 N
15. (a) 158.8 lb; (b) 66.1 lb-ft; (c) 13.2 hp
16. (a) 85.1 lb-ft; (b) 16.3 hp
17. 362 A
18. 394 A
19. Yes
20. 85.6%
21. 82.3%
22. 3450 W, 84.4%
23. (a) 2410 W; (b) 2870 W; (c) 69.5%
24. (a) 85.1%; (b) 2050 W, 1750 W
25. (a) 42.6 A; (b) 15.7 hp; (c) 10 kW; (d) 41.7 A

CHAPTER 2

1. 273.2 V
2. 50 kW
3. 101 V
4. (a) 240 V; (b) 21.4 kW
5. (a) 9; (b) 141.5 V; (c) 166.5 deg
6. (a) 7, 14; (b) 267 V
7. 8, 16, 184 deg
8. 430 At, 4600 At
9. 580 At, 5230 At

10. (a) 2.32 lb; (b) 404 lb-ft
11. 404 lb-ft, 371 lb-ft
12. (a) 192 lb-ft; (b) 222 A
13. (a) 690 lb-ft; (b) 630 lb-ft; (c) 935 rpm
14. (a) 335 lb-ft; (b) 2.82×10^6 lines; (c) 2.52 lines/in.^{2}; (d) 0.735 lb; (e) 240.8 V
15. (a) 249.2 V; (b) 346 lb-ft; (c) 42.5 kW
16. 1027 lb-ft
17. 885 W, 2065 W
18. 85%
19. (a) 227 A; (b) 57.9 kW; (c) 85.75%
20. 3395, 3450, 3500, 3550, 3605 W
21. 1656, 2447, 3635, 5225, 7215 W
22. 75.5%, 80.9%, 83.7%, 85.0%, 85.3%
23. 81.7%, 584 W, 1086 W
24. 3280, 3240, 3190, 3150, 3100 W
25. 2031, 2835, 4055, 5728, 7836 W
26. 66.4%, 80.7%, 84.7%, 85.8%, 85.9%
27. 75 hp — 85.8%, 91.3 hp — 85.9%

CHAPTER 3

1. (a) 4.8%; (b) 16.67 kW
2. (a) 231.2 V; (b) 23.33 kW
3. 16.67 kW
4. (a) 6.0 kW; (b) 11.4 kW
5. 1237 rpm; (b) 7.55%
6. (a) 2145 rpm; (b) 22.6%
7. (a) 1152 rpm; (b) 19.75%
8. 120%
9. (a) 584 rpm; (b) 12.7 hp; (c) 0.78 Ω
10. (a) 703 rpm; (b) 581 rpm
11. (a) 850 rpm; (b) 2.85 hp; (c) 0.89 Ω
12. 922 rpm
13. 0.663 Ω
14. (a) 2115 rpm; (b) 0.868 Ω
15. (a) 1010 rpm; (b) 196.5 lb-ft; (c) 37.8 hp

16. 1050 W
17. (a) 0.165 Ω; (b) 1380 rpm; (c) 100.3 lb-ft
18. (a) 124 lb-ft, 1180 rpm; (b) 649 lb-ft, 508 rpm
19. (a) 0.72 Ω; (b) 99.2 lb-ft; (c) 1310 rpm
20. 2000 W
21. (a) 8.22 Ω; (b) 2.84 Ω
22. 1.58 Ω, 0.90 Ω, 0.87 Ω
23. 101 V, 42.9%, 159 V, 67.7%, 204.2 V, 86.8%
24. 3.64 Ω
25. 2.94 Ω, 0.32 Ω

CHAPTER 4

1. Table
2. 160 V
3. 0.689×10^6 lines
4. 4.14 turns/coil, 1.035×10^6 lines
5. 1580 V
6. 4720 V
7. (a) 50 Hz; (b) 2.52×10^6 lines
8. 3.05×10^6 lines
9. (a) 8220 V; (b) 882 kW
10. 80 poles, 10 poles
11. 25 Hz, 40 Hz
12. (a) 1640 V; (b) 1975 V; (c) 1183 V
13. (a) 1780 V; (b) 2220 V; (c) 1192 V
14. 3590 V, 2230 V
15. 3.37 Ω, 3320 V, 2230 V
16. (a) 3.36 Ω; (b) 1660 V; (c) 1175 V
17. (a) 28.5%; (b) 59.5%; (c) −7.0%
18. (a) 19.8%; (b) 45.8%; (c) −13.9%
19. (a) 45.1%; (b) −24.2%
20. (a) 8.3%; (b) 24.7%
21. 91.7%, 90.0%
22. 88.75%
23. (a) 662*R* kVA; (b) 300*R* kVA
24. (a) 1106 kVA; (b) 0.905 lagging
25. (a) 2490 kVA; (b) 0.966 lagging

CHAPTER 5

1. (a) 4.58%; (b) 4.45%
2. (a) 3600 rpm; (b) 2; (c) 3590 rpm; (d) 0.278%
3. (a) 720 rpm; (b) 35 rpm; (c) 685 rpm
4. (a) 715 rpm; (b) 4.67%; (c) 8; (d) 35 rpm
5. (a) 66.3 V; (b) 3.24 V; (c) 2.93 Hz
6. (a) 299 V; (b) 14.95 V; (c) 486 V
7. (a) 584 A; (b) 122 A; (c) 720 lb-ft; (d) 136 hp; (e) 627 lb-ft
8. (a) 423 A; (b) 29.6 A; (c) 1570 lb-ft; (d) 33.4 hp; (e) 154 lb-ft
9. (a) 105 A; (b) 39.5 lb-ft; (c) 18.8 A; (d) 7.55 hp; (e) 23 lb-ft
10. (c) 20.5 A; (d) 8.27 hp; (e) 25.4 lb-ft
11. (a) 106.5 A; (b) 38.3 lb-ft; (c) 19.6 A; (d) 7.44 hp; (e) 22.7 lb-ft
12. 1263 rpm, 63.8 lb-ft
13. (a) 0.331 Ω; (b) 72.9 A
14. (a) 0.0555, 0.0444, 0.0333, 0.0250, 0.0111, 0.00444, 0.00277; (b) 6.2, 5.01, 3.76, 2.60, 1.30, 0.43, 0 hp; (c) 0.915, 0.878, 0.811, 0.724, 0.538, 0.311, 0.167; (d) 84.7%, 85.4%, 85.9%, 85.8%, 74.0%, 47.0%, 0
15. (a) 5.0 hp; (b) 85.3%; (c) 15.27 lb-ft
16. (a) 10.0 hp; (b) 60.4 lb-ft; (c) 85.12%; (d) 0.772
17. (a) 9.87 hp; (b) 59.6 lb-ft; (c) 85.05%; (d) 0.770
18. (a) 51.3 hp; (b) 230 lb-ft; (c) 88.62%; (d) 94.8%
19. (a) 144 A; (b) 81.8 lb-ft; (c) 177 lb-ft; (d) 702 rpm
20. (a) 353 A; (b) 289 lb-ft; (c) 1040 rpm; (d) 628 lb-ft
21. (a) 1.503 hp, 73.6%, 4.59 lb-ft, 0.786; (b) 2.7 hp, 68.3%, 11.0 lb-ft, 0.806; (c) 50 A, 594%, 15.57 lb-ft, 339%, 0.474
22. (a) 696 A, 557 A, 450 lb-ft; (b) 566 A, 368 A, 190 lb-ft; (c) 435 A, 217.5 A, 90 lb-ft
23. (a) 0.0655 Ω, 3.18 kW; (b) 0.0881 Ω, 2.8 kW; (c) 0.1402 Ω, 2.65 kW
24. (a) 1.875 kVA, 0.1033 mH; (b) 2.26 kVA, 0.1875 mH; (c) 2.12 kVA, 0.297 mH
25. (a) 290 A, 187%; (b) 150 lb-ft, 33.3%
26. (a) 514, 720, 1200, 1800 rpm

CHAPTER 6

7. (a) 417/1; (b) 12.5/1
8. (a) 5210/1; (b) 50/1
9. (a) 436/1; (b) 375
10. (a) 8000/1; (b) 500

Index